THEORY AND DECISION LIBRARY

General Editors: W. Leinfellner (*Vienna*) and G. Eberlein (*Munich*)

Series A: Philosophy and Methodology of the Social Sciences

Series B: Mathematical and Statistical Methods

Series C: Game Theory, Mathematical Programming and Operations Research

Series D: System Theory, Knowledge Engineering and Problem Solving

SERIES C: GAME THEORY, MATHEMATICAL PROGRAMMING AND OPERATIONS RESEARCH

VOLUME 41

Scope: Particular attention is paid in this series to game theory and operations research, their formal aspects and their applications to economic, political and social sciences as well as to sociobiology. It will encourage high standards in the application of game-theoretical methods to individual and social decision making.

The titles published in this series are listed at the end of this volume.

Mohammed Abdellaoui · R. Duncan Luce
Mark J. Machina · Bertrand Munier
(Editors)

Uncertainty and Risk

Mental, Formal, Experimental Representations

With 45 Figures and 49 Tables

Springer

Prof. Mohammed Abdellaoui
Maison de la recherche de l'ESTP
GRID
30 ave du Président Wilson
94230 Cachan
France
abdellaoui@grid.ensam.estp.fr

Prof. Mark J. Machina
University of California
Department of Economics
San Diego
La Jolla, CA 92093-0508
USA
mmachina@ucsd.edu

Prof. R. Duncan Luce
University of California
2133 Social Sciences Plaza A
5100 Irvine, CA 92697
USA
rdluce@uci.edu

Prof. Bertrand Munier
Maison de la recherche de l'ESTP
GRID
30 ave du Président Wilson
94230 Cachan
France
munier@grid.ensam.estp.fr

Library of Congress Control Number: 2007928747

ISSN 0924-6126
ISBN 978-3-540-48934-4 Springer Berlin Heidelberg New York

Springer is a part of Springer Science+Business Media

springer.com

Typesetting and production: LE-TEX Jelonek, Schmidt & Vöckler GbR, Leipzig, Germany
Cover design: WMX Design GmbH, Heidelberg

Spin 11924630 Printed on acid-free paper 43/3180/YL - 5 4 3 2 1 0

Contents

Introduction

The concept of uncertainty has much evolved since F. Knight wrote his seminal book on *Risk, Uncertainty and Profit*. Economists have generally reduced the concept to the idea that no probability was available, as opposed to the case of risk. What Knight meant might have been sensibly different: Was not uncertainty the case where probability could not be defined with precision, where there was no consensus measure? In the 1920s, such an imprecision was often sufficient to make any corresponding amounts at stake uninsurable.

Insurance companies, fortunately for us, have since then widely changed their minds as to what can be insured and what cannot be. It would have been amazing and detrimental if researchers had not changed their minds. Already in the 1930s, J. M. Keynes felt the need to deal with another sort of uncertainty. In chapter XII of *The General Theory*, *uncertainty* is defined in a more radical way: it is the situation where *we just don't know*.

The literature about uncertainty deals with different levels of investigation: the neuronal level, giving an account of brain activity; the cognitive level, assessing the role of mental procedures; and finally the choice representation level. At each level uncertainty is modelled in its own way, and that defines rationality in specific ways.

The neuronal level, considering the system of neurons in the brain, as in the newly emerging field of neuroeconomics – which has come to be called *neuronomics* – though promising, has not yet generated enough formal output to be considered in this volume. The reader of this book will rather explore, in virtually each part of the book, the cognitive level, where sources and examinations of *mental* procedures, conjectures and solutions are dealt with, much like in traditional cognitive psychology. Other authors of this book have selected the level of choices, as is more traditional in economics and management science. This last level is the most widely dealt with here. Of the three known levels of study – *neuronal, mental* and *choice pragmatic* – this book focuses on the last two.

At each level of analysis, a researcher has several *tools* at his or her disposal. It is legitimate to rely on previously accepted logical schemes and to develop their consequences formally or to explore the possibilities that they open to rational behavior. At a less general level, decision theoretic and game theoretic tools offer similar ways to proceed. Both types of tools are here characterized as *formal.* A more and more frequent way to proceed, both in psychology and more recently in economics, consists in designing and performing various types of experiments. Such approaches also appear in the present book.

As a result of the newer approaches, today we recognize that uncertainty can be either deeper than the situation described by F. Knight or, in the opposite direction, less radically ignorant of what might happen. Between complete ignorance of the possible futures and mere ambiguity about the probabilities, there is a wide array of different types of uncertainty.

What form do these types take, under which type of conditions, and how can we manage to reach decisions in each type of such really difficult situations? This is the very the topic of the present book.

The papers assembled were given some at the FUR XI conference, organized at GRID (Cachan and Paris, France) in the second half of 2004. They have been selected from some 175 papers, refereed once again, further revised and selected again to form the present set of contributions.

The book is organized in *four parts*: foundational, representational tools, alternative decision rules, and risk attitude modelling.

Part One: Foundations

S. Grant and J. Quiggin propose to explore uncertainty using the language of logic and decision trees, as is often done in artificial intelligence. Needless to say, such an approach opens new avenues of impressive research: The link between their representation and the seminal Max-Min utility model of I. Gilboa and D. Schmeidler (section 9 of the paper) is one of the most fascinating ones.

M. Amarante and F. Maccheroni, in a formal mathematical development, show that a connection to the same seminal model can be found in the very idea that several probability measures have to be considered simultaneously.

J.V. Howard connects the mental and the choice pragmatic levels through the formal representation of finite event trees, thus yielding a new foundation to Bayesian statistics, which assumes, as is well known, the less realistic assumption of countable additivity over a σ-field of events.

Finally, E. Borgonovo and L. Peccati show that, under some relatively modest hypotheses about the structure of the set of the possible states of the world, one can use Sobol's theorem to determine the impact of what they call "parameter uncertainty" on the level of performance of the decision, as evaluated through the relevant utility functional. They have in mind the epistemic notion of uncertainty. They recommend using simulation as a way to let the decision maker concentrate on the quest for information that appears the most important to acquire as a result of this procedure.

Part Two: The Importance of Representational Tools in Understanding Behavior Under Uncertainty and Risk.

A. Guerdjikova argues, on the basis of an experiment, that when diversifying portfolios, the issue is less about whether or not EU or non-EU type of rational behavior obtains than about the use of similarity considerations. The paper offers a generic connection to another tradition derived from artificial intelligence, namely case-based reasoning, specifically linking to the model derived by I. Gilboa and D. Schmeidler within this tradition.

H. Haller and Sh. Mousavi, in a formal development, show that uncertainty can, under given hypotheses, improve the welfare reached in a Second Best situation such as generated by adverse selection market equilibrium. In the insurance market, the Rothschild and Stiglitz model with adverse selection is used to establish the claim of the authors.

E. Camacho-Cuena and Ch. Seidl investigate experimentally the violation of Lorenz relations in the treatment of an income distribution or of an individual multiple outcome lottery. They show that the nature of responses that are requested from the subjects is the key variable. Merely invoking a framing effect either provides an insufficient explanation, or a quite imprecise one.

B. Sopher and A. Sheth also investigate experimentally inter-temporal choice rationality. By using their design in a variety of cases that have different initial periods, levels of discounting, types of discounting, number of periods, etc. they show that exponential discounting is the clear modal choice pattern of behavior in virtually all cases, even though the tendency toward hyperbolic discounting increases when the compounding rate increases. Their investigation thus confirms the latter point, which has already been found in other samples with other protocols, and seems therefore a rather robust result.

Part Three: The Assessment of Several Alternative Decision Rules

The first alternative rule, the focus of R.M. Hogarth and N. Karelaia, concerns 'simple heuristics that make us smart', paraphrasing the title of the book by Berlin psychologists G. Gigerenzer and P.M. Todd. They examine rules of choice between binary cues (which they emphasize as a limitation of their work). They argue that decision rules succeed according to two factors – aside from error – which they identify as characteristics of choice sets: one is the number of binary cues in the set and the presence or not of a dominance situation; and the other is the way that cues are weighted. The structure of choice sets (also called 'the environment' by the authors) may be separable or compensatory: the more separable, the more effective the simplest heuristics (like "take the best"), the more compensatory the environment, the better performing are the more complex models, like hybrids of different simple cues. But error in the environment makes the predictive ability of any model less and less satisfactory.

J.N. Bearden and R.O. Murphy examine rather sophisticated decision rules that may govern search behavior in the well-known "secretary problem", which they generalize under the name of GSP ("Generalized Secretary Problem"). They show that the existence of suboptimal search behavior (too few rounds of investigation) can be accounted for by a stochastic component in the search policy. To this bias, they oppose the optimal search induced by a dynamic programming procedure which they define and present. They manage to give tentative psychological explanations for the possible biased stopping rules, although they admit that working on what has been called here the – observable – *choice pragmatic* level of investigation is not easy to interpret in terms of the – unobservable – *mental* level.

N.P. Thomas offers an interesting contribution to the literature on collective choice. Using Monte-Carlo simulations he tests two alternative MCDM procedures:

(a) Either each individual evaluates the alternatives at hand; some voting procedure being then started, based on the global scoring of each individual,
(b) Or a voting procedure is organized first on the relevance of each attribute and one can then design a group preference ordering using the attributes selected by the vote.

The paper shows that the second type of procedure can be superior to the first type in cases where value conflicts have emerged in the group of decision makers, whereas, in the other cases, the first procedure leads to a higher welfare.

Part Four: Models of Risk Attitudes Modelling and Methodological Issues

E. Paté-Cornell examines the relationships between the methods of probabilistic risk analysis (PRA), derived from engineering, and those of decision analysis (DA), mainly the expected utility tradition derived from economics and the social sciences. The interest of this paper lies in the experience the author has in both domains, especially PRA. Two PRA cases, taken as benchmarks, make the comparison possible. One is the case of the shuttle's PRA, the other is the case of terrorism prevention. The frequentist and Bayesian concepts of probability are examined. The main conclusions she draws from her analysis are that the risk analyst has to be less specific than the analyst helping the decision maker to actually make a decision, because PRA is, in general, developed before the final decision maker, who will be relying on the PRA model, has been precisely determined. This discussion is quite fascinating and might be echoed in various environments.

H. Grossmann, M. Brocke and H. Holling design a procedure to induce preferences for multi-attribute options. The paper relates finite conjoint measurement and multi-attribute utility analysis. The idea is to set up a computer-based procedure leading the participants gradually to order a given finite set of (multiattribute) alternatives using a single (weak) order. The paper does two things: it shows first that the qualitative information represented by the weak order is sufficient to determine a unique set of numerical utilities; it shows then that the procedure, described in the paper is effective, i.e. that preferences are effectively changed w.r.t. their previous state by the computer interactive program. Two experiments lead to that conclusion.

Odilo W. Huber builds on a line of thought to which he has contributed since the mid-Nineties. The idea is that probabilities are not always actively searched for by corporate executives. Z. Shapira already pointed out some time ago that managers tend to believe that they can change the odds and "get around" issues of risk, although that statement needs some qualification. Precisely, Huber reports some experiments on the topic. They lead to the conclusion that mental representations differ among subjects and among tasks, on one hand; and on another hand, that probabilities which have been actively searched for are better recalled than is pre-existing information.

J. Sounderpandian gives first a quite original survey of the literature on the evolution of risk attitudes and on the diverse approaches which have been taken by researchers on this topic. He then introduces the idea of studying such evolution using simulations, provided that the society under investigation is not too "large" and provided, of course, that individual risk attitudes in that society are interrelated. The paper goes on to derive some understanding of the profile and the fate of a society from this perspective.

Oswald Huber completes part four, as well as the volume, by focusing on the concept of risk-defusing operators (RDOs). He asks whether the practical existence of such RDO's can impact on the decision process and in particular on the search for information. The paper summarizes several experiments indicating that there is, indeed, such an impact. Although extreme interpretations given to the phenomenon are to be questioned (based on the fact that RDO does not get rid of all risk), there are in these findings some undoubtedly important and interesting topics to be further studied by future research.

In summary, this set of papers finds sources of ideas in quite a few disciplines contributing to decision theory. We hope that the reader will also agree that these ideas exhibit a rather unusual degree of originality. May readers enjoy reading this volume.

Part I

Foundations

Conjectures, Refutations and Discoveries: Incorporating New Knowledge in Models of Belief and Choice under Uncertainty

Simon Grant[1] and John Quiggin[2]

[1] Department of Economics, Rice University and Australian National University
e-mail: sgrant@rice.edu
[2] School of Economics, University of Queensland, Australia
e-mail: j.quiggin@uq.edu

Abstract. The purpose of this paper is to develop a model of choice under uncertainty in which individuals do not possess a complete description of the space of states of the world, and in which this description evolves over time. The crucial analytical tool is the description of knowledge in terms of a finite set of propositions.

Keywords: unforeseen contingencies, incomplete state-space, propostions

1 Conjectures, Refutations and Discoveries: Incorporating New Knowledge in Models of Belief and Decision under Uncertainty

In any complex decision problem, the usefulness of formal decision procedures is limited by the knowledge that, decisions commonly proved unsatisfactory because of the occurrence of contingencies that were unforeseen, and perhaps unforeseeable, at the time the decisions were taken. This problem is particularly severe in relation to complex environmental problems such as global warming and the sustainable management of large ecosystems.

One popular answer to the question of sustainable design is the 'precautionary principle', namely, that where there is a serious, but unproven, possibility of environmental damage arising from some action or inaction, policy should be designed on the assumption that the risk is in fact real. The opposite position, which may be described as the 'permissive principle', is one which suggests that, in the absence of conclusive proof of danger, the proposed activities of firms and individuals should be given the benefit of any doubt.

Under standard approaches to decision theory, both of these alternatives are rejected in favour of a model in which all possible events (or 'states of nature') are described in advance and assigned a subjective probability. The preferred course of action is the one that maximises the expected return, which may be expressed in monetary terms, or, more generally, in terms of expected utility.

From the decision-theoretic perspective, a strong interpretation of the precautionary principle leads to adoption of a maximin rule, which is excessively conservative and leads to poor average outcomes. Weak interpretations do no more than assert that action should not require full scientific proof of dangers, which is the same as the standard decision-theoretic view. Hence, from the usual decision-theoretic viewpoint, the precautionary principle is either wrong or a restatement of the obvious.

Although highly effective in many contexts, the standard decision-theoretic model has long been criticised for its inability to deal with events for which well-defined probabilities are not available and, even more, for problems where not all possible outcomes can be foreseen. The difficulties associated with the first of these problems were described by Ellsberg (1961) and remained unresolved for many years. Recent work, including that of Epstein (1999), Ghirardato and Marinacci (1999) and Grant and Quiggin (2002d) has resulted in the development of improved characterisations and analytical tools.

The purpose of this paper is to develop a model of choice under uncertainty in which individuals do not possess a complete description of the space of states of the world, and in which this description evolves over time. The crucial analytical tool is the description of knowledge in terms of a finite set of propositions.

2 Epistemology

The problem posed for theories of choice under uncertain by the existence of unforeseen contingencies has been widely recognised. The term 'unknown unknowns', recently used in (widely-derided) remarks by US Defense Secretary Donald Rumsfeld, is commonly used to describe such contingencies. Since all formal theories of decision under uncertainty in widespread use at present rely, implicitly or explicitly, on the availability of a complete description of the state-contingent consequences of actions under consideration, the existence of unforeseen contingencies represents a serious difficulty.

Two main approaches have been adopted. The first has been to choose some form of maximin rule. Such rules have been proposed in a wide range of contexts and arise as a polar case in many different models. In expected utility theory, for example, maximin arises as the polar case for the class of concave utility functions (normally referred to, in this context, as risk-aversion). In rank-dependent models, it is the polar case for a convex probability weighting function (pessimism). Since maximin also arises as a polar case for other models of choice under uncertainty, it is, therefore, not always clear whether a maximin rule is being proposed as a response to lack of knowledge or reflecting an extreme aversion to risk.

The second approach, more directly related to the problem of unforeseen contingency, has been to augment the state space with some form of residual event, aimed at capturing the existence of radically incomplete knowledge. However, it turns out to be quite difficult to implement this approach in such a way that uncertainty about the residual event is non-trivially different from uncertainty about the known set of states of nature.

Both approaches share with expected utility the fact that normative and positive models of choice are derived simultaneously. In the most common approach, the functional form of the model are derived from a set of axioms that are held to be both normatively compelling and descriptively realistic.

It is not strictly necessary, in this approaches, to combine normative and positive models. Some modellers are interested only in description, or only in prescription. Others hold that while some particular set of axioms, typically those of expected utility, is normatively compelling, other axioms yield a more realistic model of observed behavior. Nevertheless, descriptive models are typically constructed in such a way that they can, if desired, be treated as normative models for use by decision-makers. In particular, such models typically refer only to information that is available to decision-makers.

The feature of the state-act model that yields the close fit between normative and positive models is, in the terminology of the rational expectations literature, model-consistency. The information on which individuals are assumed to base decisions is, broadly speaking, the same as that used to model those decisions.

The model-consistent act-state approach has yielded important insights. However, it is inherently unsatisfactory for the case of incomplete knowledge. An adequate external description of the behavior of an individual with radically incomplete knowledge must employ some notion of complete (or at least more extensive) knowledge than that possessed by the individual being described.

Closely related to this is the problem of learning. The concept of learning that has been analyzed most extensively in the literature on decision theory is that of Bayesian updating which is, in a crucial sense, a negative form of learning. The Bayesian decision-maker begins with a prior probability distribution over all states of the world. The occurrence of a particular event amounts to news that a particular subset of states, those making up the complementary event, are no longer possible. Hence the probabilities of these states must be set to zero, while the probabilities of the states that make up the observed event are replaced by their event-conditional probabilities.

If learning in the ordinary sense of the term is to be modelled, it must be possible to represent additions to, as well as subtractions from, the set of possibilities considered by the individual. A natural way of doing this, once the postulate of model-consistency is dropped, is to suppose that, at any given time, the individual has access to a proper subset of some global set of possibilities.

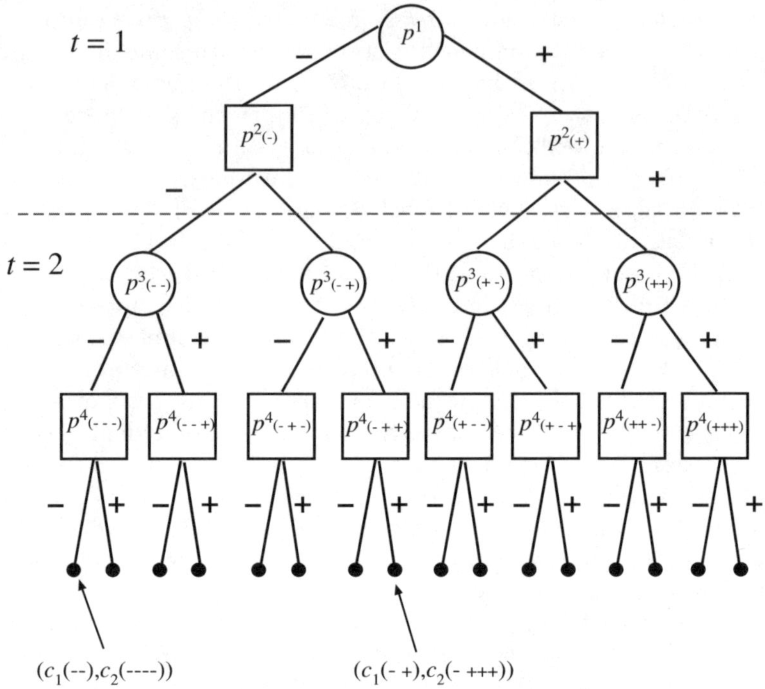

Fig. 1. Decision tree with full information

In this paper, we will argue that, rather then seeking to work directly with a generalized concept of the state space, it is preferable to consider a specification of knowledge in terms of propositions, and to postulate that, at any given time, individuals are equipped to consider only a finite subset of a potentially infinite set of such propositions. The propositional approach may be related back to the state space approach, since any state of nature is characterised by the set of propositions that are true in that state.

3 Example

Consider a standard decision tree with two decision nodes and two chance nodes, as illustrated in Fig. 1. As illustrated there are chance nodes (decisions by Nature) at $t = 1$ and $t = 3$, and decision nodes at $t = 2$ and $t = 4$. Thus there are a total of $2^4 = 16$ possible terminal nodes. We will assume that payoffs are received at times $t = 2$ and $t = 4$ after decisions are made. At each node, we will denote a move to the left by 0 $(-)$ and a move to the right by 1 $(+)$.

This representation is natural for a fully informed decision-makers or outside observers. In decision theory, however, we are normally concerned with decision-makers who are not fully informed. In Fig. 2 we illustrate the case

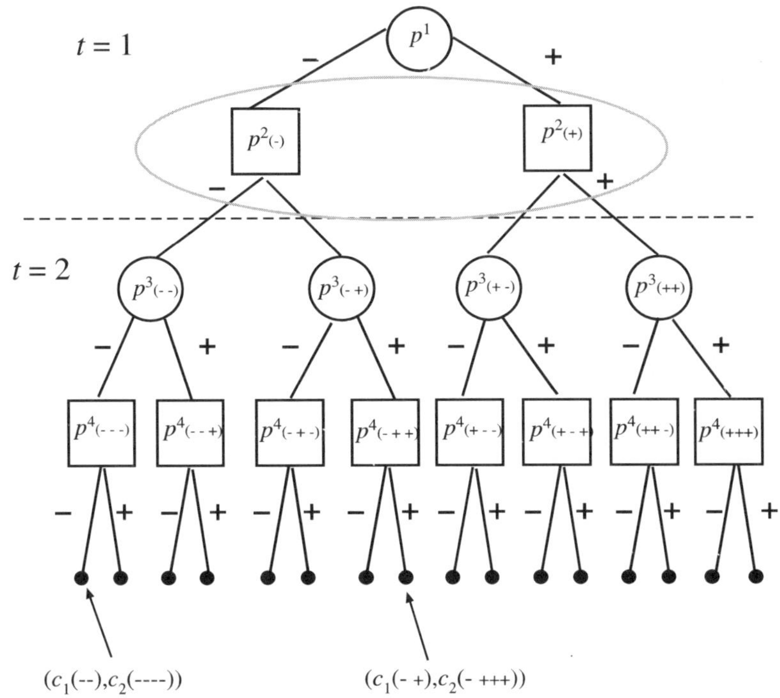

Fig. 2. Decision tree with state contingent uncertainty

where the decision at $t = 2$ must be taken without knowledge of the act of Nature at $t = 1$ ...

We now consider a more radical form of uncertainty, in which the decision at $t = 4$ is taken without the decision-maker even being aware of the chance node at $t = 3$. Thus, the decision-maker cannot distinguish between the nodes 000 and 001, 010 and 011.

In these circumstances, the decision-maker not only does not know what outcome will arise if, say, decision 1 is taken at $t = 4$, but also does not possess a complete state-contingent description of the possible outcomes.

In considering problems of this general kind, the main focus of attention has been on the question of whether, given sufficient information on preferences over decisions, it may be possible to infer a state-contingent model consistent with those preferences. The most promising approach begins with the work of Kreps (1992) and has been developed by Dekel, Lipman and Rusticchini (2001) and Epstein and Marinacci (2006). For these purposes it is more convenient to adopt the logically equivalent representation in Fig. 3 in which the choice node at $t = 3$ is eliminated, and attention is focused on the coarsely specified consequences of the decision at $t = 4$.

The focus of this paper is very different. We are primarily concerned with describing the structure of beliefs like those illustrated in Fig. 4 and the

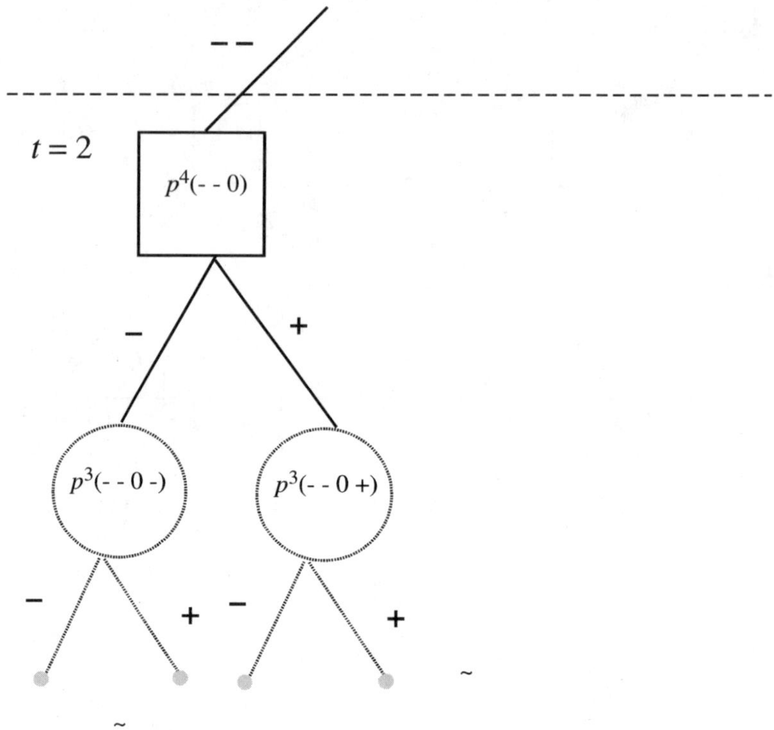

Fig. 3. Decision tree with unreconised contingencies

way in which such beliefs may evolve over time, in the light of both acts of nature and decisions taken by individuals. The natural way to represent both decisions and beliefs, we claim, is in terms of binary propositions. That is, in terms of Figs. 1–4 we impose the (previously implicit) restriction that both chance and decision nodes should have exactly two branches.

4 Propositions

Let the set of states of the world be Ω. We focus on the representation

$$\Omega = 2^{\mathbf{N}},$$

where $\mathbf{N} = \{1, 2, \ldots n, \ldots\}$ is supposed to be a finite or countably infinite set, indexing a family of 'elementary' propositions $p^1, p^2 \ldots p^n \ldots$ about the world. Each proposition is a statement such as 'The winner of the 2008 US Presidential election is Hillary Clinton'. An exhaustive description of the state of the world therefore consists of an evaluation of each of the propositions $p^n, n \in N$. As will be shown in more detail below, the elementary propositions may be used to generate a larger set of propositions $\mathbf{P}$.

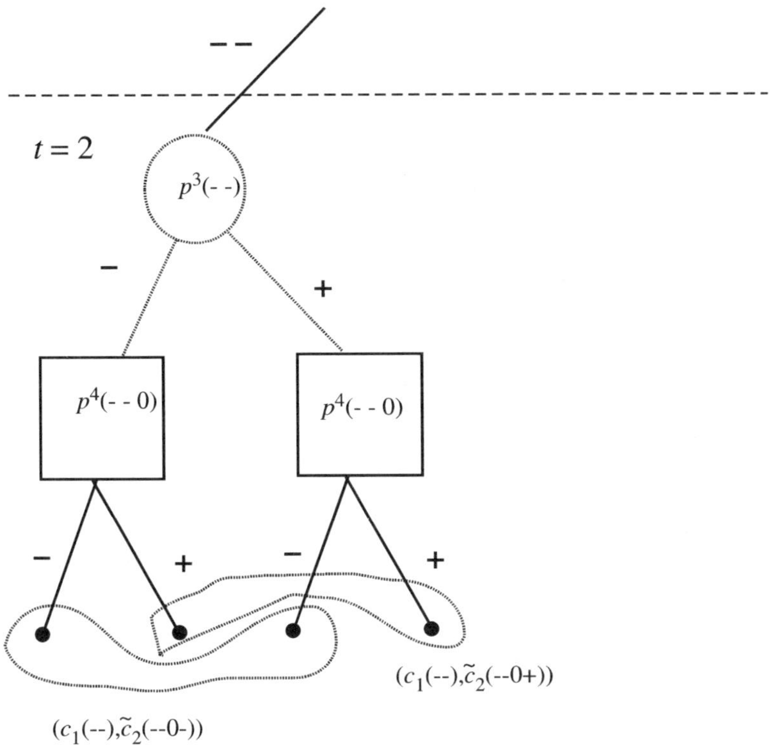

Fig. 4. Decision tree with coarse contingencies

With each proposition and each possible state of the world, a fully informed observer can associate a truth value t^n, which will be denoted 1 (True) or -1 (False). From the viewpoint of a fully informed observer, any state of the world can therefore be described by a real number $\omega \in \Omega \subseteq [0,1]$ [1], given by

$$\omega = \sum_{n \in \mathbf{N}} 2^{-(n+1)} \left(t^n + 1\right) .$$

An elementary proposition p^n is true in state ω if and only if $\omega_n = 1$, where $\omega_n \in \{0,1\}$ is the nth element in the *binary* expansion of ω. Note that, since the mapping $p^n(\omega) = \omega_n$ is defined from the viewpoint of a fully informed observer, the truth value $p^n(\omega)$ does not vary over time.

From this external viewpoint of the model any proposition p^n corresponds to a event $E_n \subseteq \Omega$. More precisely we have

[1] If some propositions may be true in all states of the world, Ω may be a proper subset of $[0,1]$. Alternatively, Ω may be set equal to $[0,1]$ with some states having zero probability in all evaluations.

$$E_n = \{\omega \in [0,1] : \omega_n = 1\} \subset \Omega\,.$$

4.1 Decision-Makers and Decisions

Decision-makers are finitely rational individuals who are not, in general, able to formulate all the propositions in $\mathbf{P}$, or even the elementary propositions $p^n, n \in \mathbf{N}$ and therefore not able to give an exhaustive specification of the state space. We will assume more concretely that, at time t, each individual i is able to conceive a finite set of propositions $\mathbf{P}_t^i$, all of which are generated by a set of elementary propositions $p^n, n \in \mathbf{N}_t^i$ which will be derived below. Note that the elements of the set E_n are not in general, accessible to a decision-maker, even if the proposition p^n is accessible. More generally, proper subsets of E_n are not in general, accessible to a decision-maker.

Example 1. Suppose that the elements of $\mathbf{N}$ are two statements about possible winners of the Melbourne Cup which is a horse race that is run in Melbourne, Australia on the first Tuesday after the first of November. [The winner in 1861 was Archer. The defeated favourite in 1931 was Phar Lap.]

p^1: The winner of the 1861 Melbourne Cup is Archer
p^2: The winner of the 1931 Melbourne Cup is Phar Lap

A decision-maker in October 1861 might be expected to have beliefs about p^1 but not about p^2. However, from the external viewpoint, we have

$$E_1 = \{10, 11\}$$

so that any state of the world consistent with p^1 gives a truth value to p^2.

Decisions are modelled by allowing the decision-maker to control (at time t) the truth value of some proposition. A decision is, therefore, the act of determining the truth value of a proposition. In the example above, we might consider elementary propositions such as p^3: Decision-maker i bets on Archer, and p^4 : Decision-maker i bets against Archer. We will denote by $\mathbf{\Delta}_t^i \subseteq \mathbf{N}_t^i$ the set of elementary propositions decidable by decision-maker i at time t.

Note: We need to consider whether a decision-maker can fail to decide on an element of $\mathbf{\Delta}_t^i$ at time t and if so how to represent this.

4.2 Compound Propositions

The individual can also consider compound propositions p. A compound proposition is derived by assigning truth values of 1 or -1 to all p^n where n is a member of some (possibly empty) subset $\mathbf{N}(p) \subseteq \mathbf{N}$, leaving all p^n, $n \subseteq \mathbf{N}(p)$ unconsidered. The set $\mathbf{N}(p)$ is referred to as the *scope* of p, and is the disjoint union of $\mathbf{N}_-(p)$, the set of elementary propositions false under p, and $\mathbf{N}_+(p)$, the set of elementary propositions true under p. The simple proposition p^n has scope $\mathbf{N}(p^n) = \{n\}$. We define the null proposition $p^{\emptyset}$ such that $p^{\emptyset}_n = 0, \forall n$ and do not assign a truth value to $p^{\emptyset}$.

Any (non-null) compound proposition p corresponds, from the external viewpoint, to an event

$$E_p = \{\omega \in [0,1] : \omega_n = 0, \forall n \in \mathbf{N}_-(p)\,;\ \omega_n = 1, \forall n \in \mathbf{N}_+(p)\} \subset \Omega\,.$$

We set

$$E_{p^{\emptyset}} = \emptyset\,.$$

A compound proposition p is true in state ω if $\omega \in E_p$ (that is, if $\omega_n = 0, \forall n \in N_-(p)$; $\omega_n = 1, \forall n \in N_+(p)$) and false otherwise. We denote the truth value of proposition p in state ω by $t(p;\omega)$. That is,

$$t(p;\omega) = \left\{ \begin{array}{l} 1 \text{ if } \omega_n = 0, \forall n \in \mathbf{N}_-(p)\,; \omega_n = 1, \forall n \in \mathbf{N}_+(p) \\ 0 \text{ otherwise} \end{array} \right\}.$$

A numerical representation of compound propositions is possible using ternary numbers, where the value 0 denotes 'not considered'. Denote the truth value of proposition p^n under p by $p_n \in \{-1, 0, 1\}$.

As already noted, certain propositions are under the control of decision-makers. The set of all decisions available to decision-maker i at time t is denoted $\mathbf{D}^i_t$. Without loss of generality, we will assume that all elements of $\mathbf{D}^i_t$ are compound propositions derived from elementary decisions, that is,$\mathbf{D}^i_t \subseteq \{-1,0,1\}^{\mathbf{\Delta}^i_t}$. Since some combinations of elementary decisions may be inconsistent or unconsidered, we do not assume that $\mathbf{D}^i_t = \{-1,0,1\}^{\mathbf{\Delta}^i_t}$.

A given decision/action may jointly determine the value of a number of propositions - most obviously if the value of a compound proposition p is set to 1, this determines the truth value of all the elementary propositions in $N(p)$, and of any compound propositions derived from these elementary propositions. Not all of these compound propositions are necessarily accessible to the decision-maker. So we want a category of 'conscious action', roughly, a decision-maker i consciously acts to determine proposition p at time t if $p \in P_{it}$ and the action of decision-maker i at time t determines the truth value of p.

4.3 Classes of Propositions

The class of all propositions in the model is denoted by $\mathbf{P} = \{-1,0,1\}^{|\mathbf{N}|}$. It is useful to consider more general classes of propositions $P \subseteq \mathbf{P}$. To any class of propositions P, given state ω, we assign the truth value

$$t(P;\omega) = \sup_{p \in P} \{t(p;\omega)\}\ .$$

That is, P is true if any $p \in P$ is true and false if all $p \in P$ are false. In terms of the logical operations defined below, the set P has the truth value derived by applying the OR operation to its members.

5 Logical Operations from the External Viewpoint

From the external viewpoint, the usual logical operations are available with the standard set-theoretic interpretation. It is usual in decision theory to focus on the set theoretic interpretation and, from the external viewpoint the two are isomorphic. But the propositional interpretation is more satisfactory when describing a decision-maker with only partial awareness.

5.1 Implication

The implication relationship $p \rightarrow p'$ holds if and only if

$$p'_n \in \{-1,1\} \Rightarrow p'_n = p_n \, .$$

That is, $p \rightarrow p'$ if and only if any elementary proposition p^n that is true (false) under p' is also true (false) under p.

The implication relationship is

(i) reflexive $p \rightarrow p$,
(ii) transitive $p \rightarrow p'$ & $p' \rightarrow p'' \Rightarrow p \rightarrow p''$,
(iii) anti-symmetric $p \rightarrow p'$ & $p' \rightarrow p \Rightarrow p = p'$.

Observe that $p \rightarrow p^{\emptyset}, \forall p$ and that $p \rightarrow p'$ if and only if $E_p \subseteq E_{p'}$.

With each proposition p, we can associate the class of propositions

$$[p] = \{p' : p' \rightarrow p\} \, .$$

That is, $[p]$ is the class of propositions stronger than p. For an elementary proposition p^n,

$$E_{[p^n]} = E_{p^n} \, .$$

More generally, for any class P of propositions we define $[P]$,

$$[P] = \{p' : \exists p \in P, p' \rightarrow p\} \, .$$

Observe that

$$E_P = E_{[P]} \, .$$

We refer to $[P]$ as the *completion* of P and say that P is *complete* if $P = [P]$.

5.2 Consistency and Logical Independence

Two propositions p and p' are consistent, denoted $p \sim p'$ if there exists p'', $p'' \rightarrow p$ and $p'' \rightarrow p'$. The consistency relationship is reflexive and symmetric, but not transitive. To illustrate the latter point informally, note that the proposition 'Hillary Clinton is the winner of the US Presidential election in 2008' is consistent with 'George Bush is the winner of the US Presidential election in 2004' which in turn is consistent with 'Hillary Clinton is not the winner of the US Presidential election in 2008', but the first and third propositions are inconsistent.

The following lemma (proof left to the reader) characterises consistency in terms of the ternary representation used above:

Lemma 1. *For any p, p', $p \sim p'$ if and only if for all n, such that $p_n \in \{-1, 1\}$ either $p'_n = p_n$ or $p'_n = 0$.*

With each proposition p, we can associate the class of propositions

$$\langle p \rangle = \{p' : p' \sim p\} .$$

More generally, for any class P of propositions, we define

$$\langle P \rangle = \{p' : \exists p \in P, p \sim p'\} .$$

Observing that $[P] \subseteq \langle P \rangle$, we define the set of propositions logically independent of p as

$$\rangle P \langle = \langle P \rangle - [P] .$$

Conjecture $\langle\langle P \rangle\rangle = \langle P \rangle$, $\rangle\rangle P \langle\langle = [P]$.

5.3 OR and AND

For any two classes of propositions, P and P', define

$$P \vee P' = [P] \cup [P'] ,$$
$$P \wedge P' = [P] \cap [P'] .$$

Observe that

$$E_{P \vee P'} = E_P \cup E_{P'} ,$$
$$E_{P \wedge P'} = E_P \cap E_{P'} .$$

The distributive laws apply to $\vee$ and $\wedge$. Moreover, for the set of complete classes of propositions $\vee$ and $\wedge$ define a lattice structure.

5.4 Negation

The final logical operation to be considered is that of negation. Define:

$$\neg P = \mathbf{P} - \langle P \rangle .$$

That is, the negation of P is the set of propositions inconsistent with all elements of P.

Conjecture $[\neg P] = \neg P$.

For any elementary proposition p^n,

$$\neg [p^n] = [\neg p^n] \ ,$$

where $\neg p^n$ is true if and only if p^n is false. More formally, $\neg p^n$ is a proposition having the value $\neg p^n_m = -1, m = n, \neg p^n_m = 0$ otherwise.

We can see this by observing that

$$\begin{aligned}
[p^n] &= \{p : p_n = 1\} \ , \\
\langle [p^n] \rangle &= \{p : p_n = 1\} \cup \{p : p_n = 0\} \ , \\
[\neg p^n] &= \{p : p_n = -1\} \\
&= \mathbf{P} - \langle [p^n] \rangle \\
&= \neg [p^n] \ .
\end{aligned}$$

More generally, for any p

$$\neg [p] = [\neg p] \ ,$$

where

$$\neg p = \{p' : \exists n, \text{ s.t. } p_n p'_n = -1\} \ .$$

The following lemma (proof left to the reader) characterises consistency.

Lemma 2. *The negation operation has the following properties*

$$\begin{aligned}
E_{[p^n]} \cup E_{[\neg p^n]} &= \Omega \, , \\
[P] &= \neg [\neg P] \ , \\
\rangle P \langle &= \langle P \rangle \cap \langle \neg P \rangle \ , \\
\rangle P \langle &= \rangle \neg P \langle \ , \\
\mathbf{P} &= [P] \cup [\neg P] \cup \rangle P \langle \ .
\end{aligned}$$

Note that the sets making up the union in the last line are mutually disjoint.

6 The Decision-Maker's Viewpoint

The class of all propositions considered by individual i at time t is denoted P^i_t. The scope of the individual's proposition set is given by

$$\mathbf{N}^i_t = \cup_{p \in P^i_t} \mathbf{N}(p) \ .$$

For a given set P^i_t, the *definitely false set* is given by

$$\mathbf{N}^{i-}_t = \cap_{p \in P^i_t} \mathbf{N}_-(p)$$

and the *definitely true set* by

$$\mathbf{N}_t^{i+} = \cap_{p \in P_t^i} \mathbf{N}_+ (p) \ .$$

These sets characterise the elementary propositions that are true (false) for every element $p \in P_t^i$. Combining these yields the *characterising proposition* $\underline{p}_{it}$

$$\underline{p}_{tn}^i = \begin{cases} -1 & n \in N_t^{i-} \\ 1 & n \in N_t^{i+} \\ 0 & \text{otherwise} \end{cases} \ .$$

We assume that $\underline{p}_{tn}^i \in P_t^i$.

The *set of active possibilities* is given by

$$N_t^{i*} = N_t^i - \left(N_t^{i-} \cup N_t^{i+}\right) \ .$$

Thus, $n \in N_t^{i*}$ if and only if there exist $p, p' \in N_t^i$ with $p_n \neq p'_n$. Recall that p_n can take the three values $0, 1, -1$.

6.1 Logical Operations for the Decision-Maker

Logical operations for the decision-maker are applied with respect to the set P_t^i and may be derived with reference only to propositions $p \in P_t^i$[2]. Thus, for any $P, P' \subseteq P_t^i$

$$\begin{aligned} [P]_t^i &= \left\{p' \in P_t^i : \exists p \in P_t^i, p' \to p\right\} , \\ P \vee_t^i P' &= [P]_t^i \cup [P']_t^i , \\ P \wedge_t^i P' &= [P]_t^i \cap [P']_t^i , \\ \langle P \rangle_t^i &= \langle P \rangle \cap P_t^i , \\ \neg P_t^i &= P_t^i - \langle P \rangle_t^i \ . \end{aligned}$$

7 Changes in Knowledge

In the model set out above, there are four possible states of knowledge for individual i at time t about an elementary proposition p^n

[2] Spelling out the definition of $\langle P \rangle_{it}$ given blow in terms of the definition of $\langle P \rangle$ would require references to propositions that are not in P_{it} which seems unsatisfactory. However, using the direct characterisation of consistency in terms of the ternary truth values, it is possible to derive $\langle P \rangle_{it}$ without reference to unconsidered propositions.

(i) (believed to be) impossible, $n \in \mathbf{N}_t^{i-}$,
(ii) (believed to be) certain, $n \in \mathbf{N}_t^{i+}$,
(iii) active possibility, $n \in \mathbf{N}_t^{i*}$,
(iv) not under consideration $n \in \mathbf{N}_t^{i0} = \mathbf{N} - \mathbf{N}_t^i$.

A crucial feature of the model proposed here is that knowledge can change over time, say from period t to $t+1$ in several different ways. First, some elementary proposition p^n, under consideration at time t, may be verified or falsified by experience at time $t+1$. For the case when p^n is an active possibility at time t, this is analogous to the observation of data in a Bayesian model. However, we allow for the possibility that a proposition treated by the decision-maker as impossible may be verified in period $t+1$ or vice versa.

Next, the state of knowledge may change as a result of inference. For example, the truth value of compound propositions may change as a result of information about elementary propositions. In addition, as will be discussed below, beliefs about active possibilities may be updated in the light of changes in knowledge. The canonical example of such updating is the Bayesian inference procedure in which a posterior distribution is derived from a prior distribution following the observation of data.

The most important, and novel, case treated in the model proposed here is that when a proposition that was previously not under consideration is either verified by experience or becomes an active possibility as a result of inference. Informally, at least, we may distinguish several processes by which this may take place. Surprises arise when an unanticipated event occurs, independently of the actions of the decision-maker, so that some previously unconsidered proposition is verified or falsified. Discoveries are similar, but arise from events that are not fully anticipated, but result from purposive thought and experiment on the part of the decision-maker[3]. Conjectures arise when a previously unconsidered proposition becomes active, typically as a result of formal or intuitive inference.

Symmetrical with the process by which new propositions come under consideration are processes of forgetting, by which propositions previously under consideration cease to be so. Given the finite capacity of human minds, it is reasonable to suppose that, on some appropriate measure of information content, the size of the set of propositions under consideration by any individual remains roughly constant over time. If this measure is approximately equal to the number of elementary propositions under consideration, then the number of propositions forgotten should be equal, on average to the number acquired through discovery and related processes.

[3] At this stage in the project, actions have not been modelled explicitly, and therefore the distinction between surprises and discoveries must remain informal.

8 Inference, Conjecture and Refutation

8.1 Inference

One standard form of discovering new propositions, first considered in formal terms by the ancient Greeks, is that of logical inference. If $p, p' \in P_t^i$ it is natural, in a normative framework, to postulate that $p \vee p'$ and $p \wedge p'$ should be available for inclusion in $P_{(t+1)}^i$ and, further that, if p and p' both have belief values 1 or 0, standard truth-table techniques should apply to determine the belief values of derived propositions such as $p \vee p'$. In a descriptive framework, we must be more cautious. It is well-known that individuals commonly fail to derive all the logical consequences of their beliefs. Furthermore, as the ancient Greek logicians observed when they created lists of common fallacies, individuals frequently attribute incorrect or unjustified beliefs to derived propositions. Nevertheless, the formulation of Π_t^i should give a high probability to the derivation of logical inferences, at least for intelligent individuals with some formal or informal training in logical reasoning.

8.2 Popper and Lakatos on Conjectures and Refutations

Until the early 20th century, most discussion of new knowledge, particularly scientific knowledge, relied either on observation (induction) or inference. The work of Karl Popper, along with the reports of Poincare and others on the process of mathematical and scientific discovery, drew attention to the importance of processes such as conjecture and refutation. Popper drew a sharp distinction between the context of discovery (conjectures) and the context of justification (potential refutation). Whereas previous philosophers of science, and particularly the Vienna school of logical positivism, with which Popper was associated, had focused their attention on evidence that confirmed scientific hypotheses, Popper made the point that the crucial property of a scientific hypothesis was potential refutation. In our terms, the simplest statement of the Popperian model may be stated as one in which no hypothesis can ever be definitely proved (so that no positive proposition can ever be an element of N_t^{i+}) but any non-trivial hypothesis p can be refuted by the observation of some element of $\neg p$, with the result that $p \in N_t^{i-}$ [4]. Subsequent work in the Popperian, such as that of Lakatos has presented a more complex and nuanced view, but has retained a central focus on potential refutation.

Popper's most important contributions to the understanding of conjectures were negative. In the pre-Popperian picture, scientific hypotheses were derived from observed regularities derived from the patient accumulation of observations. This leads naturally to the confirmationist view of justification rejected by Popper. The most useful work on the generation of conjectures

[4] Popper's main point was to deny scientific status to theories like Marxism and Freudian psychology for which, he claimed, there was no possible refutation by evidence.

from previous knowledge is that of Lakatos who shows how a concern with deriving testable implications from a model under challenge leads naturally to the generation of new conjectures.

9 Research Agenda

Thus far, the discussion has been concerned solely with beliefs, to the exclusion of preferences and actions. This is in sharp contrast with the expected-utility approach, where probability beliefs are derived from preferences over actions, considered as mappings from the state space to some outcome space. Other models of choice under uncertainty provide more of a separate role for beliefs, without wholly separating beliefs, preferences and actions. It is clear that a satisfactory account of problems involving uncertainty must encompass preferences and actions.

It is not, as yet clear, how this should best be undertaken within the framework set out. Given the absence of a state-space accessible to the decision-maker, it is not clear that maintaining the separation between the state space and the outcome space, crucial in standard Bayesian decision theory, is appropriate here. It may be more desirable to consider partly or completely probablized subsets of P_t^i as 'possible worlds', each with their own associated outcome space.

As far as preferences are concerned, the most promising approach appears to involve adaptation of ideas developed by Gilboa and Schmeidler. Within possible worlds, preferences may be described by some version of the 'multiple priors' model. When considering actions that have consequences that appear to depend importantly on unforeseeable events, some version of the 'case-based decision theory' model, also proposed by Gilboa and Schmeidler, may be appropriate.

10 Concluding Comments

The problem of unforeseeable events is critical in decision theory. This paper has set out a framework within which this problem can be addressed.

References

Dekel, E., Lipman, B. and Rusticchini, A. (2001), 'A unique subjective state space for unforeseen contingencies', *Econometrica*, 69, 891–934.

Ellsberg, D. (1961), 'Risk, ambiguity and the Savage axioms', *Quarterly Journal of Economics*, 75(4), 643–669.

Epstein, L. (1999), 'A definition of uncertainty aversion', *Review of Economic Studies*, 66(3), 579–608.

Epstein and Marinacci (2006), "Coarse contingencies", mimeo.

Ghirardato, P., (2001), "Coping with ignorance, unforeseen contingencies and non-additive uncertainty", *Economic Theory*, **17**, 247–276.

Ghirardato, P. and Marinacci, M. (1999), 'Ambiguity made precise: a comparative foundation', *Working Paper*.

Gilboa, I. and Schmeidler, D. (1989), 'Maxmin expected utility with non-unique prior', *Journal of Mathematical Economics*, 18(2), 141–153.

Gilboa, I. and Schmeidler, D. (1995), 'Case-based decision theory', *Quarterly Journal of Economics*, 110, 605–639.

Grant, S. and Quiggin, J. (2002), 'A model-free definition of increasing uncertainty', Risk, Uncertainty and Decision Conference, Paris.

Gul, F. and Pesendorfer, W. (2004), 'Self-control and the theory of consumption', *Econometrica*, 72, 119–158.

Heifetz, A., M. Meier, and B. Schipper, (2004), "Interative Unawareness", forthcoming *Journal of Economic Theory*.

Kreps, D. (1992), "Static choice and unforeseen contingencies", in Economic Analysis of Markets and Games: Essays in Honor of Frank Hahn, ed. by Partha Dasgupta, Douglas Gale, Oliver Hart and Eric Maskin. Cambridge, MA: MIT Press, 259–281.

Mukerji, S., (1997), "Understanding the nonadditive probability decision model", *Economic Theory* 9, 23–46.

Nehring, K. (1999), 'Preference for flexibility in a Savage framework', *Econometrica*, 67(1), 101–119.

When an Event Makes a Difference

Massimiliano Amarante[1] and Fabio Maccheroni[2]

[1] Department of Economics, Columbia University, NY, USA
[2] Istituto di Metodi Quantitativi and IGIER, Università Bocconi, Milan, Italy
e-mail: ma734@columbia.edu, fabio.maccheroni@unibocconi.it

Abstract. For (S, Σ) a measurable space, let $\mathcal{C}_1$ and $\mathcal{C}_2$ be convex, weak* closed sets of probability measures on Σ. We show that if $\mathcal{C}_1 \cup \mathcal{C}_2$ satisfies the Lyapunov property, then there exists a set $A \in \Sigma$ such that $\min_{\mu_1 \in \mathcal{C}_1} \mu_1(A) > \max_{\mu_2 \in \mathcal{C}_2} \mu_2(A)$. We give applications to maxmin expected utility and to the core of a lower probability.

Keywords: Lyapunov theorem, maximin expected utility, lower probability

1 Introduction

In the theory of decision making under uncertainty as well as in the theory of cooperative games several questions can be reduced to the problem of whether or not two distinct sets of measures disagree on a set. For instance, if two maxmin expected utility preferences have the same utility on the prize space and the same willingness to bet, are they necessarily the same? Under which conditions, does the core of a lower probability coincide with the weak* closed and convex hull of any set of measures defining it? Both questions are answered in the affirmative if and only if one knows that there exists a set A such that $\min_{\mu_1 \in \mathcal{C}_1} \mu_1(A) > \max_{\mu_2 \in \mathcal{C}_2} \mu_2(A)$, whenever $\mathcal{C}_1$ and $\mathcal{C}_2$ are two (convex, weak* closed) disjoint sets of measures. This is our main result, which we prove in the next section under the conditions stated therein. In Sect. 3, we provide a quick sample of the usefulness of Theorem 1, by answering the two questions stated above. We do not discuss, however, the full range of applications of Theorem 1. For another less immediate application, we refer the reader to [2], where our Theorem 1 turns out to be a key tool to characterize those events which are unambiguous either in the sense of [14] or of [5]. In general, we expect Theorem 1 to be widely applicable in areas different from the ones we consider such as in Quasi-Bayesian Statistics (due to the central role played by upper probabilities; see, for instance [13]) or in social choice theory.

2 Main Result

If μ_1 and μ_2 are two probability measures on a σ-algebra Σ, then (by definition) $\mu_1 \neq \mu_2$ means that there exists a set $A \in \Sigma$ such that $\mu_1(A) > \mu_2(A)$. Equivalently, the two disjoint sets $\{\mu_1\}$ and $\{\mu_2\}$ can be separated by means of a linear functional having an especially simple form, namely one that is defined by an indicator function. Here, we are concerned with extending this property to sets of measures which are not singletons.

Let (S, Σ) be a measurable space and let $\Delta(\Sigma)$ denote the set of all (countably additive) probability measures on Σ. $\Delta(\Sigma)$ is a subset of the norm dual of the Banach space of bounded, Σ-measurable functions.

Definition 1. *Let $\mathcal{C} = \{\mu_i\}_{i\in I} \subset \Delta(\Sigma)$. We say that $\mathcal{C}$ has the Lyapunov property if the range of the vector measure $(\mu_i)_{i\in I}$ on E is a convex and compact subset of $\mathbb{R}^I$ (equipped with the product topology), for all $E \in \Sigma$.*

Notice that if $\mathcal{C}$ has the Lyapunov property and all of its elements are absolutely continuous with respect to a given probability measure, then any subset of $\mathcal{C}$ has the Lyapunov property, and that finite dimensional sets of nonatomic measures have this property (see [10]). Sets of measures with the Lyapunov property have special importance in the theory of decision making under uncertainty. For a decision maker described by a set of priors like in [8] or in [7], the Lyapunov property corresponds to the demand that the class of unambiguous events in the sense of [12] or [7] be "rich" (see Sect. 2).

Theorem 1. *Let $\mathcal{C}_1$ and $\mathcal{C}_2$ be convex, weak* closed subsets of $\Delta(\Sigma)$ such that $\mathcal{C}_1 \cup \mathcal{C}_2$ has the Lyapunov property. Then $\mathcal{C}_1 \cap \mathcal{C}_2 = \emptyset$ if and only if there exists A in Σ such that*

$$\min_{\mu_1 \in \mathcal{C}_1} \mu_1(A) > \max_{\mu_2 \in \mathcal{C}_2} \mu_2(A) .$$

Proof. Since each $\mathcal{C}_i$ is weak* compact, then it is weak compact, and convexity of $\mathcal{C}_i$ implies that there exist a measure $\lambda_i \in \mathcal{C}_i$ such that $\mu_i \ll \lambda_i$ for all $\mu_i \in \mathcal{C}_i$, $i = 1, 2$ (see, for instance [3]). Hence, all the measures in $\mathcal{C}_1 \cup \mathcal{C}_2$ are absolutely continuous with respect to $\lambda = \frac{1}{2}\lambda_1 + \frac{1}{2}\lambda_2$, and the sets $\mathcal{C}_1'$ and $\mathcal{C}_2'$ of all Radon–Nikodym derivatives of elements of $\mathcal{C}_1$ and $\mathcal{C}_2$ are disjoint, weakly compact, and convex subsets of $\mathcal{L}^1(\lambda)$. The Separating Hyperplane Theorem (see, [4], V.2.10) guarantees that there exist $g_0 \in \mathcal{L}^\infty(\lambda) - \{0\}$ such that

$$\min_{f_1 \in \mathcal{C}_1'} \int g_0 f_1 \, \mathrm{d}\lambda > \max_{f_2 \in \mathcal{C}_2'} \int g_0 f_2 \, \mathrm{d}\lambda . \tag{1}$$

W.l.o.g. $0 \leq g_0(s) \leq 1$ for λ-almost all $s \in S$ (otherwise take $\frac{g_0 - \mathrm{essinf} g_0}{\|g_0 - \mathrm{essinf} g_0\|_\infty}$).

By assumption, $\mathcal{C}_1 \cup \mathcal{C}_2$ has the Lyapunov property. Hence, $\mathcal{C}_1' \cup \mathcal{C}_2'$ is thin in the sense of [10]. By Lemma 1 in [10], $g_0 = \chi_A + h$ where $A \in \Sigma$ and

$h \in \mathcal{L}^{\infty}(\lambda)$ is such that $\int hf \, \mathrm{d}\lambda = 0$ for all $f \in \mathcal{C}_1' \cup \mathcal{C}_2'$. For all $\mu \in \mathcal{C}_1 \cup \mathcal{C}_2$, setting $f = d\mu/\mathrm{d}\lambda$ we have

$$\mu(A) = \int_A f \, \mathrm{d}\lambda = \int \chi_A f \, \mathrm{d}\lambda = \int (\chi_A + h) f \, \mathrm{d}\lambda = \int g_0 f \, \mathrm{d}\lambda$$

and Eq. (1) becomes

$$\min_{\mu_1 \in \mathcal{C}_1} \mu_1(A) > \max_{\mu_2 \in \mathcal{C}_2} \mu_2(A) .$$

The converse is obvious. □

Corollary 1. *Under the assumptions of Theorem 1, $\mathcal{C}_1 \subseteq \mathcal{C}_2$ if and only if $\min_{\mu_1 \in \mathcal{C}_1} \mu_1(A) \geq \min_{\mu_2 \in \mathcal{C}_2} \mu_2(A)$ for all $A \in \Sigma$.*

Proof. Let $\min_{\mu_1 \in \mathcal{C}_1} \mu_1(A) \geq \min_{\mu_2 \in \mathcal{C}_2} \mu_2(A)$ for all $A \in \Sigma$. Assume that $\mathcal{C}_1$ is not contained in $\mathcal{C}_2$. Then there exists $\bar{\mu} \in \mathcal{C}_1 - \mathcal{C}_2$. Since $\mathcal{C}_2 \cup \{\bar{\mu}\}$ is thin, Theorem 1 yields that there exists $B \in \Sigma$ such that $\bar{\mu}(B) < \mu_2(B)$ for all $\mu_2 \in \mathcal{C}_2$. Therefore $\min_{\mu_1 \in \mathcal{C}_1} \mu_1(B) \leq \bar{\mu}(B) < \min_{\mu_2 \in \mathcal{C}_2} \mu_2(B)$, which is absurd. The converse is trivial. □

We conclude this section, by proving another separation result. This extends an obvious property of two nonatomic measures: if $\mu_1 \neq \mu_2$, there exist $A, B \in \Sigma$, $A \cap B = \emptyset$ such that $\mu_1(A) > \mu_1(B)$ and $\mu_2(A) < \mu_2(B)$. Notice that this is no longer true if the nonatomicity assumption is removed. In this form, the separation theorem turns out to be a basic tool in study unambiguous events in the sense of [14] and [5] (see [2]).

Corollary 2. *Under the assumptions in Theorem 1, there exist $A, B \in \Sigma$, $A \cap B = \emptyset$, such that $\mu_1(A) - \mu_1(B) > 0 > \mu_2(A) - \mu_2(B)$ for any $\mu_1 \in \mathcal{C}_1$ and any $\mu_2 \in \mathcal{C}_2$.*

Proof. Let $A \in \Sigma$ be such that $\mu_1(A) > \mu_2(A)$ for any $\mu_i \in \mathcal{C}_i$, $i = 1, 2$. Since $\mathcal{C}_1 \cup \mathcal{C}_2$ has the Lyapunov property, the range on S of the vector measure defined by $\mathcal{C}_1 \cup \mathcal{C}_2$ is compact and convex. Hence, for any $\alpha \in [0, 1]$ there exists $B \in \Sigma$ such that $\mu_1(B) = \mu_2(B) = \alpha$ for all $\mu_i \in \mathcal{C}_i$, $i = 1, 2$. Pick one such a B so that $2\mu(B) = \min_{\mu_1 \in \mathcal{C}_1} \mu_1(A) + \max_{\mu_2 \in \mathcal{C}_2} \mu_2(A)$ for all $\mu \in \mathcal{C}_1 \cup \mathcal{C}_2$. Then, $\mu_1(A) - \mu_1(B) > 0$ and $\mu_2(A) - \mu_2(B) < 0$ for all $\mu_i \in \mathcal{C}_i$, $i = 1, 2$.

If $A \cap B \neq \emptyset$, write $B = (A \cap B) \cup B'$ and $A = (A \cap B) \cup A'$. Then, for any $\mu_i \in \mathcal{C}_i$, $i = 1, 2$,

$$\begin{aligned} \mu_1(A) - \mu_1(B) &= \mu_1(A \cap B) + \mu_1(A') - \mu_1(A \cap B) - \mu_1(B') \\ &= \mu_1(A') - \mu_1(B') , \\ \mu_2(A) - \mu_2(B) &= \mu_2(A') - \mu_2(B') , \end{aligned}$$

and A' and B' do the job. □

3 Application: MEU Preferences and Lower Probabilities

In the theory of decision making under uncertainty, one is concerned with a decision maker ranking the elements of a set $\mathcal{A}$ of mappings $a : S \to X$, where S is the state space and X the prize space. For the sake of simplicity, let X be a convex subset of a vector space and $\mathcal{A}$ be the set of all simple and measurable functions from S to X. The decision maker's ranking $\succeq$, is said to satisfy the maxmin expected utility (MEU) criterion if and only if for $a, b \in \mathcal{A}$

$$a \succeq b \Leftrightarrow \min_{\mu \in \mathcal{C}} \int (u \circ a)\, \mathrm{d}\mu \geq \min_{\mu \in \mathcal{C}} \int (u \circ b)\, \mathrm{d}\mu\,,$$

where $u : X \to \mathbb{R}$ is a nonconstant and affine utility function on the prize space, and $\mathcal{C}$ is a weak* closed and convex set of finitely additive probability measures on (S, Σ). The willingness to bet of a MEU decision maker is the lower probability

$$\rho(A) = \min_{\mu \in \mathcal{C}} \mu(A), \quad \forall A \in \Sigma\,.$$

The core of a lower probability ρ is the set $core(\rho)$ of all finitely additive probability measures ν on (S, Σ) such that $\nu \geq \rho$.

Preferences satisfying the MEU criterion have been axiomatized in [8]. In [11] and [3] necessary and sufficient conditions on $\succeq$ are given that guarantee that all the measures in $\mathcal{C}$ be countably additive. An event $A \in \Sigma$ is unambiguous in the sense of Nehring [12] or Ghirardato, Maccheroni and Marinacci [7] if $\mu(A) = \mu'(A)$ for all $\mu, \mu' \in \mathcal{C}$. In [1] (Proposition 4), it was shown that (i) the class of unambiguous events is "rich", that is there exist unambiguous events of measure α for every $\alpha \in [0, 1]$, and (ii) there exists a countably additive, nonatomic probability measure on the class of unambiguous events if and only if $\mathcal{C}$ has the Lyapunov property.

In the context of maxmin expected utility, a natural question is whether or not two MEU preferences with the same utility on the prize space and the same willingness to bet are necessarily the same preference. A related question in the theory of lower probabilities is whether or not the weak* closed and convex set $\mathcal{C}$ defining a lower probability ρ coincides with its core. The following example, due to Huber and Strassen [9], answers negatively to both questions.

Example 1. Let $S = \{1, 2, 3\}$, $X = \mathbb{R}$, $\mu = (\frac{1}{2}, \frac{1}{2}, 0)$, $\nu = (\frac{4}{6}, \frac{1}{6}, \frac{1}{6})$. Consider two MEU preferences, $\succsim_1$ and $\succsim_2$, with $u_1(x) = u_2(x) = x$ for any $x \in \mathbb{R}$ and sets of priors

$$\mathcal{C}_1 = co\{\mu, \nu\} \quad \text{and} \quad \mathcal{C}_2 = \left\{ \left(\frac{3+t}{6}, \frac{3-t-s}{6}, \frac{s}{6} \right) : 0 \leq s, t \leq 1 \right\}.$$

It is readily checked that:

- $\rho_1(A) = \min_{\mu_1 \in \mathcal{C}_1} \mu_1(A) = \min_{\mu_2 \in \mathcal{C}_2} \mu_2(A) = \rho_2(A)$ for all $A \subset S$, but $\succsim_1$ is different from $\succsim_2$;
- $\mathcal{C}_1$ is a weak* closed and convex set defining the lower probability ρ_1, and it is strictly included in *core* (ρ_1) (which coincides with $\mathcal{C}_2$).

Both conclusions are reverted under the assumptions of Theorem 1 as the next two corollaries show. In reading Corollary 3, notice that point 1. amounts to say that $\succsim_1$ is more ambiguity averse than $\succsim_2$ (see Ghirardato and Marinacci [6]), and remember that xAy is the mapping from S to X taking value x on A and y on A^c.

Corollary 3. *Let $\succsim_1$ and $\succsim_2$ be two MEU preferences with (weak* closed and convex) sets of priors $\mathcal{C}_1$ and $\mathcal{C}_2$ contained in $\Delta(\Sigma)$ and such that $\mathcal{C}_1 \cup \mathcal{C}_2$ has the Lyapunov property. Then the following conditions are equivalent:*

1. *For all $a \in \mathcal{A}$ and $z \in X$,*
$$a \succsim_1 z \Rightarrow a \succsim_2 z\,. \tag{2}$$
2. *For all $x, y, z \in X$ such that $x \succsim_i y$ for $i = 1, 2$, and all $A \in \Sigma$,*
$$xAy \succsim_1 z \Rightarrow xAy \succsim_2 z\,.$$
3. *u_1 is a positive affine transformation of u_2 and $\rho_1 \leq \rho_2$.*

In particular, if $u_1 = u_2$ and $\rho_1 = \rho_2$, then $\succsim_1$ coincides with $\succsim_2$.

Corollary 4. *Let ρ be a lower probability such that core$(\rho) \subset \Delta(\Sigma)$ and core(ρ) has the Lyapunov property. Then core(ρ) is the weak* closed and convex hull of any subset $\mathcal{K}$ of $\Delta(\Sigma)$ such that*

$$\rho(A) = \inf_{\nu \in \mathcal{K}} \nu(A), \quad \forall A \in \Sigma\,.$$

The easy proofs are omitted.

Acknowledgement. We thank Mohammed Abdellaoui and an anonymous referee for helpful suggestions. The financial support of the *Ministero dell'Istruzione, dell'Università e della Ricerca* is gratefully acknowledged.

References

1. Amarante, M. (2005), "Ambiguity, measurability and multiple priors", *Economic Theory* **26**, 995–1006.
2. Amarante, E., Filiz, M. (2004), Ambiguous events and maxmin expected utility, Columbia University discussion paper # 0405–09.
3. Chateaneuf, A., Maccheroni, F., Marinacci, M., Tallon, J.-M. (2005), "Monotone continuous multiple priors", *Economic Theory* **26**, 973–982.

4. Dunford, N., Schwartz, J.T. (1954) Linear Operators, Part I: General Theory. Wiley-Interscience, London.
5. Epstein, L.G., Zhang, J. (2001), "Subjective probabilities on subjectively unambiguous events", *Econometrica* **69**, 265–306.
6. Ghirardato, P., Marinacci, M. (2002), "Ambiguity made precise: A comparative foundation", *Journal of Economic Theory* **102**, 251–289.
7. Ghirardato, P., Maccheroni, F., Marinacci, M. (2003), "Differentiating ambiguity and ambiguity attitude", *Journal of Economic Theory* **118** (2004), 133–173.
8. Gilboa, I., Schmeidler, D. (1989), "Maxmin expected utility with a non-unique prior", *Journal of Mathematical Economics* **18**, 141–153.
9. Huber, P.J., Strassen, V. (1973) "Minimax tests and the Neyman–Pearson lemma for capacities", *Annals of Statistics* **1**, 251–263.
10. Kingman, J.F.C., Robertson, A.P. (1968), "On a theorem of Lyapunov", *Journal of the London Mathematical Society* **43**, 347–351.
11. Marinacci, M. (2002), "Probabilistic sophistication and multiple priors", *Econometrica* **70**, 755–764.
12. Nehring, K. (1999), "Capacities and probabilistic beliefs: A precarious coexistence", *Mathematical Social Sciences* **38**, 197–213.
13. Wasserman, L.A., Kadane, J.B. (1990), "Bayes' theorem for Choquet capacities", *Annals of Statistics* **18**, 1328–1339.
14. Zhang, J. (2002), "Subjective ambiguity, expected utility and Choquet expected utility", *Economic Theory* **20**, 159–181.

Countable Additivity and the Foundations of Bayesian Statistics

John V. Howard

London School of Economics, Houghton Street, London WC2A 2AE, UK
e-mail: j.v.howard@lse.ac.uk

Abstract. At a very fundamental level an individual (or a computer) can process only a finite amount of information in a finite time. We can therefore model the possibilities facing such an observer by a tree with only finitely many arcs leaving each node. There is a natural field of events associated with this tree, and we show that any finitely additive probability measure on this field will also be countably additive. Hence when considering the foundations of Bayesian statistics we may as well assume countable additivity over a σ-field of events.

Keywords: Bayesian statistics, foundations, countable additivity, finite additivity

1 Introduction

When laying the foundations for Bayesian statistics we obviously seek neither to rule out possibilities in an arbitrary way nor to make unnecessary restrictive assumptions. These considerations have led some of the pioneers of Bayesianism to avoid the assumption of countable additivity, and instead to assume only finite additivity of the probability measure. Good, for example, in [6] wished to make as few assumptions as possible, so he declined to strengthen his axiom system by adding countable additivity. This approach we will term *minimalism*. It deliberately leaves open the question whether finitely but not countably additive probability measures exist.

However, the standard assumption nowadays is to strengthen the axiom to insist on countable additivity. This immediately allows all the theorems of mathematical probability theory to be used in Bayesian statistics. But alternative strengthenings of the axiom system are possible. de Finetti [4], for example, wished to retain the possibility that one might choose a natural number at random, each number being equally likely to be selected. Let us call this the requirement of *uniform choice*. Such probability measures cannot be countably additive, so to assume they exist means we must forego countable additivity as an axiom. We can have minimalism plus countable additivity or minimalism plus uniform choice, but not both. However, we shall try to show that there is a third alternative. If minimalism is combined with a new requirement, *realism*, it leads to the conclusion that countable additivity is in fact an innocuous addition to our axiom system. (Realism will basically be the requirement that we can observe only a finite amount of information in a finite time.)

The tension between the three desiderata showed clearly in Hill's discussion of the Fraser–Monette–Ng example in [1] (an interesting modification of this example is given in [5]). Hill supports the usefulness of allowing finitely additive uniform distributions over the natural numbers (i.e. uniform choice), which in this example would lead him into incoherency. However, adding the *realistic* assumption that only numbers less than some bound N could be reported to the participants in the game allows him to escape from the trap. So realism immediately rules out the paradoxes that uniform choice allows in. (Another example often cited as showing problems with uniform choice is the "two envelope paradox". However, Brams and Kilgour [3] show that this paradox can also occur with proper probability measures. They suggest resolving it as a St. Petersburg type paradox.)

Berger and Wolpert, commenting on Hill's argument, say "Do examples of the type we are discussing exist for finite sample spaces? If so, such would seem to provide a counter-example to Professor Hill's argument. If not, one could indeed not object, philosophically, to the use of finitely additive measures." We shall in fact work with a restricted sample space: although infinite, the range of possibilities observable in a finite time is always finite. Within this framework we will assume only finite additivity, but we will show that essentially this gives no extra generality over assuming countable additivity (in terms of making probability forecasts over events which can be observed in a finite time). Hence any paradoxes which might arise (and which can be observed) would also occur with a countably additive probability measure. Consequently, we will argue that one may as well always assume countable additivity, in a sense "without loss of generality".

2 A Simple Model of a Bayesian Statistician

We assume, firstly, that all data can be digitized without losing any essential information, provided that enough digits are allowed. Secondly, we assume that only a finite amount of information can be absorbed per unit time. Clearly there is no loss of generality in restricting the digits to 0's and 1's. Then our model of any observer (whether human or computer) is that he, she, or it has a sample space consisting of the set of all binary sequences (we used this idea in [7]). This is our interpretation of "realism". Note that our observers may have infinite lives, so the sample space can be (uncountably) infinite. The restriction is simply to the amount they can observe in a finite

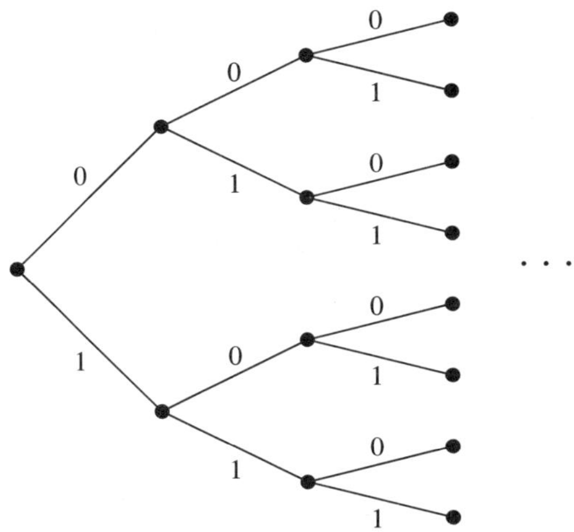

Fig. 1. Binary tree of possible observations

time. Effectively Nature chooses a path for them through the binary tree shown in Fig. 1 below.

In particular cases some or all of the branches of the tree may be finite. This framework also harmonises well with a decision theory approach to Statistics, which would allow some of the nodes in the tree to be choice nodes. For our purposes it suffices to consider trees with chance nodes only.

Sometimes we are almost constrained to think of information as arriving in a stream of bits. Whenever we do not make direct measurements ourselves, but instead use a robotic observer (for example, on a space probe to the moon or Mars, or inside a nuclear reactor), the information will come to us as a datastream of this form. In general, we do not feel that any significant information is necessarily lost when we do this: if the probe has sufficiently good cameras, environmental monitors, and other sensing devices, we think that we are gathering as much information as if we were there ourselves (even though without the excitement of walking on another world).

The assumption of realism imposes a strong constraint. It means that sometimes observation must be thought of fundamentally as a sequential process rather than as a simple act. We shall show that the assumption leads naturally to the adoption of countable additivity and the rejection of uniform choice. However, it is likely that some will prefer to regard observation as an act rather than a process. This allows them to build models using improper priors, and to try to develop objective Bayesianism. Unfortunately, it is also liable to generate paradoxes and contradictions. The realism assumption automatically stops the paradoxes.

After a finite time has elapsed the observer will have observed some finite sequence s of 0's and 1's. We can define a natural field $\mathcal{F}$ of events on the sample (sequence) space consisting of finite unions of sets of the form

$$S_s = \{\omega : \omega \text{ an infinite sequence of 0's and 1's having } s \text{ as an initial finite subsequence}\}\,.$$

This conforms to the requirement that we restrict ourselves to events that could be observed. We assume our (Bayesian) statistician has a finitely additive probability measure P on sets in $\mathcal{F}$ defined by the numbers $P(S_s)$ (which we will write as $P(s)$), for all finite s. So the probability that the sequence starts with a "0" is $P(0)$, with a "1" followed by a "0" is $P(10)$, etc. P must satisfy

$$P(s) = P(s0) + P(s1) \text{ for all } s\,.$$

($s0$ is the sequence consisting of the digits in s followed by 0: similarly $s1$.) We also require $P(s) \geq 0$ for all s and $P(0) + P(1) = 1$. However, in keeping with "minimalism", we do not require that P be countably additive.

But although we do not require it, we still get it, albeit in a vacuous sense. Suppose an event S_s is a union of countably many disjoint sets of the form S_{s_i}. Then the union is in fact finite. Hence $P(S_s) = \sum_i P(S_{s_i})$ and countable additivity is established.

To show the union is finite we could appeal to Tychonoff's theorem (the compactness of the product of compact spaces). A direct argument would be as follows. If the union is infinite, at least one of S_{s0} and S_{s1} has an infinite cover. Say it is S_{s0}. Then we can extend to $s00$ or $s01$, and so on. Continuing in this way we define an infinite sequence extending s which cannot lie in any of the S_{s_i}. This contradiction establishes the result. (Note that although in general Tychonoff's theorem requires the Axiom of Choice, we do not need it for this particular case. If at some stage both $t0$ and $t1$ are possible extensions of t we stipulate that $t0$ will be chosen.)

Hence we can by the extension theorem (see, for example, [2] Theorem 3.1) extend P to a unique countably additive measure on the σ-field $\mathcal{S}$ generated by $\mathcal{F}$. There is no reason why we should not do this, even if we believe that ultimately only events in $\mathcal{F}$ and their probabilities are basic. We have in no way restricted the finitely additive measures we allowed on $\mathcal{F}$. It is true that there may be other extensions of P to $\mathcal{S}$ which are only finitely additive, but why should we ever prefer these. On any observable event the two extensions agree, so it is a matter of convenience which we choose – and for convenience σ-additivity always wins. In any case, whenever we have two extensions of P (say Q and R) no operational test will ever be able to distinguish them. Whatever is observed in a finite time will be assigned the same probability by Q and R. Hence nothing fundamental can depend on the choice: we are not restricting ourselves by always working with the countably additive Q and ignoring the finitely additive R.

Of course the mathematical results are well known (see [2] or [7]). What we are claiming is that their significance for the foundations of Bayesian inference has not been fully appreciated.

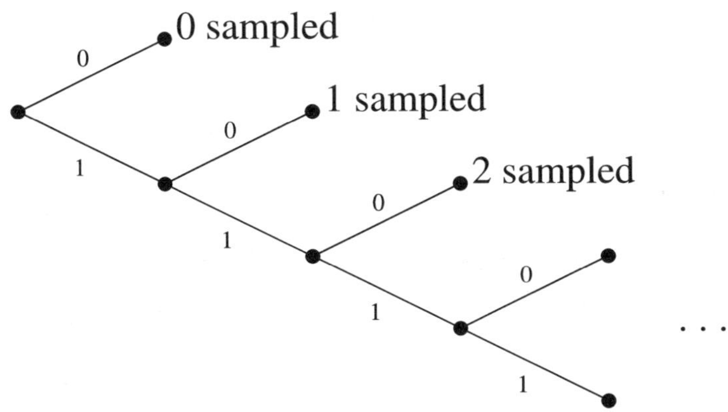

Fig. 2. Uniform sampling of a natural number

3 An Example

Suppose we try, within our structure of a binary tree, to choose a natural number n at random, all numbers being equally likely. We can imagine the number is to be given to us as a sequence of n 1's followed by a 0. (The zero serves as an end marker.) The tree of possibilities is given in Fig. 2.

We will imagine a machine which prints out a digit every second, and which has printed nothing but 1's since it was started a long time ago. We can see that if the machine ever does print a "0" it must then stop. The machine is being watched by Ursula (who believes in uniform choice) and Charles (who favours countable additivity).

At each node the probability of observing another 1 is one, and we have a finitely (and countably) additive probability measure P on events in $\mathcal{F}$. The unique countably additive extension Q of P to $\mathcal{S}$ gives probability one to the infinite sequence 111... and zero to all other sequences.

There are other extensions (say R) which give weight zero to 111... and which are finitely additive on $\mathcal{S}$. (To prove they exist we need the Axiom of Choice, which hardly seems in the spirit of making minimal assumptions.) But why bother with R? Q assigns exactly the same probabilities to anything that can actually be observed.

Ursula views the sample space as $\mathbb{N} = \{0, 1, 2, \ldots\}$, each element having probability zero. Charles thinks the sample space is $\mathbb{N} \cup \{\infty\}$ with $Q(\{\infty\}) = 1$. Ursula is certain that the machine will eventually stop. Charles believes (with probability 1) that it will continue to print 1's for ever. In one sense, their views could hardly be more different. (However, they are in total agreement that the machine will not stop within any particular stated time T.) If in fact at some point the machine does print a 0 and stop, Ursula will feel vindicated, but Charles will now think his model of the situation was wrong. (He is very unlikely to be converted to a belief in uniform choice.) However Charles can cover himself against this eventuality by making a Dutch book against Ursula (see Williamson [8]). She should be prepared to offer odds of $2^{n+2} : 1$ against n occurring for every n. Hence if Charles stakes $\frac{1}{2^{n+1}}$ on n for every n, he will make a profit of at least 1 unit if the machine ever stops. If the machine does go on for ever, Ursula is wrong (this possibility was not even in her sample space), but she cannot make a Dutch book against Charles, who would not accept any bets on $\{\infty\}$.

The above problem might be compared to a situation where Ursula thinks a real number will be sampled uniformly in $[0, 1]$, whilst Charles believes it will be a specific real number outside this range (say π) with certainty. We have a similar complete disagreement, but neither can now make a Dutch book against the other. Ursula could offer odds of $1 : 1$ against the number being in $[0, 0.5]$, of $3 : 1$ against the number being in $[0.5, 0.75]$, and so on, but this does not give rise to a Dutch book.

Often problems which start "a number n is chosen at random ..." continue by instructing us to sample another number, x say, depending on the value of n. (See, for example, the Fraser–Monette–Ng paradox.) But in practice would we continue to accept Q or R after an event of probability zero had been observed? The alternative hypothesis (say) that $P_n = \frac{1}{2^{n+1}}$ has now got a likelihood ratio of ∞ in its favour. And similar arguments could be made against the use of uniform priors for (say) the mean of a Normal distribution.

We can still within our framework choose a real number uniformly between 0 and 1 to any given level of precision. (Take the finite sequence s to represent all reals whose binary expansion starts with the digits of s.) However, if we try to construct a tree corresponding to any of the finitely but not countably additive probability measures on [0,1] which give probability zero to all singletons, it will be equivalent (for all sets in $\mathcal{F}$) to the standard uniform σ-additive measure on $[0, 1]$.

4 Conclusions

We have worked within a framework which allows an infinite sample space, permits the sampling of real numbers to arbitrary accuracy, and is consistent with a Decision Theory approach to statistics. Within this framework it appears that there is no loss in generality involved in assuming countable additivity as a universal axiom. Moving outside the framework implies a loss of realism, because we are then assuming that an infinite amount of information can be observed in a finite time.

It is therefore suggested that the choice of finite or countable additivity cannot be regarded as a fundamental philosophical problem for Bayesian statistics. It should rather be regarded as a matter of making a convenient assumption. Many statisticians will find countable additivity both simpler and more convenient.

References

1. Berger, J.O., Wolpert, R.L. (1988), *The Likelihood Principle.* Institute of Mathematical Statistics.
2. Billingsley, P. (1986), *Probability and Measure.* Wiley.
3. Brams, S.J., Kilgour, D.M. (1995), The Box problem: To switch or not to switch, *Mathematics Magazine* **68**, 27–34.
4. de Finetti, B. (1974), *Theory of Probability (Volume 1).* Wiley.
5. Goldstein, M., Howard, J.V. (1991), A likelihood paradox, *J. R. Statist. Soc. B* **53**, 619–628.
6. Good, I.J. (1950), *Probability and the Weighing of Evidence.* Griffin.
7. Howard, J.V. (1975), Computable Explanations, *Zeitschr. f. math. Logik und Grundlagen d. Math,* **21**, 215–224.
8. Williamson, J. (1999), Countable additivity and subjective probability, *Brit. J. Phil. Sci.* **50**, 401–416.

On the Quantification and Decomposition of Uncertainty

Emanuele Borgonovo and Lorenzo Peccati

IMQ – Bocconi University, 20135 Milano, Viale Isonzo 25, Italy
e-mail: emanuele.borgonovo@unibocconi.it

Abstract. In this work we deal with the quantitative assessment and decomposition of uncertainty. The decision making process is often accompanied by an uncertainty propagation exercise in the practice. We first analyze the meaning of uncertainty propagation from a subjective decision-making point of view. We show that, in order to quantify uncertainty, one has to resort to the distribution of the expected utility (U) originated from parameter uncertainty. We undertake the analytical determination of the moments U. We show that, if one considers the uncertain parameter space as subdivided in alternative preference regions delimited by indifference hypersurfaces, the moments of U are the sum of the moments of the expected utility of alternatives in the regions alternatives are preferred. As a consequence, if an alternative is never preferable, it does not contribute to uncertainty. In order to decompose uncertainty, we focus on the variance of U. By stating of Sobol' variance decomposition theorem in the Decision-Theory framework, we show that the variance of U can be expressed as sum of the variances brought by uncertain parameters individually and/or in groups. We then determine and discuss the meaning of global importance of parameters. Since parameters associated with the highest value of the global importance are the most effective in reducing uncertainty, gathering information on these parameters would reduce uncertainty in the most effective way. We illustrate the moment calculation and variance decomposition procedures by means of an analytical example. The application to the uncertainty analysis of an industrial investment decision-making problem concludes the paper.

Keywords: variance decomposition, global sensitivity analysis, industrial decision making

1 Introduction

This paper presents a method for the quantification and decomposition of uncertainty in decision-making (DM). The problem of uncertainty is and has been widely debated in the literature, with authors proposing a formal distinction between risk and uncertainty, other ones rejecting it and coming to a unifying theoretical framework [1, 3, 4, 8–10, 12–16, 20–22, 25, 27–29, 31–33].

The general settings of this paper is that of subjective uncertainty as dealt with in [12, 16, 22, 30]. de Finetti's theorem paves the way to the use of Bayesian inference as a normative tool for DM under uncertainty [22, 30]. Bayesian inference allows to include in the subjective framework uncertainty generated by lack of knowledge in technical aspects of the problem – an uncertain failure probability of a piece of equipment or an uncertain proportion of goods that customers will buy [1, 22] – and subjective/expert judgment [11, 19]. The term "Epistemic" uncertainty [1], and [16] is used (for a discussion on model uncertainty we refer to [1]). In the remainder of the paper, the symbol π is used to denote an uncertain quantity (*parameter)*; the uncertainty regarding π will be deemed as *parametric* or *epistemic* (see also [1, 16]). If there is more than one uncertain quantity in the DM problem, $\pi = \{\pi_1, \pi_2, \ldots, \pi_n\}$ will be the vector of parameters, and $\mathbf{F} = \{F_1(\pi_1), F_2(\pi_2), \ldots, F_n(\pi_n)\}$ the corresponding vector of epistemic distributions. The symbol π^* denotes a realization of π. This work has the purpose of dealing with the quantitative structure of uncertainty, determining how and how much the uncertainty in alternatives and parameters contributes to the decision-maker (DMr) uncertainty. In order to state the problem in a quantitative fashion, we start with the analysis of the DM process as usually implemented in the industry [11,30]. The process begins with the identification of the random events and consequences [11,30]. A decision-support model in the form of an influence diagram or a decision tree is usually built to represent the problem [11,30]. The utility assessment for each consequence is performed next. Subjective probability elicitation is the next step [11,30]. Assuming exchangeability [5, 12, 22], and that the DMr does not apply a loss function to the parameters [5][1], then the subjective probability p_i for event i, is:

$$p_i = \int \pi_i \, \mathrm{d}F(\pi_i) \,, \tag{1}$$

where π_i is one of the possible values p_i can be assigned and $F(\cdot)$ is the subjective probability measure for π_i. The preferred alternative is the one that maximizes expected utility with Eq. (1) determining the DMr subjective probabilities [22, 30]. A numerical uncertainty propagation exercise (*uncertainty analysis*[2]) concludes the process, with the purpose of displaying the DMr uncertainty in the problem [11,19]. A vector of parameters ($\pi^{*\prime}$) is sampled from $\mathbf{F}$. The expected utility (U) of the alternatives is evaluated in correspondence of the sampled values. Thus, let us say that in this first sampling, the DMr would choose alternative I corresponding to an expected utility of U'. In

[1] The issue of uncertainty aversion is part of future work of the authors.

[2] Most of the standard Decision-Making Software (DATAPRO by the Treeage Corporation, or Precision Tree by Palisade) are equipped with Monte Carlo based subroutines for Uncertainty Analysis [41].

a next sampling, the probabilities would assume a different value ($\pi^{*\prime\prime}$) and the DMr preferred act would be, say, II with an expected utility of U''. Now, if the process is repeated a M times, with M a suitably large sample size for inspecting the uncertainty space, one gets the distribution of U provoked by the uncertain parameters. Sometimes also the frequency with which alternatives are selected is registered and called strategy selection frequency [11]. In a subjective DM framework, epistemic uncertainty propagation is equivalent to the DMr assigning values π^* to the parameters according to his subjective distribution $\mathbf{F}$, and inspecting how he/she would act consequently.

From the above discussion, it is clear that in order to quantify uncertainty one has to consider the dependence of U on parameters. U becomes a function of random variable characterized by what we call an epistemic distribution. As a first task of the paper, we undertake the analytical calculation of the moments of U. To do so, we use the fact that the uncertain parameter space (Ω) can be thought of as divided into preferred alternative regions delimited by indifference hypersurfaces. We show that any lth order epistemic moment of U is the sum of the moments of the expected utility of alternatives over the uncertainty space regions where alternatives are preferred. As a consequence, if an alternative is never preferred, it does not contribute to any of the moments of U, *i.e.* it does not contribute to uncertainty. The extension of these results to the central moments of U concludes our analysis on the quantification of U. We then focus on the analysis of variance of U ($V_{\mathbf{F}}[U]$), to determine the contributions of the parameters to uncertainty. We first state Sobol' variance decomposition in the DM framework. This enables us to: 1) state that the alternative contributions to $V_{\mathbf{F}}[U]$ equals the sum of the contributions of the uncertain parameters taken individually and in groups;

Table 1. List of the symbols used in this work

Symbol	**Meaning**
Ω	Parameter uncertainty space
π	Vector of the uncertain parameters
π^*	One of the possible values of π
n	Number of uncertain parameters
N	Number of alternatives
$\mathbf{F}(\cdot)$	Epistemic uncertainty distribution of π
a^j	Alternative j
$a^*(\pi)$	Preferred alternative at π
$u^j(\pi)$	Utility of alternative j at π
$u^*(\pi)$	Preferred alternative utility at π
U	Expected utility
Ω_s	Region in which a_s is preferred

2) Introduce and compute the global importance of parameters [38,40]. This last step can have a direct application to uncertainty management. In fact, global importance was defined so that the parameters associated with the highest value of their global importance turn out to be the most effective ones in reducing uncertainty [2,6,7,17,18,23,24,26,34–39]. The knowledge of which parameter reduces uncertainty in the most effective way is of help to the DM process, especially in presence of limited resources – time or money –, since gathering information on the most important parameters would mean to reduce uncertainty in the most effective way. In Sect. 2, we present an analytical approach to the uncertainty analysis of U and investigate how the moments of U can be expressed as functions of the moments of preferred alternatives. In Sect. 3 we propose a variance decomposition method to connect the uncertainty contributions of alternatives to the uncertainty contributions of probabilities. In Sect. 4 we illustrate the method by means of a simple example. In Sect. 5 we present the application of the method to the uncertainty analysis of an industrial decision-making problem. Sect. 6 offers some conclusions and future work perspectives.

2 Uncertainty Quantification

The purpose of this Section is the computation of the epistemic moments of U. Table 1 illustrates the notation and symbols used throughout the work.

Let $\Omega \subseteq^n$, where n is the number of uncertain parameters and $\pi \in \mathbf{\Omega}$. In general, one can write the expected utility as a function of the uncertain parameters as:

$$U = u(\pi)\,. \tag{2}$$

We note that U is defined $\forall \pi \in \Omega$, i.e. $U : \Omega \subset \mathbb{R}^n \to \mathbb{R}$. We illustrate this with an example that should also help in clarifying the notation.

Example 1. Suppose that a decision maker faces the choice among two lotteries, I and II, that depend on the outcome of random events 1 and 2 respectively, each of which has two possible outcomes. Let epistemic uncertainty on the probabilities of the two events be present, and let

$$F_1(\pi_1) = \begin{cases} 1 & \text{if } 0 \le \pi_1 \le 1 \\ 0 & \text{otherwise} \end{cases} \text{ and } F_2(\pi_2) = \begin{cases} 1 & \text{if } 0 \le \pi_2 \le 1 \\ 0 & \text{otherwise} \end{cases}$$

be the respective subjective density functions[3]. Let $u_i, i = 1 \ldots 4$, the DMr utility on each of the consequences. Then, the utility of alternatives I and II are, respectively:

[3] A structure of this type emerges in a DM problem of the type discussed in [16]. Suppose the DMr is undecided to bet on team A or B in the following lottery. He suffers a consequence of u_1 or u_2 respectively, if team A of a certain sport

$$\begin{aligned} U_I &= u^I(\pi) = \pi_1 u_1 + (1-\pi_1)u_2\,, \\ U_{II} &= u^{II}(\pi) = \pi_2 u_3 + (1-\pi_2)u_4\,. \end{aligned} \tag{3}$$

At any π, U is equal to the value of the utility of the preferred alternative. Thus, U as a function of the parameters is written as:

$$U = u^*(\pi) = \max\left[\pi_1 u_1 + (1-\pi_1)u_2, \pi_2 u_3 + (1-\pi_2)u_4\right]\,. \tag{4}$$

In order to quantify the effect of the propagation of the uncertainty in the parameters, one needs to compute the epistemic moments of U. To do so, let us introduce the following definitions:

Definition 1. *We call preference region ($\Omega_s \subseteq \Omega$) the region of the uncertainty space where alternative s is preferred. We call indifference epistemic hypersurface, an hypersurface delimiting an uncertainty region.*

Let us illustrate the two above definitions by means of example 1. In this case, the indifference hypersurface is the straight line given by:

$$\pi_1 = \pi_2\left(\frac{u_3-u_4}{u_1-u_2}\right) + \frac{u_4-u_2}{u_1-u_2} \tag{5}$$

and Ω is subdivided into two regions, $\Omega_I = \left\{\pi_1, \pi_2 : \pi_1 \geq \pi_2\left(\frac{u_3-u_4}{u_1-u_2}\right) + \frac{u_4-u_2}{u_1-u_2}\right\}$ and $\Omega_{II} = \Omega \setminus \Omega_I$. For the following values for the utilities: $u_1 = 1$, $u_2 = 0$, $u_3 = 2/3$, and $u_4 = 1/3$ the indifference line is represented in Fig. 1.

We note that the number of regions is $1 < s < N$. In fact, let us say that there are N alternatives. Suppose that one of the alternatives, namely a^k is preferred $\forall \pi \in \mathbf{\Omega}$. Then $U = \max\left[u^1(\pi), u^2(\pi), \ldots, u^k(\pi), \ldots, u^N(\pi)\right] = u^k(\pi)$ and there is only one region, i.e. $\Omega_s = \Omega$. If there are two alternatives that dominate the others, there will be two regions, and so on so forth.

Before coming to an expression for the moments of U, we need the following definitions.

Definition 2. *We call the quantity*

$$\mu^s_{l,\mathbf{F}}[U] = \int_{\Omega_s} \left[u^s(\pi)\right]^l \,\mathrm{d}\mathbf{F} \tag{6}$$

alternative s contribution to the lth order moment of U.

Definition 3. *We call the quantity*

$$M^s_{l,\mathbf{F}}[U] = \int_{\Omega_s} \left(u^s(\pi) - E_F[U]\right)^l \,\mathrm{d}\mathbf{F} \tag{7}$$

contribution of alternative s to the lth order central moment of U.

---

wins or looses the next game. Similarly, a consequence of u_3 or u_4 is experienced if team B of the same or of another sport wins or looses the next game. A is not playing against B, obviously. If the DMr is not an expert of the sports, he can be uncertain on the victory probability of the two teams.

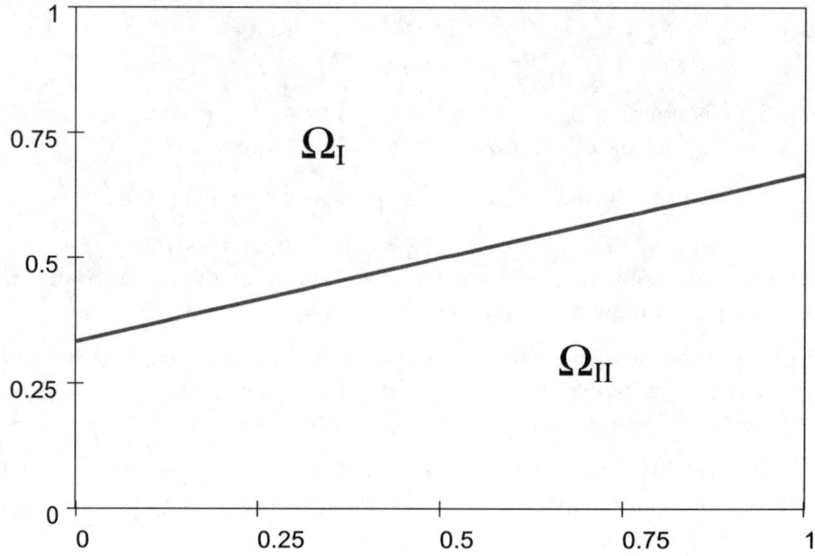

Fig. 1. Ω_I and Ω_{II} for the example

We are now ready to show that:

Theorem 1.

$$\mu_{l,\mathbf{F}}[U] = \sum_{i=1}^{N} \mu^i_{l,\mathbf{F}}[U] \,, \tag{8}$$

$$M_{l,\mathbf{F}}[U] = \sum_{i=1}^{N} M^i_{l,\mathbf{F}}[U] \tag{9}$$

i.e. any l^{th} order moment (central or non-central) of U equals the sum of the contributions to that moment of all the alternatives.

Proof. We carry on first the proof for non-central moments [Eq. (7)]. We note that at every point π:

$$U = u^*(\pi) = \max_j \left[u^j(\pi)\right] \;\; j = 1, \ldots, N \,, \tag{10}$$

that is

$$u^*(\pi) = u^j(\pi) \text{ if } u^j(\pi) > u^k(\pi) \; \forall k = 1, \ldots, N, \; k \neq j \,. \tag{11}$$

Thus, at every point π, $u^*(\pi) = u^j(\pi)$, where $a^j(\pi)$ is the preferred alternative. Let Ω_s the region in which alternative s is preferred. Clearly if alternative s is never preferred $\Omega_s = \emptyset$. Then, utilizing the fact that, by construction, $\Omega = \Omega_1 \cup \Omega_2 \cup \ldots \cup \Omega_N$ and the linearity of the integration operator, one gets:

$$\mu_{l,\mathbf{F}}[U] = \int_{\Omega} [u^*(\pi)]^l \, \mathrm{d}F = \int_{\Omega_1 \cup \Omega_2 \cup \ldots \cup \Omega_N} [u^*(\pi)]^l \, \mathrm{d}F = \sum_{s=1}^{N} \int_{\Omega_s} [u^j(\pi)]^l \, \mathrm{d}\mathbf{F} \,. \tag{12}$$

The extension of the proof to central moments is obtained just noticing that $E_F[U]$ is a constant with respect to integration. □

Let us then express $V_{\mathbf{F}}[U]$ as a function of the alternative contributions.

Corollary 1.

$$E_{\mathbf{F}}[U] = \sum_{s=1}^{N} \int_{\Omega_s} u^s(\pi) \mathrm{d}\mathbf{F} = \sum_{i=1}^{N} M^i_{1,\mathbf{F}}[U] \tag{13}$$

and

$$V_{\mathbf{F}}[U] = \sum_{s=1}^{N} \int_{\Omega_s} [u^s(\pi)]^2 \, \mathrm{d}\mathbf{F} - \left[\sum_{s=1}^{N} \int_{\Omega_s} u^s(\pi) \mathrm{d}\mathbf{F}\right]^2 = \sum_{i=1}^{N} M^i_{2,\mathbf{F}}[U] \,, \tag{14}$$

where N is the number of alternatives.

Proof. Equation (13) is immediately proven by Eq. (7), setting $l = 1$. Equation (14) is a consequence of the properties of variance and of Eqs. (7) and (13). □

Equation (13) states that $E_{\mathbf{F}}[U]$ is the sum of the expected value (according to **F**) of the utility of alternatives in the regions they are preferred.

Equation (14) states that the epistemic variance of U is the sum of the variance of the utility of alternatives in the regions they are preferred.

3 Variance Decomposition and Global Importance

In order to introduce the notion of parameter global importance [2, 17, 18, 23, 24, 34–36], we need focus on the variance of U as a function of the uncertain parameter, $V_{\mathbf{F}}[U]$. In particular, we make use of the following.

Theorem 2 (Sobol', 1990). *Let $\mathbf{x} \in [0,1]^n$ be a set of random independent variables uniformly distributed in the unitary hypercube*[4], *and $Y = f(\mathbf{x})$ an*

[4] We are reporting Sobol' theorem following the notation originally utilized by the author in Sobol' (1993). The choice of proving the theorem is the $[0,1]$ space is classical in the numerical integration and Monte Carlo literature. It is a consequence of the following theorem at the basis of the Monte Carlo method:

Theorem 3. *Let $F_X(x)$ the distribution function of X and consider the random variable $y = g(x)$ with $g(x) = F_X(x)$. Then y is uniformly distributed in $[0,1]$.*

integrable function. Then the variance of Y can be uniquely decomposed in the following sum:

$$V_x[Y] = \sum_{i=1}^{n} V_i + \sum_{i<j} V_{i,j} + \sum_{i<j<m} V_{i,j,m} \ldots + V_{1,2,\ldots n} , \tag{16}$$

where

$$V_{i,j,\ldots,m} = \int \cdots \int [f_{i,j\ldots,m}(x_i, x_j, \ldots, x_m)]^2 \prod_{k=i,j,\ldots,m} \mathrm{d}x_k , \tag{17}$$

with

$$f(\mathbf{x}) = f_0 + \sum_{i=1}^{n} f_i(x_i) + \sum_{i<j} f_{i,j}(x_i, x_j) + \ldots + f(x_i, x_j, \ldots, x_n) , \tag{18}$$

with the $f_{i,j}(x_i, x_j)$ calculated as follows:

$$\begin{aligned} f_0 = E_{\mathbf{x}}[Y] &= \textstyle\int \cdots \int f(\mathbf{x}) \prod_k \mathrm{d}x_{\mathbf{k}}{}^5 \\ f_0 + f_i(x_i) &= \textstyle\int \cdots \int f(\mathbf{x}) \prod_{k \neq i} \mathrm{d}x_k , \\ f_0 + f_i(x_i) + f_{i,j}(x_i, x_j) &= \textstyle\int \cdots \int f(\mathbf{x}) \prod_{k \neq i,j} \mathrm{d}x_k \\ &\ldots \end{aligned} \tag{19}$$

$V_{i,j,\ldots,m}$ are deemed as interaction terms of order r, where r is the number of parameters they involve. $\sum_{i=1}^{n} V_i$ represents the portion of the variance of Y explained by the individual parameter uncertainty. Similarly, $\sum_{i<j} V_{i,j}$ is the portion of V explained by terms containing parameter pairs and so on.

Then, one defines the parameter global importance as [6, 7, 36–38, 40]:

$$\Phi_i = \frac{V_i + \sum_{j \neq i} V_{i,j} + \ldots + V_{1,2,\ldots n}}{V_x[Y]} , \tag{20}$$

There follows that the random numbers X:

$$x = F_X^{-1}(u) \tag{15}$$

have distribution $F_X(x)$ if and only if random variable U is uniformly distributed in the $[0,1]$ interval.

The above theorem states that any distribution of random variables can be obtained by a deterministic transformation ($F_X^{-1}(u)$) from numbers uniformly generated in the $[0,1]$ space. The fact that once specified $F_X(u)$, the operation $x = F_X^{-1}(u)$ is deterministic explains the reason why Sobol' theorem, proven in the $[0,1]^n$ space holds for generic distributions.

[5] The notation follows the one of Sobol' (1993).

i.e. the ratio of all individual and interaction terms involving x_i and $V_x[Y]$. Φ_i is the fraction of $V_x[Y]$ associated with x_i . Φ_i is used in the practice to provide guidance in data and information collection [2, 23, 24, 26]. In the literature, Φ_i were introduced as uncertainty importance measures, and built in such a way that the parameter with the highest Φ_i are the most effective in reducing $V_x[Y]$. One says that they are the most effective in reducing uncertainty [2, 6, 7, 17, 18, 23, 24, 26, 34–39]. This is related to the traditional choice of $V_x[Y]$, in the practice, as the privileged indicator of uncertainty, and is due to the fact that if $V_x[Y]$ was reduced to 0, then all the higher order moment would collapse to 0. In the realm of Decision-Theory we recall that standard deviation was utilized by Fishburn as a measure of epistemic uncertainty aversion [16].

In the previous section, we have stated the decomposition of $V_{\mathbf{F}}[U]$ in terms of alternative contributions to uncertainty. We now want to state the relationship between alternative contributions and uncertain parameter contributions. Making use of Sobol' theorem, we are ready to relate the alternative contributions to uncertainty and the parameter contributions to uncertainty. We can, in fact, prove the following:

Theorem 4. *Let $\Omega_i \subset \Omega$ denote the subspace obtained fixing π_i at a certain value, $\Omega_{i,j}$ the subspace obtained fixing π_i and π_j and so on so forth, and let the symbol dF/i mean the integration all the parameters but π_i, $dF/i,j$ mean the integration on all the parameters but i and j, etc. Then it is true that:*

$$
\begin{aligned}
&\sum_{s=1}^{N} \int_{\Omega_s} [u^s(\pi)]^2 \,\mathrm{d}F - \left[\sum_{s=1}^{N} \int_{\Omega_s} u^s(\pi)\,\mathrm{d}F\right]^2 = \\
&\sum_{i=1}^{n} \int \mathrm{d}F_i \left[\sum_{s=1}^{N} \int_{\Omega_i \cap \Omega_s} u^s(\pi)\,\mathrm{d}F/i\right]^2 \\
&+ \sum_{i<j} \int \mathrm{d}F_i\,\mathrm{d}F_j \left[\sum_{s=1}^{N} \int_{\Omega_{i,j} \cap \Omega_s} u^s(\pi)\,\mathrm{d}F/i,j \right.\\
&\left. - \int_{\Omega_i \cap \Omega_s} u^s(\pi)\,\mathrm{d}F/i - \int_{\Omega_j \cap \Omega_s} u^s(\pi)\,\mathrm{d}F/j + E_{\mathbf{F}}[U]\right]^2 + \ldots
\end{aligned}
\tag{21}
$$

Proof. It is enough to combine Corollary 1 [Eq. (14)] and Sobol' Theorem [Eqs. (16)–(19)]. In fact, both sides of Eq. (21) equal $V_{\mathbf{F}}[U]$. The right hand side of Eq. (21) is Sobol' theorem applied to U, where each of the terms inside the square is one of the terms in Eq. (19), and the same logic as in Theorem 1 proof has been applied to identify the integration regions. □

The meaning of Eq. (21) is as follows. The sum of the contributions to uncertainty of the preferred alternatives equals the sum of the contributions to uncertainty of the uncertain parameters.

4 An Illustrative Example

In this section we apply the results discussed in Sect. 1, 2 and 3 for the computation of the moments of U and for its variance decomposition to Example 1 discussed in Sect. 2.

U as a function of the parameters becomes [Eq. (4)]:

$$U = \begin{cases} u_I(\pi) & \text{if } \pi \in \Omega_I \\ u_{II}(\pi) & \text{if } \pi \in \Omega_{II} \end{cases} . \tag{22}$$

Let us then start applying Theorem 1 and Corollary 1 (Sect. 2). We have [Eq. (13)]:

$$E_{\mathbf{F}}[U] = \int_{\Omega_I} (\pi_1 u_1 + (1-\pi_1)u_2)\, d\pi_1\, d\pi_2 + \int_{\Omega_{II}} (\pi_2 u_3 + (1-\pi_2)u_4)\, d\pi_1\, d\pi_2 \tag{23}$$

$$= \int_0^1 \int_{\pi_2(\frac{u_3-u_4}{u_1-u_2})+\frac{u_4-u_2}{u_1-u_2}}^1 (\pi_1 u_1 + (1-\pi_1)u_2)\, d\pi_1\, d\pi_2 + \int_0^1 \int_0^{\pi_2(\frac{u_3-u_4}{u_1-u_2})+\frac{u_4-u_2}{u_1-u_2}} (\pi_2 u_3 + (1-\pi_2)u_4)\, d\pi_1\, d\pi_2 \tag{24}$$

Thus:

$$E_{\mathbf{F}}[U] = \frac{1}{6} \frac{u_3^2 - 3u_2u_4 + u_3u_4 + u_4^2 - 3u_2u_3 + 3u_1^2}{u_1 - u_2} . \tag{25}$$

Substituting for the numbers, one gets: $E_{\mathbf{F}}[U] = \frac{17}{27}$. For the calculation of $V_{\mathbf{F}}[U]$, Eq. (14) becomes:

$$V_{\mathbf{F}}[U] = \int_{\Omega_I} \left\{u^I(\pi) - E_{\mathbf{F}}[U]\right\}^2 d\pi_1\, d\pi_2 + \int_{\Omega_{II}} \left\{u^{II}(\pi) - E_{\mathbf{F}}[U]\right\}^2 d\pi_1\, d\pi_2 = 2.95 \times 10^{-2} . \tag{26}$$

The contributions to the variance of alternatives I and II are:

$$\int_0^1 \int_{\pi_2(\frac{u_3-u_4}{u_1-u_2})+\frac{u_4-u_2}{u_1-u_2}}^1 \left\{(\pi_1 u_1 + (1-\pi_1)u_2) - E_{\mathbf{F}}[U]\right\}^2 d\pi_1\, d\pi_2 = 1.89 \times 10^{-2} \tag{27}$$

$$\int_0^1 \int_0^{\pi_2(\frac{u_3-u_4}{u_1-u_2})+\frac{u_4-u_2}{u_1-u_2}} \left\{(\pi_2 u_3 + (1-\pi_2)u_4)^2 - E_{\mathbf{F}}[U]\right\}^2 d\pi_1\, d\pi_2 = 1.06 \times 10^{-2} . \tag{28}$$

Theorem 1 allows to compute the remaining central moments of U through a similar approach. The moments up to order six and their alternative contributions are displayed in Table 2.

Table 2 shows that I is the alternative that provides the highest contribution to the central moments of even order of U. The odd-order central moments of U see a positive contribution from I and a negative one from II, with the contributions of I having a higher absolute value.

Let us now turn to the assessment of variance contribution of the parameters – the probabilities in this case – [Eq. (21)]. The left hand side, in accordance with Sobol' variance decomposition method, will contain three terms: V_1, V_2, the two individual terms, and $V_{1,2}$, the interaction term. We have:

$$V_1 = \int_0^1 (\int_0^1 u(\pi)\,\mathrm{d}\pi_2 - E_F[U])^2\,\mathrm{d}\pi_1 = 2.52 \times 10^{-2} \tag{29}$$

$$V_2 = \int_0^1 (\int_0^1 u(\pi)\,\mathrm{d}\pi_1 - E_F[U])^2\,\mathrm{d}\pi_2 = 2.3 \times 10^{-3} \tag{30}$$

and

$$\begin{aligned} V_{1,2} &= \int_0^1 \int_0^1 \left\{ u(\pi) - \int_0^1 u(\pi)\,\mathrm{d}\pi_1 - \int_0^1 u(\pi)\,\mathrm{d}\pi_2 + E_F[U] \right\}^2 \mathrm{d}\pi_1\,\mathrm{d}\pi_2 \\ &= 2.0 \times 10^{-3}\,. \end{aligned}$$

The global importance of π_1 and π_2 is, then, respectively:

$$\Phi_1 = \frac{V_1 + V_{12}}{V} = 0.92\,, \tag{31}$$

$$\Phi_2 = \frac{V_2 + V_{12}}{V} = 0.14\,. \tag{32}$$

The results in Eqs. (31) and (32) indicate that π_1 is the most important parameter, while π_2 and the interaction term play a less relevant role. We recall that both π_1 and π_2 were assigned uniformly in [0, 1]. Thus, although the DMr is, in principle, "equally uncertain" on the parameters, the above indication means that the most effective way for the DMr to reduce His/Her uncertainty would be to reduce uncertainty in π_1.

Table 2. Expected value and central moments of U

l	$M_{l,F}[U]$	$M^{I}_{l,F}[U]$	$M^{II}_{l,F}[U]$
2	2.95×10^{-2}	1.89×10^{-2}	1.06×10^{-2}
3	2.10×10^{-3}	4.36×10^{-3}	-2.27×10^{-3}
4	1.98×10^{-3}	1.14×10^{-3}	5.24×10^{-4}
5	0.290×10^{-3}	0.42×10^{-3}	-1.27×10^{-3}
6	0.17×10^{-3}	0.14×10^{-3}	0.031×10^{-3}

Let us now compare this result to the contribution of the probabilities to the uncertainty of each of the alternatives. The probability global importance on I, $\mathbf{\Phi}^I$, is immediately computed as:

$$\Phi_1^I = 1, \, \Phi_2^I = 0 \,. \tag{33}$$

That is, π_1 has global importance equal to unity on alternative I. In fact, uncertainty in π_2 is non-influential on the uncertainty in I since π_2 is non-influential on $u_I(\pi)$. As a consequence, there are also no interaction terms. For alternative II, symmetrically:

$$\Phi_1^{II} = 0, \, \Phi_2^{II} = 1 \,. \tag{34}$$

These results reflect the fact that uncertainty in lottery I is provoked by π_1 alone and uncertainty in lottery II is provoked by π_2 alone. However, when one turns to the uncertainty of the expected utility U of the optimal act, both probabilities play a role (Sect. 1). Mathematically, the coupling effect is attributable to the presence of the max function [Eq. (10)]. Based on the above numbers, it is easy to see that $V_{\mathbf{F}}[u^I] > V_{\mathbf{F}}[u^{II}]$. Utilizing $V_{\mathbf{F}}[u^I]$ as a measure of dispersion or uncertainty, one would say that I is the alternative characterized by the highest uncertainty. I is also the alternative that contributes to uncertainty the most (Table 2). This effect is also consistent with the fact $u^I(\pi)$ depends on π_1, which is the most influential uncertain parameter. However, while this can provide some insights on the structure of uncertainty, and particularly in the structure of the uncertainty space, it may not be a generalizable result, i.e. the most uncertain alternative is not necessarily the most influential one on the overall uncertainty. This happens if, for example, an uncertain alternative is dominated by a certain one. Then no uncertainty in the decision would even be present.

5 Case Study Application: Conflict-of-Interest Uncertainty Analysis

We apply the previous framework to the uncertainty analysis of the following industrial investment decision problem. The contractual structure is displayed

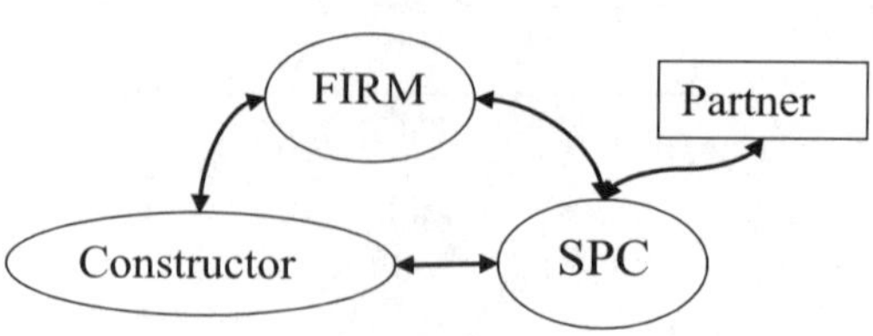

Fig. 2. The contractual structure evidences the conflict of interest generating the DM problem

in Fig. 2. A company is constructor of a production facility. At the same time, the company has won the sale concession bid, and in order to operate the facility and exploit the concession a special purpose company (SPC) has been founded, owned at 95% by the company. The remaining 5% is in the hands of a partner that, at the moment of the decision, has the option of stepping in at 50% of the project.

The partner option lasts one year. After the start of construction, a change in some conditions allows a reduction in the construction price. Management has to make a decision on whether to allocate the benefit to the company as a constructor or leave it to the SPC in the form of an investment cost discount. Leaving the benefit to SPC gives the highest benefit from a financial point of view. However, the company would be exposed to the partner option, i.e., the possibility of the partner stepping back into the project and thus reducing the benefit by 45%. Besides, there is a regulatory risk related to the possibility of the Regulator reducing the sale price, in view of the reduced construction cost. In this case, the financial benefit would decrease to zero. Furthermore, since leaving the benefit to the SPC would mean to recover it along the installation life, the company assesses the probability that an unfavorable event stopped the life of the project (Operation Risk).

The problem can be represented by the decision tree illustrated in Fig. 3.

Figure 3 also shows the utilities assessed for each consequence. The uncertain events and the corresponding probabilities are illustrated in Table 3.

The parameters are the probabilities of the three events ($\pi = \pi_1, \pi_2, \pi_3$) which the DMr assigned uniformly in $[0, 1]$. Hence: $\Omega = [0, 1] \times [0, 1] \times [0, 1]$.

Let us now discuss the uncertainty quantification and decomposition in the framework of Eqs. (12), (13) and (14). As a first task, we have to determine

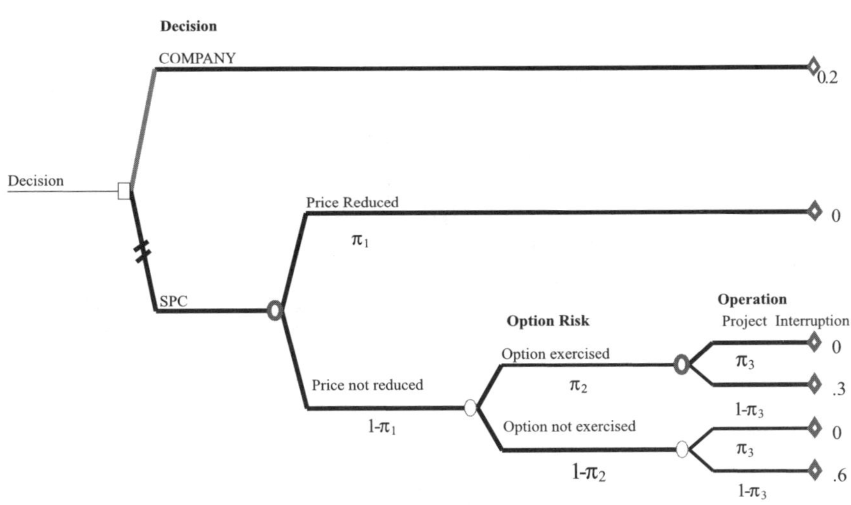

Fig. 3. The DM model for the case study

Table 3. Uncertain events and probabilities for the case study

Event	Outcomes	Probabilities
Regulatory intervention	Price reduced\Price Not Reduced	$\pi_1\backslash(1-\pi_1)$
Partner option	Option exercised\Option not exercised	$\pi_2\backslash(1-\pi_2)$
Unfavorable event	Project interrupted\ Project not interrupted	$\pi_3\backslash(1-\pi_3)$

the indifference hypersurface. The condition $u_A(\pi) = u_B(\pi)$ leads to the surface:

$$\pi_3 > 1 - \frac{2}{3}\frac{1}{(1-\pi_1)(2-\pi_2)} \tag{35}$$

The two preference regions are, then:

$$\Omega_A = \left\{\pi : \pi_3 > 1 - \frac{2}{3}\frac{1}{(1-\pi_1)(2-\pi_2)}\right\} \text{ and } \Omega_B = \Omega - \Omega_A \,.$$

The indifference surface is plotted in Fig. 4.

The value of $V_{\mathbf{F}}[U]$ is 2.13×10^{-3}. The first 6 central moments of the epistemic distribution of U are shown in Table 4, which also illustrates the alternative contributions.

Table 4 shows that uncertainty is driven by alternative B, whose contributions to the moments are constantly much higher than the contributions of A.

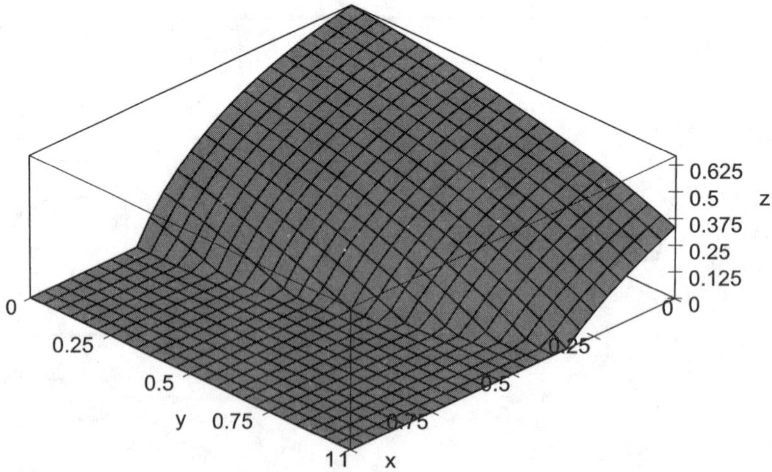

Fig. 4. Indifference surface and alternative preference regions for the case study

We now turn to the analysis of the contributions of the parameters (the three event probabilities in this case) to $V_{\mathbf{F}}[U]$. The left hand side of Theorem 3 [Eq. (21)], leads to the results for the variance decomposition on the probabilities shown in Table 5.

The corresponding parameter importance is reported in Table 6.

The results shows that π_1 and π_3 are the most influential parameters on the DMr uncertainty, with a less relevant role played by π_2.

Figure 5 compares the global importance of the parameters to their individual term contributions.

Table 4. Alternative contributions to uncertainty: central moments up to $l = 6$

l	$M_{l,F}[U]$	$M^{A}_{l,F}[U]$	$M^{B}_{l,F}[U]$
2	2.13×10^{-3}	2.23×10^{-4}	1.91×10^{-3}
3	3.39×10^{-4}	-3.70×10^{-6}	3.42×10^{-4}
4	7.16×10^{-05}	6.14×10^{-8}	7.16×10^{-5}
5	1.66×10^{-05}	-1.02×10^{-9}	1.66×10^{-5}
6	4.12×10^{-06}	1.69×10^{-11}	4.12×10^{-6}

Table 5. Results of the uncertainty decomposition for the case study

Term	**Value**
V_1	4.62×10^{-4}
V_2	1.05×10^{-4}
V_3	4.62×10^{-4}
$V_{1,2}$	1.33×10^{-4}
$V_{1,3}$	6.79×10^{-4}
$V_{2,3}$	1.33×10^{-4}
$V_{1,2,3}$	1.6×10^{-4}

Table 6. Global Importance of the uncertain probabilities in the case study

π	$\mathbf{\Phi}$
π_1	0.67
π_2	0.25
π_3	0.67

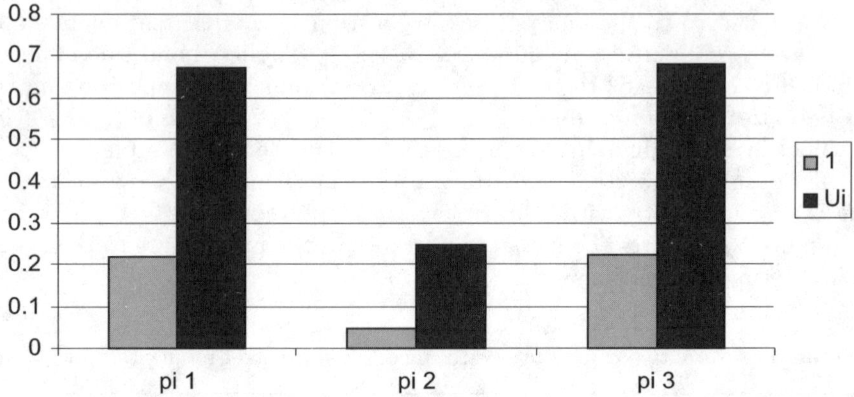

Fig. 5. Individual (*left* of each pair of rectangles) parameter contribution to Φ_i

Figure 5 shows that π_1 and π_3 are more relevant than π_2 also when taken individually. However, it is immediate to note that interaction terms play a relevant role. In particular, it is significant the interaction between π_1 and π_3, with $V_{1,3}/V_{\mathbf{F}}[U] = 0.32$. π_1 and π_3 individually, and through their interaction, explain more than 80% of the DM uncertainty. The practical implication of this result is as follows (see our discussion in Sect. 3): collecting information on the probability of a regulatory intervention (π_1) and of an unfavorable event during operation (π_3) would be the most effective way for reducing uncertainty in the decision.

A numerical propagation of uncertainty was also performed, with 10000 Monte Carlo trials. The results obtained numerically confirmed the ones utilizing the methodology proposed above.

6 Conclusions

In this work, we have dealt with a framework for the quantification and decomposition of uncertainty in DM problems. We have seen that, in order to quantify uncertainty, the decision expected utility (U) must be considered a function of the uncertain parameters [Eq. (2)]. The epistemic distribution of U by propagation of the uncertainty in π has then been the target of our analysis.

We have shown that an analytical calculation of the central and non-central moments of U is feasible and that, furthermore, all the moments can be decomposed in the sum of the contributions to uncertainty of the alternatives. Instrumental to this result have been the definitions of alternative preference region and of indifference hypersurface.

We have then focused on $V_{\mathbf{F}}[U]$. We have restated Sobol' variance decomposition theorem in the DM framework and utilized it to find the decomposition of $V_{\mathbf{F}}[U]$ as a function of the parameters. The concept of global importance of the parameters probabilities has then be straightforwardly introduced. This has enabled us to provide an equality relating alternative contributions to $V_{\mathbf{F}}[U]$ and the variance contributions of the parameters.

To illustrate the previous results, we have first made use of an analytical example, highlighting the procedure for the calculation of the moments (central and non central) of U as sum of the contributions to uncertainty of the alternatives. We have computed the first 6 moments of U and analyzed the contribution to such moments of the two alternatives of the example. We have seen that, although the DMr was, in principle, equally uncertain on the probabilities, the first alternative contributed to uncertainty the most. We have then computed the probability global importance and determined the influence of the uncertain parameters on the DMr uncertainty. We have seen that although the probabilities were characterized by the same uncertainty distribution, the probability related to the first random event (π_1) influenced the uncertainty in the decision the most. We have then compared the influence of the uncertain parameters on each alternative to their influence on the overall decision. We have noted that, although π_1 influenced the sole uncertainty in I and π_2 influenced only the uncertainty in II, both π_1 and π_2 resulted influential on the overall uncertainty in the decision.

We have then applied the method to an industrial case study, and have been able to identify the alternative contributions to uncertainty, as well as the probability contributions. In particular, we have seen that the probability of regulatory intervention and of an unfavorable event during operation resulted as the most influential ones. This would mean that reducing the uncertainty on these probabilities would provide the most efficient way of reducing the uncertainty in the problem.

It is part of the future work of the authors the study of the influence of additional information on the importance of uncertain parameter, and the analysis of the relationship between parameter global importance and expected value of perfect and sample information.

Acknowledgement. We would like to thank Prof. Bertrand Munier and an anonymous referee for the very useful comments. Financial support from the Italian Ministry of University, Instruction and Research is gratefully acknowledged.

References

1. Apostolakis, G.E. (1995), "A Commentary on Model Uncertainty", *Proceedings of the Workshop on Model Uncertainty: its Characterization and Quantification*, published by Center for Reliability Engineering, University of Maryland, College Park, Maryland, USA.
2. Archer, G.E.B., Saltelli, A., Sobol, I.M. (1997), "Sensitivity measures, ANOVA-like techniques and the use of bootstrap", *Journal of Statistical Computation and Simulation*, 58, pp. 99–120.

3. Anscombe, F.J., Aumann, R. J. (1963), "A Definition of Subjective Probability", *Annual of Mathematical Statistics*, 34, pp. 199–205.
4. Allais, M. (1953), "Le comportement de l'homme rationnel devant de risque: critique de le postulats et axiommes de l'ecole Americaine", *Econometrica*, 21, pp. 503–546.
5. Berger, J. (1985), "Statistical Decision Theory and Bayesian Analysis", *Springer Series in Statistics*, Second Edition.
6. Borgonovo, E. (2001), "Importance Relations and Sensitivity Analysis in Decision Theory", *Proceedings of the Tenth International conference on Foundations and Applications of Utility, Risk and Decision Theory*, Torino, May 30-June 2 2001, pp. 20–23.
7. Borgonovo, E., Apostolakis, G.E., Tarantola, S., Saltelli, A. (2003), "Comparison of Local and Global Sensitivity Analysis Techniques in Probabilistic Safety Assessment", *Reliability Engineering and System Safety*, 79, pp. 175–185.
8. Brams (1998), "Backward Induction Is Not Robust", *Theory and Decision*, 45, pp.263–289.
9. Chenghu Ma (2001), "A No-Trade Theorem under Knightian Uncertainty with General Preferences", *Theory and Decision*, 51 (2), pp. 173–181.
10. Chow, Sarin (2002), "Known, Unknown and Unknowable Uncertainty", 52, *Theory and Decision*, pp. 127–138.
11. Celemen, R.T. (1997), "Making Hard Decisions: An Introduction to Decision Analysis", II Edition, *Duxbury Press*, Pacific Grove, Calif, USA; ISBN: 0534260349.
12. de Finetti, B. (1937), "La prevision: ses lois logiques, ses sources subjectives", *Annales de l'Institute Henri Poincarè*, Vol. 7, 1–68. Translated into English by H.E. Kyburg and reprinted in Kyburg and Smokler (1964).
13. Eichberger, J. (1997), "Dynamically Consistent Preferences with Quadratic Beliefs", *The Journal of Risk and Uncertainty*, 14 (2), 189–207.
14. Eichberger, J. (1999), "E-Capacities and the Ellsberg Paradox", *Theory and Decision*, 46 (2), pp. 107–138.
15. Ellsberg, D. (1961), "Risk, Ambiguity and the Savage Axioms", *Quarterly Journal of Economics*, 75, pp. 643–669.
16. Fishburn, P.C. (1994), "A Variational Model of Preference under Uncertainty", *The Journal of Risk and Uncertainty*, 8 (2), pp. 127–52.
17. Helton, J.C. (1993), "Uncertainty and Sensitivity Analysis Techniques for Use in Performance Assessment for Radioactive Waste Disposal", *Reliability Engineering and System Safety*, 42, 327–367.
18. Hofer, E. (1999), "Sensitivity Analysis in the Context of Uncertainty Analysis for Computational Intensive Models", *Computer Physics Communications*, 177, 21–34.
19. Keeney, R., Raiffa, H. (1993), "Decisions with Multiple Objectives", *Cambridge University Press*, Cambridge, U.K., ISBN 0-521-43883-7.
20. Knight (1921), "Risk Uncertainty and Profit", University of Chicago Press, Chicago.
21. Kobberling, V., Wakker, P.P. (2004), "A Simple Tool for Qualitatively Testing, Quantitatively Measuring, and Normatively Justifying Savages Subjective Expected Utility", *The Journal of Risk and Uncertainty*, 28 (2), pp. 135–145.
22. Kreps, D.M. (1988), "Notes on the Theory of Choice", *Underground Classics in Economics*, Westview Press Inc., Colorado, USA, ISBN 0–88133-7553–3.

23. Iman, R.L., Conover, W.J. (1987), "A Measure of Top-down Correlation", *Technometrics*, 29 (3), pp. 351–357.
24. Iman, R.L., Hora, S.C. (1990), "A Robust Measure of Uncertainty Importance for Use in Fault Tree System Analysis", *Risk Analysis*, 10 (3), pp. 401–406.
25. Machina, M. (2004), "Almost Objective Uncertainty", *Economic Theory*, 24 (1), pp. 1–54.
26. McKay, M.D. (1996), "Variance-Based Methods for Assessing Uncertainty Importance", *in NUREG-1150 Analyses*, LA-UR-96–2695, 1–27.
27. Montesano (2001), "Uncertainty with Partial Information on the Possibility of the Events", *Theory and Decision*, 51, pp. 183–195.
28. Morris (1997), "Risk, Uncertainty and Hidden Information", *Theory and Decision*, 42, pp. 235–270.
29. Ivanenko, V.I., Munier, B. (2000), "Decision Making in 'Random in a Broad Sense' Environments", *Theory and Decision*, 49 (2): 127–150.
30. Pratt, J.W., Raiffa, H., Schlaifer, R. (1995), "Introduction to Statistical Decision Theory", *The MIT Press*, 904 pages, ISBN: 0262161443.
31. Quiggin, J. (2001), "Production under Uncertainty and Choice under Uncertainty in the Emergence of Generalized Expected Utility Theory", *Theory and Decision*, 51 (2), pp. 125–144.
32. Savage, L. (1972), The Foundations of Statistics, *Dover Publications*, Second Edition, 39, pp. 31–33.
33. Schlee, E.E. (1997), "The sure thing principle and the value of information", *Theory and Decision*, 42 (1), pp. 21–36.
34. Saltelli, A. (1997), "The Role of Sensitivity Analysis in the Corroboration of Models and its Link to Model Structural and Parametric Uncertainty", *Reliability Engineering and System Safety*, 57, 1–4.
35. Saltelli, A., Tarantola, S., Chan, K.P.-S. (1999), "A Quantitative Model-Independent Method for Global Sensitivity Analysis of Model Output", *Technometrics*, 41, 1, 39–56.
36. Saltelli, A., Sobol, I.M. (1995), "About the Use of Rank Transformation in Sensitivity Analysis of Model Output", *Reliability Engineering and System Safety*, 50, 225–239.
37. Sobol, I.M. (2001), "Global sensitivity indices for nonlinear mathematical models and their Monte Carlo estimates", *Mathematics and Computers in Simulation*, 55 (1), pp. 271–280.
38. Sobol, I.M. (1993), "Sensitivity estimates for nonlinear mathematical models", *Matem. Modelirovanie*, 2 (1) (1990) 112–118 (in Russian). English Transl.: MMCE, 1 (4) (1993) 407–414.
39. Sobol, I.M. (1967), "On the Distribution of Points in a Cube and the Approximate Evaluation of Integrals", *USSR Comp. Math. Phys.*, 7, 86–112.
40. Sobol, I.M. (1993), "Sensitivity Estimates for Nonlinear Mathematical Models", *Mathematical Modeling and Computational Experiments*, 1 (4), pp. 407–414.
41. Winston, W. (1998), "Financial Models Using Simulation and Optimization", *Palisade Editor,* Newfield, NY, USA.

Part II

The Importance of Representational Tools in Understanding Behaviour Under Uncertainty and Risk

Preference for Diversification with Similarity Considerations

Ani Guerdjikova

Cornell University, Department of Economics,
Uris Hall 462, Ithaca, NY 14853, USA
e-mail: ag334@cornell.edu.

Abstract. The paper studies the connection between the form of the similarity function of a decision-maker and his willingness to diversify. It is shown that preference for diversification obtain for both high and low aspiration levels if the similarity function is convex in the Euclidean distance. However, a decision-maker with a concave similarity function and relatively high aspiration level will fail to choose diversified acts, even if his utility function is concave.

Keywords: case-based decision theory, similarity, preference for diversification

1 Introduction

The expected utility theory predicts that a decision-maker with a concave von-Neumann–Morgenstern utility function should exhibit preference for diversification. In other words,whenever he is indifferent between two acts, he should weakly prefer any linear combination of these two acts to each of them, see Dekel (1989) and Chateauneuf and Tallon (2002). Since the expected utility theory does not allow to distinguish between risk-aversion (which is equivalent to preferences for diversification) and decreasing marginal utility, both of which are implied by the concavity of the von-Neumann–Morgenstern utility, risk-aversion is commonly assumed in applications. Empirical and experimental data, however contradict this assumption, the home-bias, i.e. the unwillingness of investors to hold diversified portfolios, see Kang and Stulz (1997) and Coval and Moskowitz (1999) being the most popular example.

Chateauneuf and Tallon (2002) address the issue of preferences for diversification in the context of Choquet expected utility. They show that concavity of the utility function is neither necessary nor sufficient for preferences for diversification to obtain. They demonstrate that convexity of the capacity is necessary for preferences for diversification to obtain.

In this paper I propose a different way to separate marginal utility from risk-preferences by using the case-based decision theory. Whereas the form of the utility function determines whether the decision-maker experiences increasing or decreasing marginal utility, his preferences for diversification are determined by his aspiration level and by his similarity perceptions.

The case-based decision theory proposed by Gilboa and Schmeidler (1995, 1997, 2001) models decisions in situations of structural ignorance, in which neither states of the world, nor their probabilities can be naturally derived from the description of the problem. It is assumed that a decision-maker can only learn from experience, by evaluating an act based on its own past performance and on the performance of acts *similar* to it. Similarity between two acts a and a' can be interpreted as the likelihood that the choice of act a' leads to a utility realization identical to the one derived from the choice of a.

In this paper, I use a modified version of the model of Gilboa and Schmeidler (1996) in which a case-based decision-maker is facing an identical decision in each period of time. Whereas in Gilboa and Schmeidler (1996) two acts are similar only if they are identical, in the current paper the set of acts is assumed to be a multidimensional simplex and the similarity is a decreasing function of the Euclidean distance between acts.

Since a case-based decision-maker has very little information about the decision problem he is facing, his initial decisions might be due to chance. Hence, his preferences if elicited in the initial periods might vary significantly. To avoid this initial randomness, I use the limit evaluation of acts after the same decision has been repeated for a long period of time to define the preferences of the decision-maker. It is with respect to these limit preferences that preference for diversification is defined.

The findings of the paper show that the willingness to diversify depends on two factors: the height of the aspiration level and the curvature of the similarity function, but not on the curvature of the utility function. Especially, a decision-maker with a relatively low aspiration level will express preference for diversification independently of the form of his similarity function. However, if the aspiration level is chosen sufficiently high, preference for diversification obtains only if the similarity function is convex.

A convex similarity function implies that the greater the distance of two acts from the referential act, the less is the decision-maker able to distinguish between these two acts with respect to their similarity to the referential one. If we are ready to assume this property, then preference for diversification will obtain independently of the aspiration level of the decision-maker. Billot, Gilboa and Schmeidler (2004) have recently provided an axiomatization of a case-based rule that uses an exponential similarity function to model similarity between acts situated on the real line. However, their model does not provide an intuition of why the exponential relation might be a sensible one. This paper provides a support for the usage of an exponential similarity function. Indeed, if willingness to diversify seems to be an appealing behavioral property, then a convex similarity function would insure that this property holds in a model with case-based decision-makers. On the other hand, concave similarity functions might explain why investors fail to choose a diversified portfolio and thus provide an alternative justification of the home-bias.

The rest of the paper is structured as follows: in Sect. 2, I present the model, which is similar to the model of Gilboa and Schmeidler (1996). In Sect. 3, I discuss preference for diversification when the similarity function is convex, whereas Sect. 3 deals with the case of a concave similarity function. Sect. 5 discusses some related results from the literature and concludes. The proofs of all results are stated in the appendix.

2 The Model

I use a version of the model of Gilboa and Schmeidler (1996). A decision-maker faces an identical decision problem p in each period $t = 1, 2 \ldots$. $A \equiv [0;1]^K$ with $K \in \mathbb{N}$ denoting the set of available acts. One can think about the corner acts (the unit vectors of the K-dimensional simplex) as of projects with unknown probability distribution of returns, into which a decision-maker would like to invest his initial endowment of one unit. The set A then represents all possible allocations of his endowment among the projects available. Let $\delta^1 \ldots \delta^K$ denote the random payoffs of the corner acts and suppose that for all $i = 1 \ldots K$, the distribution of δ^i is continuous and i.i.d. over time (although δ^i and δ^j might be correlated) with finite expectation, finite variance and bounded support. Obviously, the payoff of any act in the simplex can be expressed as a linear combination of $\delta^1 \ldots \delta^K$.

If the utility function of the decision-maker is bounded and continuous, then the utility resulting from the choice of $a \in A$ is an i.i.d. random variable $\mathfrak{U}_a$ with a continuous distribution function $(\Pi_a)_{a \in A}$. The distributions $(\Pi_a)_{a \in A}$ have finite expectations μ_a, finite variance σ_a and bounded and convex supports Δ_a. μ_a is continuous with respect to a.

The decision-maker's perception of similarity is described by a function $s : A \times A \to [0;1]$:

$$\begin{aligned} s(a;a) &= 1\,, \\ s(a;a') &= s(a';a)\,, \\ s(e^i;e^j) &= 0 \end{aligned}$$

for all distinct i and $j \in \{1 \ldots K\}$, where e^i denotes the ith unit vector. s depends only on the Euclidean distance between a and a'.

The memory of the decision-maker is represented by a set of cases. A case is a triple of a problem encountered, an act chosen and a utility realization achieved. Since the problem is identical in each period of time, a case is characterized by an act and a utility realization. As in Gilboa and Schmeidler (1996), the memory M_t contains only cases actually encountered by the decision-maker until period t:

$$M_t = ((a_\tau; u_\tau))_{\tau=1,2\ldots t} \, .$$

The aspiration level of the decision-maker in period t is $\bar{u}_t$. In the present paper, I will assume that $\bar{u}_t = \bar{u} = const$ and concentrate only on the influence of $\bar{u}$ and s on the willingness to diversify, thus neglecting the effect of aspiration adaptation.

The case-based decision-rule prescribes choosing the act with maximal cumulative utility in each period of time. The cumulative utility of an act a at time t is given by:

$$U_t(a) = \sum_{\tau=1}^{t} s(a; a_\tau)(u_\tau - \bar{u}) \, .$$

The set of all possible decisions paths that can be observed can be written as

$$S_0 = \{\omega = (a_t; u_t; \bar{u})_{t=1,2\ldots} \mid a_t \in A, \; u_t \in \Delta\} \, ,$$

where $\Delta = \cup_{a \in A} \Delta_a$ denotes the set of possible utility realizations. Let S_1 be the set of those paths on which the decision-maker chooses $\arg\max_{a \in A} U_t\,(a)$ in each period:

$$S_1 = \left\{\omega \in S_0 \mid a_t = \arg\max_{a \in A} U_t(a) \text{ for all } t = 1, 2 \ldots\right\} .$$

As well as a_t and u_t all variables introduced below depend on the path ω. I neglect this dependence in the notation for simplicity of exposition.

$C_t\,(a)$ denotes the set of periods preceding t in which a has been chosen:

$$C_t(a) = \{\tau < t \mid a_\tau = a\}$$

Let P be a probability measure on S_1 consistent with $(\Pi_a)_{a \in A}$, as in Gilboa and Schmeidler (1996, p. 11).

Denote by

$$\pi\,(a) = \lim_{t \to \infty} \frac{|C_t\,(a)|}{t}$$

the frequency with which a is chosen, if the limit on the right hand side exists. In general, this frequency will be path-dependent.

Willingness to diversify is defined as in Dekel (1989) and Chateauneuf and Tallon (2002). Assume that

$$a_1 \sim a_2 \sim \ldots \sim a_n \, .$$

Then a decision maker exhibits preferences for diversification if for any $\beta_1 \dots \beta_n \geq 0$ such that $\sum_{i=1}^n \beta_i = 1$

$$\sum_{i=1}^{n} \beta_i a_i \succeq a_k \text{ for any } k = 1 \dots n \, .$$

The preferences of a case-based decision-maker are captured by the cumulative utility he assigns to different acts. Clearly, these preferences will in general vary over time. Hence, to derive meaningful statements about the willingness to diversify, it seems reasonable to consider the preferences of a decision-maker in the limit as $t \to \infty$. Then, $a_1 \sim a_2$ will correspond to

$$\lim_{t \to \infty} \frac{U_t(a_1)}{U_t(a_2)} = 1 \, .$$

Preference for diversification will obtain if for all $a_1 \dots a_n$ such that

$$\lim_{t \to \infty} \frac{U_t(a_k)}{U_t(a_l)} = 1 \text{ for all } k \text{ and } l \in \{1 \dots n\}$$

$$\frac{\lim_{t \to \infty} U_t \left(\sum_{i=1}^n \beta_i a_i \right)}{\lim_{t \to \infty} U_t(a_k)} \geq 1 \, ,$$

if both the numerator and the denominator converge to $+\infty$ and

$$\frac{\lim_{t \to \infty} U_t \left(\sum_{i=1}^n \beta_i a_i \right)}{\lim_{t \to \infty} U_t(a_k)} \leq 1 \, ,$$

if the numerator and the denominator converge to $-\infty$. Note that the definition of preference for diversification depends (through U_t) on the chosen decision path. However, it will be shown that the emergence of preferences for diversification will depend only on the aspiration level and on the form of the similarity function and not on the specific path ω.

3 Preference for Diversification with a Convex Similarity Function

Consider a decision-maker whose perception of similarity is described by

$$s(a; a') = f\left(\|a - a'\|\right) \, ,$$

with $f' < 0$. The matrix

$$f'' \mathrm{d}E \cdot (\mathrm{d}E)^T + \mathrm{d}^2 E f'$$

is assumed to be positive definite for $a \neq a'$, where E denotes the Euclidean distance functional. Note that this assumption implies that for a given a

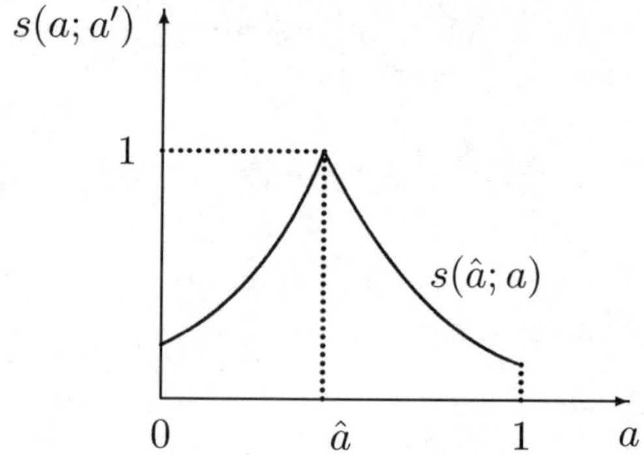

Fig. 1. Convex similarity function

$s\,(a;a')$ is convex on any set $\hat{A} \subset A$ such that $a \notin \hat{A}$. This follows from the fact that for $a \neq a'$ $s\,(a;a')$ is differentiable with:

$$\mathrm{d}^2 s = f''\mathrm{d}E \cdot (\mathrm{d}E)^T + \mathrm{d}^2 E f' \,.$$

The similarity function $s\,(\cdot;\cdot)$ is illustrated in Fig. 1 for the case $K = 2$.

Note that the similarity function itself cannot be convex over the whole set $[0;1]^K$ since it must assume a maximum at $s(a;a)$. However, I will refer to similarity functions described above as convex.

Let $a_1 = \bar{a}$ denote the act chosen in the first period and assume that $\bar{a} \in int(A)$. Ω describes the set of possible paths:

$$\Omega = \left\{ \omega \in S_1 \middle| \begin{array}{l} \bar{u}_t = \bar{u} \text{ for all } t = 1, 2, \ldots \\ a_1 = \bar{a} \end{array} \right\} .$$

Let P denote a probability measure on Ω consistent with $(\Pi_a)_{a \in A}$ as in Gilboa and Schmeidler (1996, p. 11).

Proposition 1. *Suppose that the similarity function is convex. If the aspiration level satisfies $\mu_a < \bar{u}$ for all $a \in A$ then for all a and $a' \in A$*

$$\lim_{t \to \infty} \frac{U_t(a)}{U_t(a')} = 1$$

holds P-almost surely.

Proposition 2. *Suppose that the similarity function is convex. If the aspiration level satisfies $\mu_a > \bar{u}$ for some $a \in A$, then P-almost surely an act*

$$a^* \in \tilde{A} = \{a \in A \mid \mu_a > \bar{u}\}$$

is chosen with frequency 1 in the limit.

Proposition 2 says that if the aspiration level of the decision-maker is relatively low, he will choose a single act a^* with frequency one in the limit. Moreover, $\mu_{a^*} > \bar{u}$ holds. In this case,

$$\lim_{t\to\infty} U_t(a) = +\infty$$

holds for all $a \in A$ and

$$\lim_{t\to\infty} \frac{U_t(a^*)}{U_t(a_l)} = \frac{1}{s(a^*; a_l)} \geq 1$$
$$\lim_{t\to\infty} \frac{U_t(a_k)}{U_t(a_l)} = \frac{s(a^*; a_k)}{s(a^*; a_l)} .$$

If $a^* = e^i$ for some $i \in \{1 \dots K\}$, there will be no distinct a_k and a_l such that

$$\lim_{t\to\infty} \frac{U_t(a_k)}{U_t(a_l)} = 1$$

holds. Hence, the condition of preference for diversification is trivially satisfied. If, however $a^* \in int(A)$, then

$$s(a^*; a_j) = s(a^*; a_i) ,$$

iff

$$\|a^* - a_j\| = \|a - a_i\| . \tag{1}$$

Obviously, then for any $a_1 \dots a_n$ satisfying (1) for any $i, j \in \{1 \dots n\}$,

$$\|a^* - a_k\| \geq \left\| a^* - \sum_{i=1}^{n} \beta_i a_i \right\|$$

for every $\beta_i \in [0; 1]$, $\sum_{i=1}^{n} \beta_i = 1$. Hence,

$$s\left(\sum_{i=1}^{n} \beta_i a_i; a^*\right) \geq s(a^*; a_i) = s(a^*; a_j)$$

for all i and $j \in \{1 \dots n\}$ and, therefore $\sum_{i=1}^{n} \beta_i a_i$ is (weakly) preferred to a_i in the limit for all $i = 1 \dots n$:

$$\lim_{t\to\infty} \frac{U_t\left(\sum_{i=1}^{n} \beta_i a_i\right)}{U_t(a_i)} = \frac{s\left(a^*; \sum_{i=1}^{n} \beta_i a_i\right)}{s(a^*; a_i)} \geq 1 .$$

In the case of high aspiration level (Proposition 1), since all acts fulfill

$$\lim_{t\to\infty} \frac{U_t(a)}{U_t(a')} = 1 ,$$

preference for diversification trivially obtains.

The following corollary obtains:

Corollary 1. *If the similarity function is convex, preference for diversification obtains independently of the aspiration level of the decision-maker.*

4 The Case of Concave Similarity Function

Assumption 1.

$$s(a;a') = f\left(\|a-a'\|\right) ,$$

where $f' < 0$ *and the matrix*

$$f''\mathrm{d}E \cdot (\mathrm{d}E)^T + \mathrm{d}^2 E f' ,$$

is negative definite.

Note that this assumption implies that s is concave.

s is illustrated in Fig. 2 for $K = 2$.

The concavity of s implies that the greater the distance of two acts a' and a'' from the reference act a, the more the decision-maker distinguishes between a' and a'' with respect to their similarity to a.

The next proposition shows how a decision-maker will behave if his aspiration level is higher than the mean utility of the initially chosen act.

Proposition 3. *Let the similarity function* $s(a;a')$ *of a decision-maker be concave. If* $\bar{u} > \mu_{\bar{a}}$ *and*

- $\bar{u} > \max\left\{\mu_{e^1};\ldots;\mu_{e^K}\right\}$, *then*

$$P\Bigg\{\omega \in \Omega \mid \exists \pi(a) : [0;1]^K \to [0;1] \text{ and } \frac{\pi\left(e^i\right)}{\pi\left(e^j\right)} = \frac{\mu_{e^j} - \bar{u}}{\mu_{e^i} - \bar{u}},\ \pi(a) = 0 \text{ for } a \in int(A)\Bigg\} = 1 ;$$

- $\bar{u} < \mu_{e^i}$ *for all* $i \in K' \subset \{1 \ldots K\}$, *then*

$$P\left\{\omega \in \Omega \mid \exists \pi(a) : [0;1] \to [0;1] \text{ and } \exists\, i \in K' \text{ such that } \pi\left(e^i\right) = 1\right\} = 1 .$$

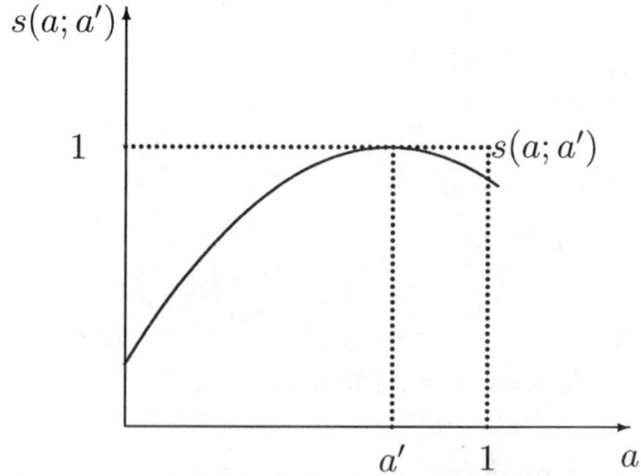

Fig. 2. Concave similarity

If the aspiration level of a decision-maker is relatively low, then the result of Corollary 1 holds and preference for diversification obtains.

However, this is not the case for relatively high aspiration levels. The following corollary obtains:

Corollary 2. *If the similarity function is concave, a decision-maker exhibits preference for diversification if and only if his aspiration level satisfies* $\bar{u} < \mu_a$ *for some* $a \in A$.

To see that the result of the corollary indeed holds, consider the case of $\bar{u} > \max_{i \in \{1 \ldots K\}} \{\mu_{e^i}\}$. An examination of the proof of Proposition 3 shows that in this case:

$$\lim_{t \to \infty} \frac{U_t\left(e^i\right)}{U_t\left(e^j\right)} = 1$$

for all i and $j \in \{1 \ldots K\}$ and

$$\lim_{t \to \infty} U_t(a) = -\infty$$

for all $a \in A$, whereas

$$\lim_{t \to \infty} \frac{U_t(a)}{U_t\left(e^i\right)} = \lim_{t \to \infty} \frac{U_t(a)}{U_t\left(e^i\right)} > 1$$

for all $i \in \{1 \ldots K\}$ and all $a \in int(A)$. Note, however that any interior act can be expressed as

$$a = \sum_{i=1}^{K} \beta_i e^i$$

for some non-negative β_i with $\sum_{i=1}^{K} \beta_i = 1$ and still the corner acts e^i are strictly preferred to $a \in int\,(A)$ in the limit. Hence, the decision-maker does not exhibit preference for diversification in this case. Note further that this result does not depend on the form of the utility function u.

5 Conclusion

The paper has shown that in the context of the case-based decision theory preferences for risk can be separated from marginal utility by means of the similarity function. A convex similarity function implies preference for diversification in the limit, independently of the form of the utility function. A concave similarity function is consistent with preference for diversification only if the aspiration level is sufficiently low.

The findings of the paper are consistent with the model of preferences for diversity proposed by Nehring and Puppe (2002, 2003). Differently from the approach of Gilboa and Schmeidler (1997), Nehring and Puppe derive a similarity function indirectly by first imposing conditions guaranteeing preference for diversity and then concluding what the similarity perceptions of a decision-maker with such a utility function might be. The similarity relation obtained is symmetric and transitive, but not necessarily complete. Nehring and Puppe (1999) compute the similarity function corresponding to preferences for diversity over acts situated on a one-dimensional simplex. They conclude that preference for diversity implies a similarity function which is convex in the Euclidean distance. Hence, despite the differences in the structure of the models, the implications of the curvature of the similarity function seem to be similar in both settings.

Up to now, few works have used the concept of similarity in case-based decisions. Gilboa and Schmeidler (2001, Chap. 19) show that positive (negative) similarity between goods can be interpreted in terms of complementarity (substitutability). Blonski (1999) proposes to model social structures using similarity functions. The similarity describes how relevant the experience of other members of the society is for the decision-maker at hand. He shows that different equilibria emerge depending on the structure of the society. However, both papers use similarity functions on finite set of acts, whereas similarity functions defined on uncountable sets are still largely unstudied, except for Gayer (2003).

Empirical evidence about similarity perceptions in economic situations is by large missing, the few exceptions being Buschena and Zilberman (1995, 1999) and Zizzo (2002). Their findings show that similarity between acts is related to the Euclidean distance between payoffs and influences decisions under risk. The present paper stresses that the exact form of the dependence on the Euclidean distance might matter at least in certain applications.

Given the difference in results obtained depending on the form of the similarity function, the question of deriving the "correct" similarity function arises. Shepard (1987) addresses the question of deriving the probability that an act a' would deliver the same outcome u as an act a actually observed to give an outcome u. This probability obviously can be interpreted as the similarity function in the model of Gilboa and Schmeidler (1997). The derivation follows the Bayesian updating rule for a given prior distribution over the set of a having identical realizations u. It can be shown, see Shepard (1987) that for an isotropic prior, e.g. a uniform prior, the resulting similarity function is convex in the Euclidean distance between the acts, see figure1 in Tenebaum and Griffiths (2001). Hence, Bayesian updating of a uniform prior can be used to rationalize a convex similarity function. A convex similarity function would on its turn imply preference for diversification in the present context. Non-isotropic priors, however, can lead to concavities in the similarity function and help explain empirically observed paradoxes such as the home-bias.

A Appendix

I start with two lemmas which will be useful in proving the results.

Lemma 1. *If $\bar{u} < \mu_{\bar{a}}$, then the expected time, during which the decision-maker will hold $\bar{a}$ is infinite. If $\bar{u} > \mu_{\bar{a}}$, then the decision-maker will almost surely switch in finite time to a corner act e^i such that*

$$a = \max_{i \in \{1 \ldots K\}} \left\| e^i - \bar{a} \right\| .$$

Proof (of Lemma 1). Suppose first that $\bar{u} < \mu_{\bar{a}}$. The cumulative utility of $\bar{a}$, as long as the investor holds it, is then a random walk with differences

$$\mu_{\bar{a}} - \bar{u} .$$

Since the expected value of the difference is $\mu_{\bar{a}} - \bar{u} > 0$ and the process starts at 0, the expected time until the first period in which the process reaches 0 is ∞. But, as long as $U_t(\bar{a}) > 0$, $U_t(a) = s(a;\bar{a})\, U_t(\bar{a}) \geq U_t(\bar{a})$, since $s(a;\bar{a}) \in [0;1]$ and, therefore, $\bar{a}$ is chosen.

Now suppose that $\bar{u} > \mu_{\bar{a}}$. Then, the expected increments of $U(\bar{a})$ are negative. Therefore, when the process starts at 0, it will cross any finite barrier below 0 in finite time. Let t be the first period, at which $U_t(\bar{a}) < 0$. Then $U_t(a) = s(a;\bar{a})\, U_t(\bar{a}) < 0$. Since $s = (a;a')$ is strictly decreasing in the distance between the acts, $U_t(a)$ has a maximum either at one of the corner acts. It follows that

$$a_{t+1} = \max_{i \in \{1 \ldots K\}} \left\| e^i - \bar{a} \right\| .$$

□

Lemma 2. *Define $V_t(a)$ as:*

$$V_t(a) = \sum_{\tau \in C_t(a)} \left[u_\tau(a) - \bar{u}_t \right] .$$

An act a is only abandoned in periods $\tilde{t}$ such that $V_{\tilde{t}}(a) < 0$.

Proof (of Lemma 2). In Lemma 1, it has already been shown that the statement of the lemma is true up to time $\bar{t}$ such that

$$\bar{t} = \min \left\{ t \mid U_t(\bar{a}) < 0 \right\} .$$

To argue by induction, suppose that the statement holds up to a period $t-1$ and consider period t. Denote by $a_1 \ldots a_l$ the acts that have been chosen up to period t in this order, $a_l = a_t$. Suppose that $V_t(a) \geq 0$. Then the cumulative utility of a_l at t can be written as:

$$U_t\,(a_l) = \sum_{\substack{i=1 \\ i \neq l}}^{l} V_t(a_i)s(a_i;a_l) + V_t(a_l)$$
$$\geq \sum_{\substack{i=1 \\ i \neq l}}^{l} V_{\bar{t}''}(a_i)s(a_i;a_l) + V_{\bar{t}''}(a_l)\,,$$

where $\bar{t}'' + 1$ denotes the last period prior to t in which the decision-maker has switched to a_l from a different act. The inequality follows from the fact that $V_{\bar{t}''}\,(a_l) \leq 0$, since either act a_l has been chosen for the first time at t'' and therefore $V_{\bar{t}''}\,(a_l) = 0$ or a_l has been abandoned for the last time at some time $t'' + 1 < \bar{t}'' + 1$ and then

$$V_{\bar{t}''}(a_l) = V_{t''}(a_l) < 0$$

must hold. Since the acts different from a_l have not been chosen after period $\bar{t}''$,

$$V_{\bar{t}''}(a_i) = V_t(a_i)$$

holds for $i \in \{1 \ldots l-1\}$.

Furthermore, since $a_{\bar{t}''+1} = a_l$, it must be that for all $a \in A$:

$$U_{\bar{t}''}(a_l) = \sum_{\substack{i=1 \\ i \neq l}}^{l} V_{\bar{t}''}(a_i)s(a_i;a_l) + V_{\bar{t}''}(a_l) \geq U_{\bar{t}''}(a)$$

holds. But then

$$U_t(a_l) - U_t(a) = \sum_{\substack{i=1 \\ i \neq l}}^{l} V_t(a_i)\left[s(a_i;a_l) - s(a_i;a)\right] + V_t(a_l)(1 - s(a_l;a))$$
$$\geq \sum_{\substack{i=1 \\ i \neq l}}^{l} V_{\bar{t}''}(a_i)\left[s(a_i;a_l) - s(a_i;a)\right] + V_{\bar{t}''}(a_l)(1 - s(a_l;a))$$
$$= U_{\bar{t}''}(a_l) - U_{\bar{t}''}(a) \geq 0\,.$$

Hence, $a_{t+1} = a_l$ if $V_t\,(a_l) \geq 0$ and hence, an act a_l can be only abandoned in a period $\tilde{t}$ such that $V_t\,(a_l) < 0$ holds. □

Proof (of Proposition 1). The proof of the proposition proceeds in two steps. First, I show that each open subset of A is chosen by the decision-maker for an infinite number of periods. This is an implication of the convexity of the similarity function and the negativity of net expected payoffs. Second, the negativity of net expected payoffs and the i.i.d. process of payoffs are used to demonstrate that the difference between the cumulative utilities of any two acts remains bounded in the limit. This implies the result of the proposition.

Lemma 3. *There is no $x \in int(A)$ such that for all acts $a \in B_x(\epsilon)$ (where $B_x(\epsilon)$ is an open ball with radius ϵ around $x \in (0;1)$), $|C_t(a)| < \infty$ holds.*

Proof (of Lemma 3). First note that no single act a can be chosen with frequency one, since then for any act $a' \neq a$, the $\mu_a - \bar{u} < 0$ would imply:

$$\lim_{t\to\infty} [U_t(a) - U_t(a')] = \lim_{t\to\infty} V_t(a)[1 - s(a';a)] \to -\infty \text{ a.s.}$$

Hence, choosing a in each period of time would contradict the case-based rule.

Suppose, therefore that only two acts a' and a'' are chosen infinitely often. Hence, there is a time T such that $a_t \in \{a';a''\}$ for all $t > T$. Denote the distinct acts chosen in periods $1\dots T$ by $a_1 \dots a_l$. The cumulative utility of act a at $t > T$ is given by:

$$U_t(a) = V_t(a')\, s(a;a') + V_t(a'')\, s(a;a'') + \sum_{i=1}^{l} V_t(a_i)\, s(a;a_i) \ .$$

By Lemma 2, a decision-maker will only switch away from an act a if $V_t(a) < 0$. Hence, $V_t(a_i) < 0$ holds for all $i = 1\dots l$. Whereas $V_t(a_i)$ are finite for all $i = 1\dots l$, $V_t(a')$ and $V_t(a'')$ a.s. tend to $-\infty$, i.e. for almost each ω there exists some time $t(\omega)$ such that $V_t(a')$ and $V_t(a'')$ are negative for all $t \geq t(\omega)$.

Since the similarity function is convex, it follows that $U_t(a)$ is strictly concave on the intervals:

$$\left(a'; \min\left\{\min_{i\in\{1\dots l\}} \{a_i \mid a_i > a'\}; a''\right\}\right)$$
$$\left(\max\left\{0; \max_{i\in\{1\dots l\}} \{a_i \mid a_i < a'\}\right\}; a'\right) ,$$

as well as on

$$\left(a''; \min\left\{\min_{i\in\{1\dots T\}} \{a_i \mid a_i > a''\}; 1\right\}\right)$$
$$\left(\max\left\{a'; \max_{i\in\{1\dots T\}} \{a_i \mid a_i < a''\}\right\}; a''\right) .$$

Hence, on almost each path, there exists a period of time $T'(\omega)$ such that there exist acts a''' and a'^{v} such that

$$U_t(a''') > U_t(a') \ ,$$
$$U_t(a'^{v}) > U_t(a'')$$

for all $t \geq T'(\omega)$ and still $a_t \in \{a';a''\}$ is chosen. This obviously contradicts the case-based rule. Clearly, the argument does not depend on the number of acts which are chosen infinitely often, as long as this number remains finite. Hence, an infinite (but countable) set of acts A' must be chosen infinitely often.

Suppose now that A' does not contain an act out of $B_x(\epsilon)$ for some $x \in int(A)$. By an argument similar to the above, we could find an element of $B_x(\epsilon)$, $\tilde{a}$ which has been chosen only for a finite number of times and show that from some point of time $T''(\omega)$, the cumulative utility of the acts in the interval

$$(\sup A' \backslash [x+\epsilon; 1]; \tilde{a})$$

is a concave function for all $t \geq T''(\omega)$. Hence, for all

$$a \in (\sup A' \backslash [x+\epsilon; 1]; \tilde{a})$$

$$U_t(a) > U_t(\sup A' \backslash [x+\epsilon; 1]) .$$

By the continuity of the cumulative utility function, there exists an act $a' \in A'$ which is chosen infinitely often and the cumulative utility of which lies below the cumulative utility of a in each period $t \geq T''(\omega)$, a contradiction. □

Remark 1. A similar argument can be used to show that every corner act in the simplex will be chosen infinitely often.

To complete the proof of the proposition, I now show that the difference between the cumulative utilities of any two acts:

$$U_t(a) - U_t(a') =: \varepsilon_t(a; a') \tag{2}$$

a.s. remains bounded over time. Since the expected mean payoffs of all acts are negative,

$$\lim_{t\to\infty} U_t(a) = -\infty$$

a.s. for all acts $a \in A$. This implies that

$$\lim_{t\to\infty} \frac{U_t(a)}{U_t(a')} = \lim_{t\to\infty} \frac{U_t(a)}{U_t(a) + \varepsilon_t(a; a')} = 1$$

holds on all paths on which $\varepsilon_t(a; a')$ remains bounded.

Hence, the proof of the following lemma would complete the proof of Proposition 1:

Lemma 4. *Define $\varepsilon_t(a; a')$ as in (2). On almost each path ω, $\varepsilon_t(a; a')$ is bounded.*

Proof (of Lemma 4). Consider first the acts in A' as defined in the proof of Lemma 3. Consider a period t in which the decision-maker switches to an act $a \in A'$ from a different act $a' \in A'$. Obviously, to satisfy the case-based rule:

$$U_{t-1}(a') \geq U_{t-1}(a)$$

and

$$U_t(a) \geq U_t(a')$$

must hold. Hence,

$$U_t(a) - U_t(a') \in \left[0; \left(\bar{u} - \min_{u \in \Delta_{a'}} u\right)(1 - s(a; a'))\right].$$

Now note, that starting from the interval

$$\left[0; \left(\bar{u} - \min_{u \in \Delta_{a'}} u\right)(1 - s(a; a'))\right],$$

the difference between the cumulative utilities of a and a' behaves as a random walk on a half-line with negative expected increments:

$$[1 - s(a; a')](\mu_a - \bar{u}) < 0,$$

as long as a is chosen. Define $\tilde{\varepsilon}_t(a; a')$ as

$$\tilde{\varepsilon}_t(a; a') = \varepsilon_t(a; a') \text{ if } \varepsilon_t(a; a') \geq 0$$
$$\tilde{\varepsilon}_t(a; a') = 0 \text{ , else}.$$

Such a random walk has an accessible atom at 0 (Meyn and Tweedie (1996, p. 105) give a definition of an accessible atom). Moreover, each set of the type $[0; c]$ is regular, see Meyn and Tweedie (1996, p. 278). This means that the state 0 is reached in finite expected time, starting from each set of the type $[0; c]$ and especially, starting from the set

$$\left[0; \left(\bar{u} - \min_{u \in \Delta_{a'}} u\right)(1 - s(a; a'))\right].$$

Denote the supremum of these expected times by $\mathcal{N}$ and observe that it is finite according to the definition of regular sets. Note that

$$\bar{u}_1 - \min_{u \in \Delta_{a'}} u$$

equals the supremum of $\varepsilon_t(a; a')$ in a period, in which the decision-maker switches from an arbitrary $\tilde{a}$ to a. Observe as well that since the probability that $\varepsilon_t(a; a') = 0$ is 0 (for atomless distributions Π_a), it follows that $\tilde{\varepsilon}_t(a; a') = 0$ a.s. coincides with $\varepsilon_t(a; a') < 0$. Hence, the decision-maker switches away from a when $\tilde{\varepsilon}_t(a; a') = 0$ is reached or in an earlier period. It follows that the expected time for which an arbitrary act a is held in a row is finite and uniformly bounded from above.

It remains to show that $\varepsilon_t(a;a')$ is bounded on almost each path of dividend realizations. At times at which a is chosen $\varepsilon_t(a;a')$ never falls below 0, since this would contradict choosing the act with highest cumulative utility in each period. Suppose, therefore that there is a sequence of periods t', $t''\dots$, such that $\varepsilon_{t'}(a;a')$, $\varepsilon_{t''}(a;a')\dots$ grows to infinity. In other words, suppose that for each $\mathcal{M}>0$ there is a k, such that $\varepsilon_{t^n}(a;a')>\mathcal{M}$ for all $n>k$. It has been shown above that each other act in A' and especially a' is chosen infinitely many times on almost each path of dividend realizations. But each time that the act a' is chosen, the difference $\varepsilon_t(a;a')$ falls below 0. If $\varepsilon_{t^n}(a;a')>\mathcal{M}$, the time needed to return to the origin is at least

$$\frac{\mathcal{M}}{(1-s(a;a'))\left[\bar{u}_1-\min_{u\in\Delta_a}u\right]},$$

which grows to infinity, as ε_{t^n} and, hence, $\mathcal{M}$ becomes very large. However, as has been explained above, the expected time for return to the origin 0 of $\tilde{\varepsilon}_t(a;a')$ is finite and uniformly bounded above by $\mathcal{N}$. The Law of Large Numbers then implies that for each $\kappa>0$ on almost each path of dividend realizations there is a period $\mathcal{K}(\omega)$, such that

$$\frac{\sum_{i=1}^{n}\tau_i}{n}\leq\mathcal{N}+\kappa$$

for all $n\geq\mathcal{K}(\omega)$, where τ_i denotes the time needed for $\tilde{\varepsilon}_t(a;a')$ to reach the origin, once a has been chosen. On the other hand, the assumption that $\varepsilon_{t^n}(a;a')\to\infty$ implies that the stopping times τ_i become infinitely large as the time grows – a contradiction. Hence, almost each sequence $\varepsilon_{t'}(a;a')$, $\varepsilon_{t''}(a;a')\dots$ (where t', $t''\dots$ denote periods at which a is chosen) is bounded from above. A symmetric argument for a' shows that $\varepsilon_t(a;a')$ is bounded from below. It follows that on almost each path $\omega\in\Omega$

$$\lim_{t\to\infty}\frac{U_t(a)}{U_t(a')}=\lim_{t\to\infty}\frac{U_t(a)}{U_t(a)+\varepsilon_t(a;a')}=1$$

holds for all acts a, $a'\in A'$.

By Lemma 3, there is no open subset of A such that A' does not contain an act out of this interval, hence for each $\epsilon>0$, and $x\in int(A)$, there is an $a\in A'\cap B_x(\epsilon)$. Moreover, for all $\epsilon>0$,

$$\lim_{t\to\infty}\frac{U_t(\tilde{a})}{U_t(a)}=1,$$

where $\tilde{a}\in A'\cap B_x(\epsilon)$ and $a\in A'$, $a\neq\tilde{a}$. Since

$$\lim_{\epsilon\to 0}A'\cap B_x(\epsilon)=x$$

and $U_t(\tilde{a})$ is continuous in $\tilde{a}$, it follows that

$$\lim_{t\to\infty} \frac{U_t(x)}{U_t(a)} = 1,$$

even if $x \notin A'$. This completes the proof of the proposition. □

Proof (of Proposition 2). The proof of Proposition 1 has shown that if there is an open subset $\tilde{A} \subset [0;1]^K$ such that

$$\mu_a - \bar{u} > 0$$

for all $a \in \tilde{A}$, then the decision-maker will eventually choose an act out of this set. By continuity of μ_a with respect to a, if $\tilde{A}$ is not an open set, it must consist of a single corner act. By Remark 1, every corner act is also eventually chosen by the decision-maker. Moreover, by the proof of proposition 2, it cannot be that only acts outside $\tilde{A}$ are chosen infinitely often. This means that at least one act $a \in \tilde{A}$ will be chosen infinitely often. Suppose to the contrary of the statement of the proposition that there are two acts from $\tilde{A}$, a and a', which are chosen with positive frequency. It is easy to show that this leads to a contradiction.

Indeed, consider the periods $z_{1a}, z_{2a}, \ldots \in \mathbb{N}$ at which the decision-maker switches to act a and denote by $z_{1a'}, z_{2a'}, \ldots \in \mathbb{N}$ the times, at which the decision-maker switches to a'. Then the proof of Lemma 2 shows that:

$$V_{z_{1a}}(a) > V_{z_{1a'}}(a) = V_{z_{2\alpha}}(a) > V_{z_{2a'}}(a) = V_{z_{3a}}(a) > \ldots$$

But these inequalities imply that $V_t(a)$, which is a random walk with positive expected increment $\mu_a - \bar{u} > 0$, crosses each of the infinitely many boundaries $V_{z_{ka}}(a)$ from above. Since, however, there is a positive probability that a random walk with positive expected increment starting from a given point, never crosses a boundary lying below this point, see Grimmet and Stirzaker (1994, p. 144), and since the stopping times are independently distributed, it follows that the probability of infinitely many switches between a and a' is 0. Hence, only one of these two acts can be chosen with positive frequency in the limit.

Alternatively, suppose that an act a' from the set $A\backslash\tilde{A}$ is chosen infinitely often with an act from $\tilde{A}$. Then, with probability 1, the cumulative utility of a will become infinitely high, whereas the cumulative utility of a' will become infinitely low, as the number of periods grows to infinity. Hence, choosing act a' infinitely often will contradict the case-based decision rule, as well. □

Proof (of Proposition 3). It has already been shown, see Lemma 1 that for $\mu_{\bar{a}} - \bar{u} < 0$, the investor switches in finite time to a corner act. Let e^i be the first corner act chosen at some time $\bar{t}$, such that $\bar{t} = \min\{t \mid U_t(\bar{a}) < 0\}$. Two cases are possible: either $\mu_{e^i} - \bar{u} < 0$ or $\mu_{e^i} - \bar{u} > 0$. Then at time $t > \bar{t}$

such that $a_\tau = e^i$ for all $\bar{t} < \tau \leq t$, the cumulative utility of an act a can be written as:

$$U_t(a) = V_{\bar{t}}(\bar{a})\, s(a;\bar{a}) + V_t\left(e^i\right) s\left(a;e^i\right) .$$

As long as $V_t\left(e^i\right) \geq 0$, e^i is chosen, according to Lemma 2. If $\mu_{e^i} - \bar{u} > 0$ holds, then $V_t\left(e^i\right) > 0$ holds infinitely long in expectation. If, however, $\mu_{e^i} - \bar{u} < 0$, then

$$V_t\left(e^i\right) < \frac{V_{\bar{t}}(\bar{a})\left(s\left(e^j;\bar{a}\right) - s\left(e^i;\bar{a}\right)\right)}{1 - s\left(e^i;1\right)} < 0$$

obtains in finite time for some $j \in \{1 \ldots K\}$. Let now $\bar{t}'$ denote

$$\bar{t}' = \min\left\{t \mid V_t(0) < \frac{V_{\bar{t}}(\bar{a})\left(s\left(e^j;\bar{a}\right) - s\left(e^i;\bar{a}\right)\right)}{1 - s(0;1)} \text{ for some } j \in \{1 \ldots K\}\right\} .$$

Note that at $\bar{t}'$ the cumulative utility of $a = e^j$ is:

$$U_{\bar{t}'}\left(e^j\right) = V_{\bar{t}}(\bar{a})\, s\left(e^j;\bar{a}\right) .$$

Moreover, since now $V_{\bar{t}}(\bar{a}) < 0$, $V_{\bar{t}'}(0) < 0$ and s is concave, it follows that at $\bar{t}'$ $U_{\bar{t}'}(a)$ is convex for every $a \in A$. Therefore, the optimal act is a corner one. Now restrain e^j to belong to the set:

$$\tilde{K} = \arg \max_{k \in \{1 \ldots K\} \setminus i} \left\{U_{\bar{t}'}\left(e^k\right)\right\} .$$

Hence,

$$U_{\bar{t}'}\left(e^j\right) = V_{\bar{t}}(\bar{a})\, s\left(e^j;\bar{a}\right) > V_{\bar{t}}(\bar{a})\, s\left(e^i;\bar{a}\right) + V_{\bar{t}'}\left(e^i\right) = U_{\bar{t}'}\left(e^i\right) ,$$

so that one of the acts e^j in set $\tilde{K}$ is chosen.

Again, if $\mu_{e^j} - \bar{u} > 0$, then $a = e^j$ will be held infinitely long in expectation, whereas if $\mu_{e^j} - \bar{u} < 0$, then the cumulative utility of e^j becomes lower than the cumulative utility of any other corner act in finite time. □

Lemma 5. *$a_t \in \left\{e^k\right\}_{k=1}^K$ for all $t > \bar{t}$.*

Proof (of Lemma 5). It has already been shown that the statement holds until period $\bar{t}'$. To argue by induction, suppose that only corner acts have been chosen up to some time t. At time t, the cumulative utility of an act a is given by:

$$U_t(a) = V_t(\bar{a})\, s(a;\bar{a}) + \sum_{k=1}^{K} V_t\left(e^k\right) s\left(e^k;a\right) .$$

Let $a_t = e^l$. As shown in Lemma 2, if $V_t\left(e^l\right) \geq 0$, then $a_{t+1} = e^l$. If, however $V_t\left(e^l\right) < 0$, then, by Lemma 2, $V_t\left(e^k\right) \leq 0$ holds for all $k = 1 \ldots K$ with strict inequality for all corner acts selected at least once in the past. Since $s(\cdot;\cdot)$ is a concave function, $U_t(a)$ becomes convex and has a corner maximum. Hence, $a_{t+1} \in \left\{e^k\right\}_{k=1}^K$. □

Consider first the case of $\mu_{e^k} < \bar{u}$ for all $k \in \{1 \dots K\}$.

Lemma 6. *Each of the corner acts* e^k, $k \in \{1 \dots K\}$ *satisfies* $\lim_{t\to\infty} \left|C_t\left(e^k\right)\right| = \infty$.

Proof (of Lemma 6). Suppose that one of the corner acts is not chosen infinitely often. Let this be act e^k Obviously, since the number of corner acts is finite, at least one of them must be chosen infinitely often. Let this be act e^j. Then, it follows that

$$\lim_{t\to\infty} U_t\left(e^j\right) = \lim_{t\to\infty} V_{\bar{t}}\left(\bar{a}\right) s\left(e^j;\bar{a}\right) + V_t\left(e^j\right) = -\infty\,,$$

since $V_{\bar{t}}\left(\bar{a}\right)$ is finite and $\mu_{e^j} - \bar{u} < 0$, whereas $U_t\left(e^k\right)$ remains finite. Hence, a.s. there is a time $T\left(\omega\right)$ such that

$$U_t\left(e^j\right) < U_t\left(e^k\right)$$

for all $t > T\left(\omega\right)$ and still $a_t = e^j$ in some of the periods in contradiction to the case-based rule. □

Now, consider the following process: let $k,\, l \in \{1 \dots K\}$, $k \neq l$ and

$$\varepsilon_{\bar{t}}\left(e^k;e^l\right) = V_{\bar{t}}\left(\bar{a}\right)\left[s\left(e^k;\bar{a}\right) - s\left(e^l;\bar{a}\right)\right]$$

$$\varepsilon_{t+1}\left(1;0\right) = \begin{cases} \varepsilon_t + u_t\left(\delta_t^k\right) - \bar{u}\,, & \text{if } \varepsilon_t \geq 0 \\ \varepsilon_t + u_t\left(\delta_t^l\right) - \bar{u}\,, & \text{if } \varepsilon_t < 0 \end{cases}.$$

$\varepsilon_t\left(e^k;e^l\right)$ represents the difference between the cumulative utilities of the acts e^k and e^l after period $\bar{t}$. To see this note that for $t \geq \bar{t}$,

$$\begin{aligned} &U_t\left(e^k\right) - U_t\left(e^l\right) \\ &= \left[V_t\left(e^k\right) + V_{\bar{t}}\left(\bar{a}\right) s\left(e^k;\bar{a}\right)\right] - \left[V_t\left(e^l\right) + V_{\bar{t}}\left(\bar{a}\right) s\left(e^l;\bar{a}\right)\right]\,, \\ &= \left[V_t\left(e^k\right) - V_t\left(e^l\right)\right] + V_{\bar{t}}\left(\bar{a}\right)\left[s\left(e^k;\bar{a}\right) - s\left(e^l;\bar{a}\right)\right] \\ &= \varepsilon_t\left(e^k;e^l\right)\,. \end{aligned}$$

An argument analogous to the one used to prove lemma 4 shows that $\varepsilon_t\left(e^k;e^l\right)$ is bounded on almost each path ω and therefore:

$$\lim_{t\to\infty} \frac{U_t\left(e^k\right)}{U_t\left(e^l\right)} = \lim_{t\to\infty} \frac{U_t\left(e^l\right) + \varepsilon_t\left(e^k;e^l\right)}{U_t\left(e^l\right)} = 1$$

with probability 1. Hence,

$$\lim_{t\to\infty} \frac{\left[V_t\left(e^k\right) + V_{\bar{t}}\left(\bar{a}\right) s\left(e^k;\bar{a}\right)\right]}{\left[V_t\left(e^l\right) + V_{\bar{t}}\left(\bar{a}\right) s\left(e^l;\bar{a}\right)\right]} = 1\,,$$

$$\lim_{t\to\infty} \frac{\left[\left|C_t\left(e^k\right)\right| \sum_{\tau\in C_t\left(e^k\right)} \frac{\left[u_\tau - \bar{u}\right]}{\left|C_t\left(e^k\right)\right|} + V_{\bar{t}}\left(\bar{a}\right) s\left(e^k;\bar{a}\right)\right]}{\left[\left|C_t\left(e^l\right)\right| \sum_{\tau\in C_t\left(e^l\right)} \frac{\left[u_\tau - \bar{u}\right]}{\left|C_t\left(e^l\right)\right|} + V_{\bar{t}}\left(\bar{a}\right) s\left(e^l;\bar{a}\right)\right]} = 1\,.$$

Since $\left|C_t\left(e^k\right)\right| \to \infty$ and $\left|C_t\left(e^l\right)\right| \to \infty$ on almost each path, it follows according to the Law of Large Numbers that

$$\lim_{t\to\infty} \frac{\sum_{\tau\in C_t(e^k)}\left[u_\tau - \bar{u}\right]}{\left|C_t\left(e^k\right)\right|} = \mu_{e^k} - \bar{u}\,,$$

$$\lim_{t\to\infty} \frac{\sum_{\tau\in C_t(e^l)}\left[u_\tau - \bar{u}\right]}{\left|C_t\left(e^l\right)\right|} = \mu_{e^l} - \bar{u}$$

obtain almost surely in the limit. Hence,

$$\lim_{t\to\infty} \frac{\left[\left|C_t\left(e^k\right)\right|\left(\mu_{e^k} - \bar{u}\right) + V_{\bar{t}}\left(\bar{a}\right)s\left(e^k;\bar{a}\right)\right]}{\left[\left|C_t\left(e^l\right)\right|\left(\mu_{e^l} - \bar{u}\right) + V_{\bar{t}}\left(\bar{a}\right)s\left(e^l;\bar{a}\right)\right]} = 1\,.$$

$$\lim_{t\to\infty} \frac{\left[\left|C_t\left(e^k\right)\right|\left(\mu_{e^k} - \bar{u}\right) + V_{\bar{t}}\left(\bar{a}\right)s\left(e^k;\bar{a}\right)\right]}{\left[\left|C_t\left(e^l\right)\right|\left(\mu_{e^l} - \bar{u}\right) + V_{\bar{t}}\left(\bar{a}\right)s\left(e^l;\bar{a}\right)\right]}$$

$$= \lim_{t\to\infty} \frac{\left[\frac{\left|C_t\left(e^k\right)\right|}{\left|C_t\left(e^l\right)\right|}\left(\mu_{e^k} - \bar{u}\right) + \frac{V_{\bar{t}}(\bar{a})s\left(e^k;\bar{a}\right)}{\left|C_t\left(e^l\right)\right|}\right]}{\left[\left(\mu_{e^l} - \bar{u}\right) + \frac{V_{\bar{t}}(\bar{a})s\left(e^l;\bar{a}\right)}{\left|C_t\left(e^l\right)\right|}\right]}$$

$$= \lim_{t\to\infty} \frac{\frac{\left|C_t\left(e^k\right)\right|}{\left|C_t\left(e^l\right)\right|}\left(\mu_{e^k} - \bar{u}\right)}{\left(\mu_{e^l} - \bar{u}\right)} = 1$$

almost surely holds (since $V_{\bar{t}}\left(\bar{a}\right)$ is finite on almost all paths, it does not influence the limit behavior). Therefore, the limit frequencies $\pi\left(e^k\right)$ and $\pi\left(e^l\right)$ satisfy

$$\frac{\pi\left(e^k\right)}{\pi\left(e^l\right)} = \lim_{t\to\infty} \frac{\left|C_t\left(e^k\right)\right|}{\left|C_t\left(e^l\right)\right|} = \frac{\mu_{e^l} - \bar{u}}{\mu_{e^k} - \bar{u}}\,.$$

If at least one of the mean utilities μ_{e^k} exceeds $\bar{u}$, then applying the argument of the proof of Proposition 2 shows that one of the acts with $\mu_{e^k} > \bar{u}$ is chosen with frequency 1 in the limit. □

Acknowledgement. I am indebted to my advisor Juergen Eichberger for his helpful guidance and to Alexander Zimper for his helpful suggestions and illuminating discussions. I would like to thank Hans Haller, Clemens Puppe, Itzhak Gilboa, Hans Gersbach, Klaus Ritzberger, Larry Blume and David Easley as well as the participants of the RUD conference in Evanston, of the FUR XI conference in Paris and of the PhD seminar at the University of Heidelberg for helpful discussions and comments. Financial support from the DFG is gratefully acknowledged.

References

Billot, A., Gilboa, I., Schmeidler, D. (2004), "Axiomatization of an Exponential Similarity Function", University of Tel Aviv, mimeo.

Blonski, M. (1999), "Social Learning with Case-Based Decision", Journal of Economic Behavior and Organization 38 (1), 59–77.

Buschena, D.E., Zilberman, D. (1995), "Performance of the Similarity Hypothesis Relative to Existing Models of Risky Choice", Journal of Risk and Uncertainty 11, 233–263.

Buschena, D.E., Zilberman, D. (1999), "Testing the Effects of Similarity on Risky Choice: Implications for Violations of Expected Utility", Theory and Decision 46 (3), 253–280.

Chateauneuf, A., Tallon, J.M. (2002), "Diversification, Convex Preferences and Non-Empty Core in the Choquet Expected Utility Model", Economic Theory 19 (3), 509–523.

Coval, J.D., Moskowitz, T.J. (1999), "Home Bias at Home: Local Equity Preference in Domestic Portfolios", Journal of Finance 54, 2045–2066.

Dekel, E. (1989), "Asset Demand without the Independence Axiom", Econometrica 57 (1), 163–169.

Gayer, G. (2003), "Perception of Probabilities in Situations of Risk. A Case Based Approach", Tel Aviv University, mimeo.

Gilboa, I., Schmeidler, D. (1995), "Case-Based Decision Theory", Quarterly Journal of Economics 110 (3), 605–639.

Gilboa, I., Schmeidler, D. (1996), "Case-Based Optimization", Games and Economic Behavior 15 (1), 1–26.

Gilboa, I., Schmeidler, D. (1997), "Act Similarity in Case-Based Decision Theory", Economic Theory 9 (1), 47–61.

Gilboa, I., Schmeidler, D. (2001), A Theory of Case-Based Decisions, Cambridge University Press, Cambridge.

Grimmet, G.R., Stirzaker, D.R. (1994), Probability and Random Processes, Oxford University Press, Oxford.

Kang, J.-K., Stulz, R.M. (1997), "Why Is There a Home Bias? An Analysis of Foreign Portfolio Equity Ownership in Japan", Journal of Financial Economics 46, 3–28.

Meyn, S.P., Tweedie, R.L. (1996), Markov Chains and Stochastic Stability, Springer Verlag, Berlin, Heidelberg.

Nehring, K., Puppe, C. (1999), "A Theory of Diversity", SFB Discussion Paper A-605, University of Bonn.

Nehring, K., Puppe, C. (2002), "A Theory of Diversity", Econometrica 70 (3), 1155–1198.

Nehring, K., Puppe, C. (2003), "Diversity and Dissimilarity in Lines and Hierarchies", Mathematical Social Sciences 45 (2), 167–183.

Shepard, R.N. (1987), "Towards a Universal Law of Generalization for Psychological Science", Science 237, 1317–1323.

Tenenbaum, J.B., Griffiths, Th.L. (2001), "Generalization, Similarity and Bayesian Inference", Behavioral and Brain Sciences 24 (4), 629–640.

Zizzo, D.J. (2002), "Preliminary Experimental Results on the Similarity Function in 2×2 and 3×3 Games", University of Oxford, mimeo.

Uncertainty Improves the Second-Best*

Hans Haller[1] and Shabnam Mousavi[2,3]

[1] Department of Economics, Virginia Polytechnic Institute and State University, Blacksburg, VA 24061, USA
e-mail: haller@vt.edu

[2] Center for Adaptive Behavior and Cognition, Max Planck Institute for Human Development, Lentzeallee 94, 14195 Berlin, Germany

[3] Department of Statistics, The Pennsylvania State University, University Park, PA 16802–2111, USA
e-mail: shabnam@stat.psu.edu

Abstract. Uncertainty, pessimism or greater risk aversion on the part of high risk consumers benefit the low risk consumers who are not fully insured in the separating equilibrium of the Rothschild–Stiglitz insurance market model.

Keywords: insurance, subjective beliefs, uncertainty

1 Introduction

Economists distinguish two kinds of asymmetric information: Moral hazard means that the action taken by an agent is not or not perfectly observable by others. The term adverse selection signifies that the type (characteristics) of an agent is not perfectly observable *ex ante* by others. The term first-best refers to the hypothetical case where everything is observable by everybody. In an adverse selection model, the first-best implies that each type can be dealt with separately.

Here we are concerned with a situation of adverse selection. The typical market solution in such models is second-best in the following sense: Whereas the "worst" types achieve their first-best outcome, the "better" types end up with less utility than in the first-best case. We shall focus on the model of a competitive insurance market introduced by Rothschild and Stiglitz (1976) which is particularly transparent and easy to analyze. Every consumer receives fair and full insurance in the first-best case of their model. In the separating equilibrium under asymmetric information, the high risk consumers still achieve their first-best outcome, that is, fair and full insurance. The low risk consumers, however, end up with a contract that offers fair but less than full insurance to them which, while preferable to their alternative choices

* Comments of a referee are much appreciated.

(full insurance at unfair odds or no insurance), is inferior to their first-best outcome. The reason is simply this: If the low risk consumers where offered fair and full insurance (at the low risk odds), then this contract would strictly dominate the contract that offers fair and full insurance at the high risk odds. Consequently, it would attract all of the consumers. Hence "self-selection" or "incentive-compatibility" can only be satisfied, if the low risk consumers are prevented from achieving their first-best outcome. Thus they suffer from the presence of the high risk consumers from whom they cannot be distinguished *ex ante*.

Now suppose that prior to insurance, the world looks grimmer to high risk consumers than it actually is. Would this do further harm to low risk consumers? We shall argue that low risk consumers can benefit from the fact that the subjective perception of the world by high risk consumers is more negative than the situation presents itself to an objective outside observer. One such instance occurs when a high risk consumer's subjective probability of an accident is higher than the objective probability. Another instance is when high risk consumers perceive some uncertainty rather than merely risk. Finally, it turns out that an increase in risk aversion on the part of high risk consumers can have a similar beneficial effect on low risk consumers.

The Rothschild–Stiglitz model assumes that all consumers are identical prior to insurance, except for the probability of an accident. It implicitly assumes that probabilities (risks) are objective or can be treated as such: Consumers and insurance providers base their decisions on the same odds. But its central findings are robust with respect to small deviations from these assumptions. Therefore, we can easily study local variations of the model. We are going to treat the original model as a benchmark case and consider its underlying probabilities as the objective ones. The local variations can be subjective in nature: uncertainty, pessimism, or change in risk aversion. Clearly, in a purely decision-theoretic context, such subjective variations would have an effect on the choices made by the individual experiencing them. They would also affect the alternatives offered to that individual by a monopolist. In a competitive insurance market environment, however, subjective perceptions become irrelevant (or of secondary importance) for the market opportunities of the high risk consumers: They still get fair and full insurance (at their objective odds) or, possibly, overinsurance. All that insurance providers care about are the objective probabilities. Market forces (free entry and exit) then assure that at the separating equilibrium only objectively fair (actuarially fair) contracts will be offered.

Since individuals often have difficulties to correctly assess small probabilities such as the probability of an accident, an investigation of the influence of subjective characteristics, especially subjective accident probabilities, on equilibrium outcomes is warranted.[1] The low risk consumers will be better

[1] We hasten to add that we assume the distortions between objective and subjective probabilities to be limited to the extent that the distinction between low

off under each of the proposed variations. The reason is that in each of these cases, the relevant part of a high risk consumer's indifference curve through his point of fair and full insurance becomes flatter and, consequently, intersects the low risk fair odds line at a higher level. Therefore, the self-selection constraint becomes binding at a point that represents more, but still not full insurance to low risk consumers. Interestingly enough, similar small subjective variations with regard to the characteristics of low risk consumers do not affect the second-best outcome at all.

Demands and attempts for more transparency in the financial services sector, in particular in commercial and central banking have intensified in recent years.[2] Charges of improper behavior against major US insurance companies could easily lead to a campaign for enhanced transparency in the insurance industry. Our findings imply that attempts to educate consumers about the objective accident probabilities, in order to reduce subjective deviations in perceptions, may prove detrimental to the welfare of some consumers without improving the welfare of others. Therefore, suggestions for more transparency in the insurance market – specifically aimed at educating consumers about accident probabilities – could be questionable and controversial.

We also study briefly the consequences of a change of objective accident probabilities and find that certain effects are strikingly different from those caused by comparable changes in subjective accident probabilities. We conclude that the distinction between objective model parameters and subjective characteristics gives rise to very insightful comparative statics.

The next section analyzes the underlying individual decision model. The third section examines equilibrium effects. The fourth section concludes.

2 The Individual Insurance Problem

Our analysis rests on the simple model of a competitive insurance market developed by Rothschild and Stiglitz. We first deal with individual demand for insurance.

There are two states of nature, $s = 1, 2$. With probability $1 - p$, where $0 < p < 1$, the good state 1 occurs in which the individual does not have an accident and has income or wealth $W > 0$. With probability p, the bad state 2 occurs where the individual does have an accident, incurs damage D, $0 < D < W$, and has reduced income $W - D$. Insurance alters the income pattern across the two states. Let W_s denote income in state s, $s = 1, 2$.

---

risk and high risk consumers persists under the subjective point of view. Let us further emphasize that insurance companies are assumed to have sufficient experience and expertise to accurately assess the objective accident probabilities of existing risk classes, notwithstanding the fact that they cannot observe the risk type of the next person walking through the door.

[2] Geraats (2002) summarizes the debate on central bank transparency.

The consumer's preferences for income patterns (W_1, W_2) are represented by the expected utility functional

$$EU(p, W_1, W_2) = (1-p)U(W_1) + pU(W_2)\,, \tag{1}$$

where $U(\cdot)$ represents the utility of money income and p is the probability of an accident. We assume $U(0) = 0$ and that U is twice differentiable with $U' > 0$ and $U'' < 0$.

The uninsured individual has income pattern $E = (E_1, E_2) = (W, W-D)$. We assume that a person can hold at most one insurance policy. If the individual insures himself, he pays a premium α_1 to an insurance company and in return will be compensated by the amount $\hat{\alpha}_2$ in case of an accident. The resulting income pattern is $(W_1, W_2) = (W-\alpha_1, W-D+\alpha_2) = (E_1-\alpha_1, E_2+\alpha_2)$ where α_1 is the premium and $\alpha_2 \equiv \hat{\alpha}_2 - \alpha_1$ is the net compensation. The pair of net trades $\alpha = (-\alpha_1, \alpha_2)$ completely describes an insurance contract. In the following diagrams, we use α as a label for the resulting income pattern, $E + \alpha$. Let $\rho = \rho(\alpha) = \alpha_2/\alpha_1$ denote the individual return on insurance so that $1/\rho$ is the relative price of insurance.

Full insurance obtains, if $W_1 = W_2$. Denote $\tau \equiv (1-p)/p$ which is the marginal rate of substitution at each point along the full insurance (45°) line. Fair insurance or fair odds means that *ex ante* the insurance company breaks even on the contract, that is, its expected profit is zero:

$$\Pi(p, \alpha) = (1-p)\alpha_1 - p\alpha_2 = 0\,. \tag{2}$$

This amounts to $\rho(\alpha) = \tau$. The company makes a positive expected profit, if the odds are unfair for the insured, i.e., $\rho(\alpha) < \tau$. It makes an expected loss, if the odds are favorable for the insured, i.e., $\rho(\alpha) > \tau$.

In the remainder of this section, we determine the individual's optimal insurance contract when he is given a choice amongst all contracts with a fixed return, $\bar{\rho}$. From now on, p is interpreted as the objective probability of an accident and is the basis for the calculation of expected profits. Thus the assumption is that insurance companies use objective probabilities whereas an individual seeking insurance may assess the odds differently.

2.1 Benchmark Case: Objective Beliefs

The benchmark case is the classical situation of Rothschild and Stiglitz where both insured and insurers base their decisions on the same (objective) accident probabilities. In this case, the individual's optimal insurance purchase is as depicted in Fig. 1.

If the odds are fair, $\bar{\rho} = \tau$, the individual purchases fair and full insurance, corresponding to the contract α. If the odds are favorable, $\bar{\rho} > \tau$, the individual overinsures himself, as in contract β. If the odds are unfair, $\bar{\rho} < \tau$, then two possibilities arise. As long as $\bar{\rho}$ exceeds the marginal rate of substitution at E, the individual under-insures himself. He purchases some, but less

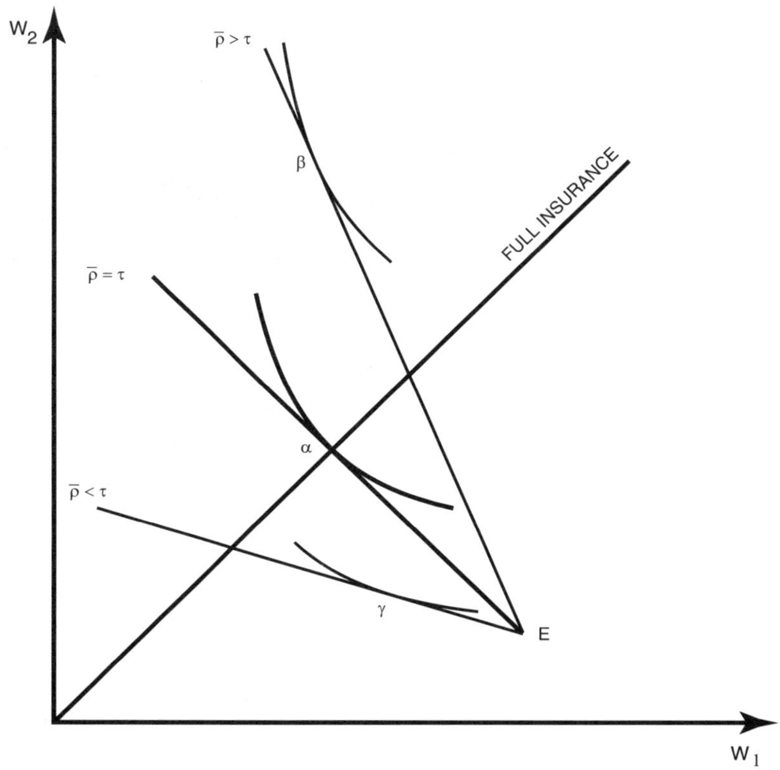

Fig. 1. Individual choices

than full insurance as in contract γ. For smaller $\bar{\rho}$, the individual does not purchase any insurance. In fact, if $\bar{\rho}$ happened to be less than the marginal rate of substitution at E and short sales were feasible, the individual would short sell insurance.

2.2 Pessimistic Beliefs

Here we assume that the individual holds subjective beliefs regarding the probability of an accident and that these beliefs are pessimistic.

The subjective probability of an accident is $q > p$. If offered objectively fair odds, the individual perceives subjectively favorable odds and chooses the over-insurance contract δ over the full insurance contract α in Fig. 2 where $\bar{U}$ and $\bar{\bar{U}}$ are indifference curves of an individual with objective and pessimistic subjective beliefs, respectively. More importantly, we observe

Fact 1. At any income pair (W_1, W_2), the indifference curve based on pessimistic beliefs is flatter than the indifference curve associated with the objective probabilities.

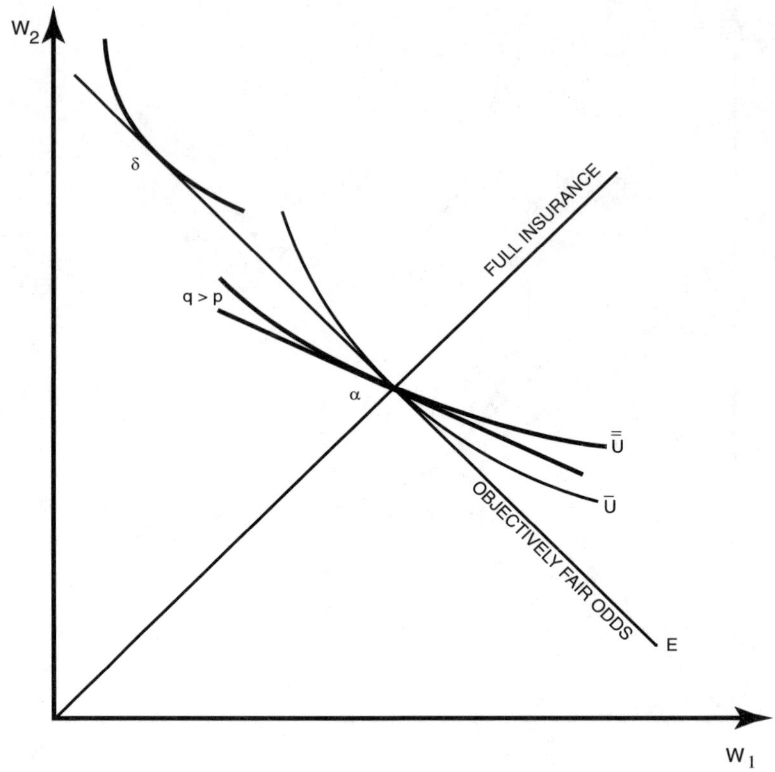

Fig. 2. Pessimism

2.3 Uncertainty

We maintain the assumption of an objective probability p of an accident. However, the individual now faces subjective uncertainty rather than risk. Subjective uncertainty (Knightian uncertainty, ambiguity) refers to situations where an individual has difficulties to assign any precise probabilities which is different from assigning the wrong probabilities. This sort of uncertainty is described by means of a capacity (non-additive probability measure, fuzzy probability measure) μ which assigns a number $\mu(T) \in [0,1]$ to every subset T of the state space $S = \{1,2\}$ with the properties

(i) $\mu(\emptyset) = 0$ and $\mu(S) = 1$;
(ii) $0 \leq \mu(\{1\}) \leq 1$ and $0 \leq \mu(\{2\}) \leq 1$.

In addition, it is assumed that the individual is uncertainty averse which in the current situation is tantamount to convexity of μ:

(iii) $\mu(\{1\}) + \mu(\{2\}) < 1$.

An ordinary (additive) probability measure π on S satisfies $\pi(\{1\})+\pi(\{2\}) = 1$. If $\mu(\{1\}) = \mu(\{2\}) = 0$, we have the case of total ignorance, described by the capacity ω with $\omega(S) = 1$ and $\omega(T) = 0$ for $T \neq S$. Since we are interested in small subjective deviations from the objective beliefs, we shall further assume

(iv) $\mu(\{1\}) + \mu(\{2\}) > 0$,

which rules out total ignorance. It turns out that with two states, a capacity μ satisfies (iii) and (iv) if and only if it is a simple capacity different from total ignorance. μ is called simple, if there exist $\lambda \in [0,1]$ and an additive probability measure π such that $\mu(S) = 1$ and $\mu(T) = \lambda \cdot \pi(T)$ for $T \neq S$. A simple capacity as a set function can also be written in the form

$$\mu = (1-\lambda) \cdot \omega + \lambda \cdot \pi \,. \tag{3}$$

The individual evaluates insurance contracts according to his Choquet Expected Utility (CEU) based on the given von-Neumann-Morgenstern utility function U[3]. For our purposes it suffices to observe that for a simple capacity of the form (3) and an income pair (W_1, W_2), CEU assumes the form

$$\begin{aligned} CEU(\lambda; \pi; W_1, W_2) = & \lambda \cdot [\pi(\{1\})U(W_1) + \pi(\{2\})U(W_2)] \\ & + (1-\lambda) \cdot \min_s U(W_s) \,. \end{aligned}$$

To isolate the effect of uncertainty, we restrict ourselves to the case

(v) $\pi(\{1\}) = 1 - p$, $\pi(\{2\}) = p$, $0 < \lambda < 1$.

Assuming $\pi(\{2\}) > p$ would reinforce the effect – which follows from the combined arguments of the current and the previous subsection. In view of (v), the expression (4) can be rewritten

$$\begin{aligned} CEU(\lambda; p; W_1, W_2) = & \lambda \cdot [(1-p)U(W_1) + pU(W_2)] \\ & + (1-\lambda) \cdot \min_s U(W_s) \,. \end{aligned}$$

Next define $p_* = \lambda \cdot p < p$ and $p^* = \lambda \cdot p + 1 - \lambda > p$. Further denote $\tau_* = (1-p_*)/p_*$ and $\tau^* = (1-p^*)/p^*$. Then (5) amounts to

$$CEU(\lambda; p; W_1, W_2) = \begin{cases} EU(p_*, W_1, W_2) & \text{if } W_1 < W_2\,; \\ EU(p, W_1, W_2) & \text{if } W_1 = W_2\,; \\ EU(p^*, W_1, W_2) & \text{if } W_1 > W_2\,. \end{cases}$$

Since $p_* < p < p^*$, this shows the following

Fact 2. In the presence of uncertainty,

---

[3] Gilboa (1987), Schmeidler (1989), and Sarin and Wakker (1992) provide a system of axioms for the representation of beliefs by capacities and of preferences by a Choquet integral of utilities with respect to these capacities.

1. at points below the full insurance line, the indifference curve is flatter than without uncertainty;
2. at points above the full insurance line, the indifference curve is steeper than without uncertainty;
3. at points on the full insurance line, the indifference curve has a kink.

This fact is illustrated in Fig. 3 where the kinked indifference curve $\bar{\bar{U}}$ reflects subjective uncertainty whereas $\bar{U}$ corresponds to objective beliefs. If the individual can choose between all contracts with fair odds, $\bar{\rho} = \tau$, then he will choose the full insurance contract α. Incidentally, a full insurance contract will be chosen not only when $\bar{\rho} = \tau$, but whenever the odds are fixed at some $\bar{\rho} \in [\tau^*, \tau_*]$. Since $\tau^* < \tau < \tau_*$, this result differs from the benchmark case. The result is the counterpart of the main theorem of Dow and Werlang (1992) regarding portfolio choice, as already mentioned by them; *ibid.*, p. 198. The result also fits into the framework of Segal and Spivak (1990) who distinguish between first and second order risk aversion. The uncertainty aversion

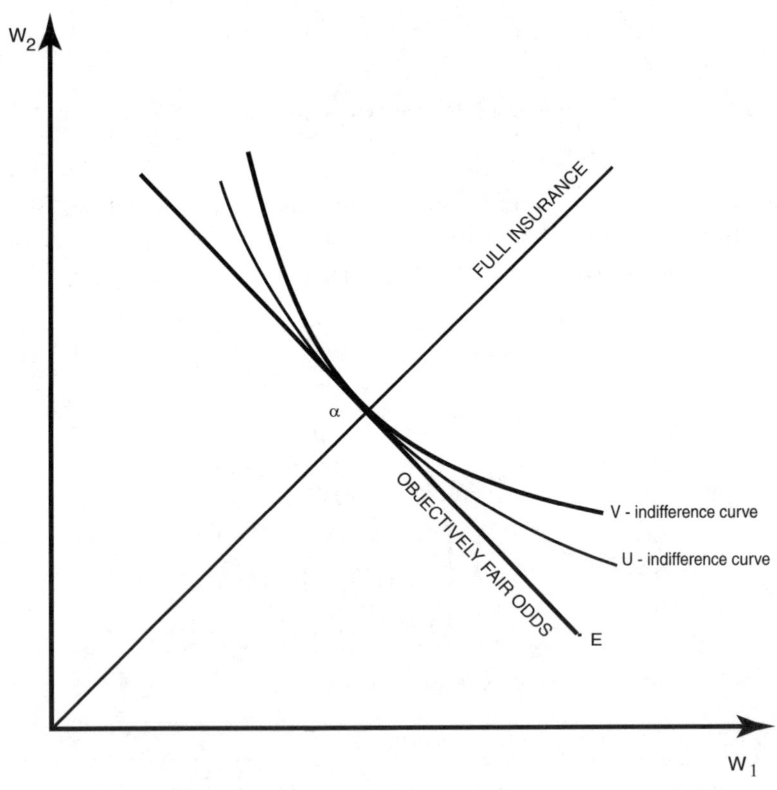

Fig. 3. Increased risk aversion

associated with simple capacities implies "first order risk aversion" in their terminology.

2.4 Increased Risk Aversion

Here we assume that the individual holds objective beliefs and consider the effect of a change in risk attitude. To be specific, let $r_A(U,x) = -U''(x)/U'(x)$ denote the Arrow-Pratt measure of absolute risk aversion at $x \geq 0$. For example, the utility function $U(x) = -e^{-ax}$ with constant $a > 0$ yields constant absolute risk aversion a. Now suppose that the individual becomes more risk averse in the sense that U is replaced by a twice differentiable von-Neumann-Morgenstern utility function V with $V(0) = 0, V' > 0, V'' < 0$, and such that

$$r_A(V,x) > r_A(U,x) \text{ for all } x \geq 0\,. \tag{4}$$

The situation of Fig. 4 obtains, that is

Fact 3. Suppose (4). Then:

1. At points on the full insurance line, U-indifference curves and V-indifference curves have the same absolute slope, τ.
2. At strictly positive points below the full insurance line, the V-indifference curve is flatter than the U-indifference curve.

The first assertion is obvious. To establish the second one, consider an income pair (W_1, W_2) with $W_1 > W_2 > 0$ and set $W_0 = (W_1 + W_2)/2 > 0$. Then $W_1 > W_0 > W_2 > 0$. (W_0, W_0) is a point on the full insurance line at which both indifference curves have absolute slope τ. The absolute slope of the U-indifference curve at (W_1, W_2) is $\tau \cdot U'(W_1)/U'(W_2)$. The absolute slope of the V-indifference curve at (W_1, W_2) is $\tau \cdot V'(W_1)/V'(W_2)$. Now

$$\ln U'(W_1) = \ln U'(W_0) + \int_{W_0}^{W_1} [U''(x)/U'(x)]\,\mathrm{d}x\,.$$

$$\ln V'(W_1) = \ln V'(W_0) + \int_{W_0}^{W_1} [V''(x)/V'(x)]\,\mathrm{d}x\,.$$

By (6), $U''/U' > V''/V'$. Hence $U'(W_1) = U'(W_0)\cdot\hat{U}_1$ and $V'(W_1) = V'(W_0)\cdot\hat{V}_1$ with $\hat{U}_1 > \hat{V}_1$. In an analogous way, we find that $U'(W_2) = U'(W_0)\cdot\hat{U}_2$ and $V'(W_2) = V'(W_0)\cdot\hat{V}_2$ with $\hat{U}_2 < \hat{V}_2$. Consequently, $U'(W_1)/U'(W_2) = \hat{U}_1/\hat{U}_2 > \hat{V}_1/\hat{V}_2 = V'(W_1)/V'(W_2)$ from where the assertion follows.

Notice that in (4), we can replace, for $x > 0$, the absolute risk aversion measures, $r_A(U,x)$ and $r_A(V,x)$, by the relative risk aversion measures, $r_R(U,x) = x r_A(U,x)$ and $r_R(V,x) = x r_A(V,x)$, respectively.

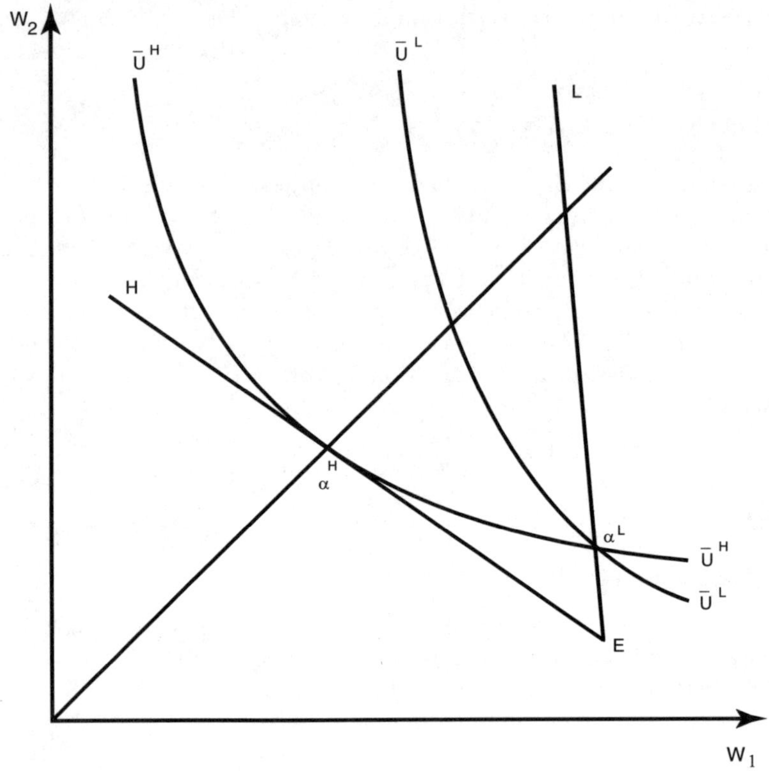

Fig. 4. Separating equilibrium

3 Equilibrium in Competitive Insurance Markets

In what follows, the Rothschild–Stiglitz model of a competitive insurance market serves as the benchmark case. There exist two classes (types) of consumers (individuals): low risk and high risk types, with respective objective accident probabilities p^L and p^H and $0 < p^L < p^H < 1$. In the first-best situation where consumer types can be observed by an insurance company, each type would receive actuarially fair and full insurance. However, an adverse selection problem exists. Under asymmetric information, only the individuals know their types while an insurance company cannot determine the type of an individual *ex ante*. The company knows, however, that there are these two types of consumers. It also knows the objective accident probability and the fraction of each consumer class.

3.1 Benchmark Case: Objective Beliefs

The benchmark case is the classical situation of Rothschild and Stiglitz where all agents base their decisions on the same (objective) probabilities and all individuals have the same von-Neumann-Morgenstern utility function U.

Rothschild and Stiglitz consider an equilibrium in contracts. They show that either there is no equilibrium or there exists a separating equilibrium. A separating equilibrium exists if there are sufficiently many high risk people. In that case, the equilibrium or second-best outcome consists of two contracts, α^L and α^H. At α^H, the high risk consumers obtain objectively (actuarially) fair and full insurance. At α^L, the low risk consumers obtain objectively (actuarially) fair, but not full insurance. Low risk types strictly prefer α^L to α^H. High risk types are indifferent between the two contracts, but they all choose α^H. This market outcome is portrayed in Fig. 5 where the respective indifference curves are denoted $\bar{U}^L$ and $\bar{U}^H$.

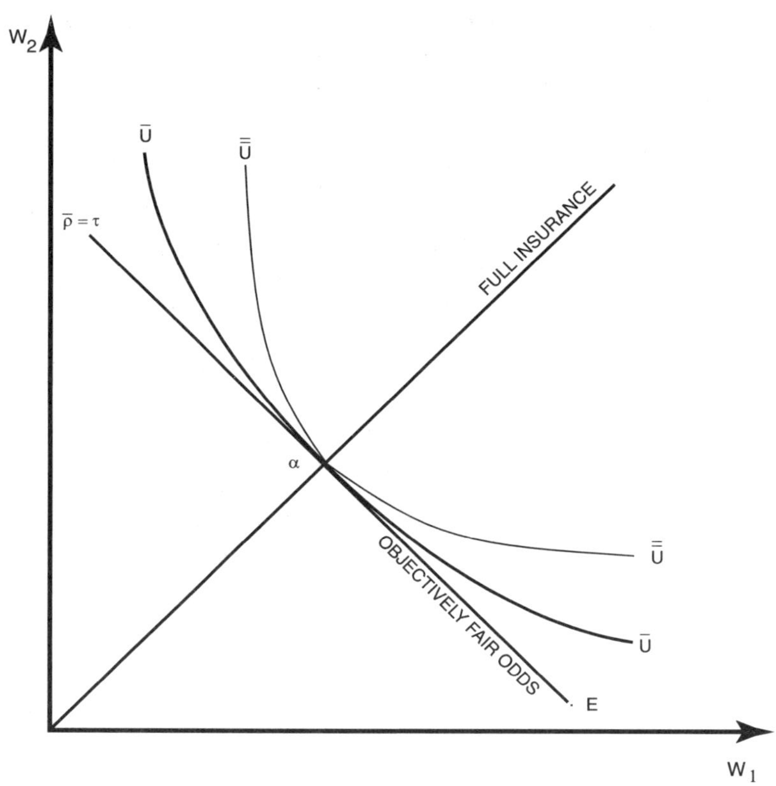

Fig. 5. Uncertainty

3.2 Subjective Variations

Now let the endowment point E, the objective accident probabilities $p^L < p^H$, and the income utility function U be given. In this and the next subsection, we assume that there are enough individuals of the high risk type so that the separating equilibrium exists for every contemplated model variation. We first consider subjective variations of the high risk consumer characteristics.

Notice that because of free entry and exit, in a separating equilibrium each contract breaks even (is objectively or actuarially fair) with respect to the corresponding consumer class. With uncertainty faced by high risk consumers, as in Sect. 2.3, α^H is still the equilibrium contract selected by high risk individuals. However, by Fact 2, preservation of incentive compatibility requires that α^L be moved further up on the fair odds line for low risk individuals – whose equilibrium contract is thus improved, but still falls short of full insurance. With increased risk aversion exhibited by high risk consumers as in Sect. 2.4, a similar effect occurs due to Fact 3.

When pessimism prevails among high risk consumers as in Sect. 2.2, two cases ought to be distinguished, depending on the criterion for entry. First, suppose that similar but not identical to a suggestion by Wilson (1977), a new contract only enters the market, if it remains profitable even after further profitable entry or consequential exit. Then α^H remains the equilibrium contract picked by high risk consumers. By Fact 2, α^L moves further up on the fair odds line for low risk individuals like in the previous cases. Second, suppose that like in the Rothschild–Stiglitz model, a new contract enters, if it can make a hit-and-run profit, that is, it is profitable in the absence of further entry or exit. Then α^H is no longer the equilibrium contract aimed at high risk individuals. It is replaced by the subjectively optimal contract along the objectively (actuarially) fair odds line for high risk individuals, analogous to the replacement of α by δ in Fig. 2. But this additional change moves α^L still further up, to the benefit of low risk consumers.

In all three instances considered (uncertainty, pessimism, increased risk aversion), we have found that a bleaker subjective perspective of high risk individuals is beneficial to low risk individuals. In contrast, minor variations of the subjective perceptions and risk attitudes of low risk individuals do not alter the separating equilibrium, since the objectively or actuarially fair odds line for low risk individuals remains unaffected by these variations.

3.3 Objective Variations

Whereas the focus of our investigation lies on the impact of variations in subjective perceptions and risk attitudes on the separating equilibrium contracts in the Rothschild–Stiglitz model of a competitive insurance market, variations of the model parameters are not necessarily subjective. Variations of objective probabilities are conceivable and potentially important. For instance, factors exogenous to the model such as added safety features may reduce the probability of a certain kind of accident over time. Conversely, the probability of specific accidents, say car accidents, could increase over time because of changing exogenous conditions, for example traffic density. In accordance with the rest of the paper, the following comparative statics deals with an increase of the objective probability of an accident for one of the consumer types. Most of the findings differ significantly from those obtained for subjective variations.

The main conclusions of this subsection are unambiguous: If *ceteris paribus* the probability of an accident increases for a consumer type, then this consumer class experiences a direct negative effect on its equilibrium contract. This finding is in stark contrast to the results obtained with respect to subjective variations. Recall that in equilibrium, both consumer types obtain objectively fair insurance, regardless of subjective variations. After a change of objective probabilities, they still obtain objectively fair insurance, but now based on the new probabilities. Let us consider first the case where p^H, the accident probability of the high risk type increases. Then at the new actuarially fair and full insurance contract, a high risk individual has less income in both states, clearly a worse outcome. Consider next the case where p^L, the accident probability of the low risk type increases to a level q^L such that $p^H > q^L > p^L$. Then this type's fair odds line becomes flatter. The new equilibrium contract is located at the intersection of the new fair odds line and the unchanged indifference curve of the high risk type. At the original contract α^L, the original indifference curve for low risk consumers is steeper than the indifference curve for high risk consumers, since $p^H > p^L$. Hence the new contract lies below the original indifference curve for low risk consumers. Moreover, the probability of an accident has increased, $q^L > p^L$, which reduces the expected utility associated with any given underinsurance contract. The cumulative equilibrium effect for low risk consumers is definitely negative.

Regarding cross-type effects, a change of p^L, the objective accident probability for low risk individuals does not affect high risk individuals, a conclusion we had also reached when considering subjective variations of the low risk consumer characteristics. The possible cross-type effects are more intriguing when p^H, the objective accident probability for high risk individuals varies. First of all, the effect of an increase of p^H on the equilibrium welfare of low risk individuals is typically non-zero, albeit hard to sign. Secondly, it can be negative which is the opposite of what happens when the high risk individuals merely become more pessimistic. The following example illustrates this possibility: An increase of p^H can reduce the equilibrium welfare of low risk consumers.

Example 1. Consider $E = (2, 1)$ and $U(x) = \sqrt{x}$ for $x \geq 0$. Let $p^H = 1/2$ and $p^L = 1/4$. An increased accident probability for high risk individuals assumes the form $p^H_\epsilon = 1/2 + \epsilon/2$, for some $\epsilon \in (0, 1)$. For convenience, we denote $p^H_0 = p^H = 1/2$. For a given $\epsilon \geq 0$, the high risk individual's fair odds line has absolute slope $\tau^H_\epsilon = (1-\epsilon)/(1+\epsilon)$. Fair and full insurance amounts to $W_1 =$

$W_2 = (3-\epsilon)/2$. We claim that for sufficiently small $\epsilon > 0$, the indifference curve through $((3-\epsilon)/2, (3-\epsilon)/2)$ based on accident probability p^H_ϵ and the indifference curve through $(3/2, 3/2)$ based on accident probability p^H do not intersect in the relevant range. Since the underlying probabilities are different, the single crossing property holds, that is, the two indifference curves intersect at most once. The respective equations for the indifference curves are:

(a) $(1-\epsilon)U(W_1) + (1+\epsilon)U(W_2) = 2U((3-\epsilon)/2)$.
(b) $U(W_1) + U(W_2) = 2U(3/2)$.

A crossing or intersection point (W_1, W_2) satisfies both equations. Set $u_1 = U(W_1)$ and $u_2 = U(W_2)$. Then from (b), $u_2 = 2\sqrt{3/2} - u_1$. Substitution for u_2 in (a) yields

$$\begin{aligned} 2\sqrt{(3-\epsilon)/2} &= (1-\epsilon)u_1 + (1+\epsilon)\cdot[2\sqrt{3/2} - u_1] \\ &= (1+\epsilon)\cdot 2\sqrt{3/2} - 2\epsilon u_1 \text{ or} \\ 2\epsilon u_1 &= (1+\epsilon)\cdot 2\sqrt{3/2} - 2\sqrt{(3-\epsilon)/2} \text{ or} \\ u_1 &= [(1+\epsilon)\cdot\sqrt{3/2} - \sqrt{(3-\epsilon)/2}]/\epsilon\,. \end{aligned}$$

Both the numerator and denominator of this fraction converge to zero as ϵ goes to zero. The rule of de l'Hospital applies:

$$\begin{aligned} \lim_{\epsilon\to 0} u_1 &= \lim_{\epsilon\to 0}\left[\sqrt{3/2} + \frac{1}{4}\cdot\frac{1}{\sqrt{(3-\epsilon)/2}}\right] \\ &= \sqrt{3/2} + \frac{1}{4}\cdot\frac{1}{\sqrt{3/2}} \\ &= 1.4288\,. \end{aligned}$$

Hence for sufficiently small $\epsilon > 0$, $u_1 > 1.4287$ and $W_1 = U^{-1}(u_1) = (u_1)^2 > (1.4287)^2 = 2.041 > 2$. This shows the claim that for sufficiently small $\epsilon > 0$, the two indifference curves do not intersect in the relevant range, that is when or before they hit the fair odds line of the low risk type. But this implies that the low risk consumers get less insurance and, hence, are worse off if the objective accident probability of the high risk type increases by a small amount.

4 Conclusion

In the separating equilibrium of the Rothschild–Stiglitz model of a competitive insurance market, low risk individuals suffer from the presence of high risk consumers when insurance companies are unable to distinguish among their customers. In the words of Rothschild and Stiglitz (1976, p. 638), "the presence of the high-risk individuals exerts a negative externality on the low-risk individuals. The externality is completely dissipative; there are losses to the low-risk individuals, but the high-risk individuals are no better off than they would be in isolation."

We address the question whether a more negative perspective on the part of high risk consumers accentuates or mitigates the negative externality they are exerting on low risk individuals. The distinction between changes of objective model parameters and changes of subjective consumer characteristics proves extremely powerful when dealing with this question. It allows us to consider a pessimistic deviation of subjective probability assessments from objectively given accident probabilities. We find that pessimistic subjective beliefs held by high risk consumers mitigate rather than accentuate the negative externality imposed on low risk consumers. In contrast, objectively higher accident probabilities of high risk consumers, can *ceteris paribus* accentuate the negative externality. We further find that variations of subjective beliefs held by low risk consumers do not affect the equilibrium outcome at all whereas a change of the objective accident probability of low risk consumers has a negative impact on their equilibrium contract.

In view of recent advances in decision theory, we have been curious to see what happens when high risk consumers perceive some uncertainty or ambiguity rather than pure risk. We find that uncertainty mitigates the negative externality exerted on low risk consumers. The idea that uncertainty might help improve equilibrium outcomes has been explored independently in a completely different context by Eichberger and Kelsey (2002). They reconsider the problem of voluntary provision of (contributions to) a public good. Because of free riding, the standard model tends to predict substantial under-provision in contradiction to empirical and experimental evidence. But as Eichberger and Kelsey show, individuals might contribute more if they are uncertain about the contributions of others.

Instead of having a different perception of risk, consumers may differ in their attitudes towards risk. We show that increased risk aversion of high risk consumers benefits low risk consumers. The idea that increased risk aversion can cause positive spill-overs has surfaced before. Safra, Zhou and Zilcha (1990) present a Nash bargaining model where a player becomes better off when the other player becomes "sufficiently" more risk averse.

References

Dow, J., Werlang, S.R.D.C. (1992), "Uncertainty Aversion, Risk Aversion, and the Optimal Choice of Portfolio," *Econometrica*, 60, 197–204.

Eichberger, J., Kelsey, D.(2002), "Strategic Complements, Substitutes and Ambiguity: The Implications for Public Goods," *Journal of Economic Theory*, 106, 436–446.

Geraats, M.P. (2002), "Central Bank Transparency," *Economic Journal*, 122, F532–F565.

Gilboa, I. (1987), "Expected Utility Theory with Purely Subjective Probabilities," *Journal of Mathematical Economics*, 16, 65–88.

Rothschild, M., Stiglitz, J.(1976), "Equilibrium in Competititve Insurance Markets: An Essay on the Economics of Imperfect Information," *Quarterly Journal of Economics*, 90, 629–650.

Safra, Z., Zhou, L., Zilcha, I.(1990), "Risk Aversion in the Nash Bargaining Problem with Risky Outcomes and Risky Disagreement Points," *Econometrica*, 58, 961–965.

Sarin, R., Wakker, P.(1992), "A Simple Axiomatization of Non-Additive Expected Utility," *Econometrica*, 60, 1255–1272.

Schmeidler, D. (1989), "Subjective Utility and Expected Utility without Additivity," *Econometrica*, 57, 571–587.

Segal, U., Spivak, A.(1990), "First Order versus Second Order Risk Aversion," *Journal of Economic Theory*, 51, 111–125.

Wilson, C. (1977), "A Model of Insurance Markets with Incomplete Information," *Journal of Economic Theory*, 16, 167–207.

Lorenz Meets Rating but Misses Valuation

Eva Camacho-Cuena[1] and Christian Seidl[2]

[1] Departemento de Analisis Economico: Teoria Economica e Historia Economica, Universidad Autónoma de Madrid, E-28049 Cantoblanco, Madrid, Spain
e-mail: eva.camacho@uam.es

[2] Institut für Volkswirtschaftslehre der Christian–Albrechts–Universität zu Kiel, Olshausenstrasse 40, D-24119 Kiel, Germany
e-mail: seidl@economics.uni-kiel.de

Abstract. Using an experiment with material incentives, this paper investigates the violation of Lorenz relations in the case of dominant and single-crossing Lorenz curves. Our experimental design consists of two treatments: an income distribution treatment and a lottery treatment. Both treatments were conducted in Italy and Spain. In each treatment, subjects were asked to judge ten multiple-outcome lotteries or ten n-dimensional income distributions, respectively, in terms of both ratings and valuations. This $2 \times 2 \times 2$ experimental design allows us to investigate the response-mode (rating versus valuation) and framing (lotteries versus income distributions) effects in subjects' perceptions concerning the two types of Lorenz relations. We found the existence of a marked response-mode effect, as only the ratings of the lotteries and income distributions confirm both Lorenz relations, whereas the valuations violate them. The framing effect is significant only for the Spanish data. For this data the sign of the framing effect depends on the type of the Lorenz relation considered. For crossing Lorenz curves, a higher conformity corresponds to the lottery frame, for Lorenz dominance a higher conformity corresponds to the income distribution frame.

Keywords: income distributions, lotteries, Lorenz curves, inequality and risk aversion, response-mode effects

1 Introduction

Consider two distributions of payoffs, say x and y, with the same mean, μ, where the probability mass of x is concentrated on the higher payoffs, while the probability mass of y is concentrated on the lower payoffs. Therefore, x provides a relatively high payoff with a high probability, and a relatively low payoff with a low probability, whereas y provides a low payoff with a high probability and a high payoff with a low probability. In this case, the Lorenz curve of x, $L(x)$, will, in most cases, either dominate or cut the Lorenz curve of y, $L(y)$, from below.

There are several methods to elicit preferences between two different distributions of payoffs: choices, ratings and valuations. The choice method refers to the observation of subjects' choices when they are asked to choose the more preferred one from a pair of distributions of payoffs. Under the rating method, subjects are asked to rate distributions on a point scale. Under the valuation method, subjects are asked for their monetary values assigned to distributions.[1] Traditional economic reasoning rules out response-mode effects, that is, subjects are assumed to express the same preferences irrespective of the mode of preference elicitation.

Moreover, when studying subjects' perception of Lorenz dominance, there are different frames to present them different distributions of payoffs. We will consider two: lotteries[2] and income distributions. In case a distribution of payoffs is presented as a lottery the payoffs represent the different prizes, whereas, when presented as an income distribution the payoffs represent the different income levels. Both frames are of outstanding economic significance. Traditional economic considerations would assume that, in case of the lotteries, risk averse subjects would prefer x to y, and, in case of income distributions, inequality averse subjects would prefer to be a member in a society in which income distribution x obtains rather than in a society in which income distribution y obtains (provided that the subjects have to make their choices under a veil of ignorance regarding their income level in a particular society).

In this paper, we investigate the response-mode effects in subjects' perceptions with respect to Lorenz dominance and single-crossing Lorenz curves. The experimental design consisted of two treatments. In the first treatment we presented subjects ten multiple-outcome lotteries, and in the second treatment ten n-dimensional income distributions whose entries corresponded exactly to the entries in the lotteries. In order to test whether response-mode effects affect the perception of the Lorenz relationships, subjects were asked in each treatment to judge each particular lottery or income distribution in terms of ratings and in terms of valuations. Material incentives were used in both treatments. This paper is a follow-up work of a study on preference reversals between lotteries and income distributions (Camacho et al. 2005). We use the data collected in this experimental study to investigate subjects' perceptions with respect to Lorenz dominance and single-crossing Lorenz curves.

In Sect. 2 we describe the experimental design. In Sect. 3 we report our results, and in Sect. 4 we summarize the main findings.

2 The Experiment

The experiment was conducted at the ESSE laboratory at the University of Bari, Italy, as well as at the LEE laboratory at the University Jaume I in Castellón, Spain. Subjects were volunteers recruited from students in different departments of these universities.

The experimental design consisted of two treatments, one concerning ten lotteries, and the other concerning ten income distributions. Each treatment encompassed two parts, a rating part, and a valuation part, and in every experimental session only one treatment was applied. We conducted a total of 21 sessions each of which lasted about one hour. Because of obviously absurd statements, we had to eliminate the data of 3 subjects. This left us with the Italian data of 52 subjects for the lottery treatment and of 56 subjects for the income distribution treatment. The Spanish data came from 51 subjects for the lottery treatment and from 50 subjects for the income distribution treatment. In order to prevent anchor effects, each subject was admitted to only one treatment and one experimental session.

We conducted the experiments before the introduction of the euro at the end of the year 2001. In this way we avoid possible money illusion effects and transitory effects due to the subjects' being poorly acquainted with a new currency. For the sake of comparability, however, in this paper we present all figures and tables in terms of euros.

For the presentation of lotteries and income distributions we checked several formats, and found the format based on the design used by Lopes (1984, 1987) and Schneider and Lopes (1986) to convey best the messages contained in the multiple-dimensional lotteries and income distributions of our experiment. The format used for the sessions conducted in Italy is displayed in Figs. 1–3. Each lottery and income distribution had the same expected value of approximately € 1800, save for differences in rates of exchange and rounding errors in order to secure decent numbers in terms of the local currencies.[3] The distributions in Fig. 1 are negatively skewed, the distributions in Fig. 2 are positively skewed, and the distributions in Fig. 3 are unimodal, rectangular, and bimodal. The ordering of the distributions in Figs. 1 to 3 was adopted for the presentation of the results in this paper. The ordering of their presentation for the Italian subjects is shown in square brackets. The ordering for the Spanish subjects was exactly opposite to the ordering for the Italian subjects.[4] The exact parameters of the distributions (mean, standard deviation, skewness, kurtosis,[5] minimum, maximum, range, and Gini coefficient) are shown in Table 1.

The lotteries and income distributions can be arranged as Lorenz curves. Two Lorenz curves either intersect or one dominates the other. We show the types of Lorenz relations of our experimental design in Fig. 4: An increasing arrow means that the Lorenz curve of the lottery or income distribution in a row cuts the Lorenz curve of the lottery or income distribution in the corresponding column from below, where intersections within two percentage points from the lower and the upper bounds were ignored. A horizontal arrow means that the Lorenz curve of the lottery or income distribution in a row dominates the Lorenz curve of the lottery or income distribution in the corresponding column. A tilde means that parts of the corresponding Lorenz curves coincide.

Table 1. Main statistics of distributions

Italy										
Main	Negatively skewed			Positively skewed				Unimodal	Rectangular	Bimodal
Statistics	1	2	3	4	5	6	7	8	9	10
Mean	1807.34	1807.49	1807.70	1806.05	1807.50	1807.39	1807.91	1807.60	1807.60	1802.54
SD	651.02	332.53	350.37	329.50	350.37	932.15	1112.54	410.83	595.61	799.50
Skewness	−0.1236	−11.742	−10.366	12.100	10.366	10.356	11.857	0.0000	0.0000	0.0048
Kurtosis	−15.302	0.6859	0.4199	0.9023	0.4199	0.4188	0.7290	−0.1537	−12.061	−17.530
Minimum	774.69	774.69	774.69	1497.72	1415.09	774.69	774.69	774.69	826.33	774.69
Maximum	2582.28	2117.47	2200.11	2840.51	2840.51	4555.15	5309.18	2840.51	2788.86	2840.51
Range	1807.60	1342.79	1425.42	1342.79	1425.42	3780.46	4534.49	2065.83	1962.53	2065.83
GINI Coef.	0.2042	0.0981	0.1063	0.0971	0.1063	0.2836	0.3277	0.1292	0.1919	0.2510

Spain										
Main	Negatively skewed			Positively skewed				Unimodal	Rectangular	Bimodal
Statistics	1	2	3	4	5	6	7	8	9	10
Mean	1745.69	1745.84	1746.04	1744.44	1745.68	1745.93	1746.24	1745.94	1745.68	1741.05
SD	628.81	321.19	338.41	318.26	338.08	900.45	1074.59	396.82	575.21	772.22
Skewness	−0.1236	−11.742	−10.366	12.100	10.344	10.356	11.857	0.0000	0.0000	0.0048
Kurtosis	−15.302	0.6859	0.4199	0.9023	0.4157	0.4181	0.7290	−0.1537	−12.061	−17.530
Minimum	748.26	748.26	748.26	1446.63	1366.82	748.34	748.26	748.26	798.02	748.26
Maximum	2494.20	2045.24	2125.06	2743.62	2743.62	4400.24	5128.07	2743.62	2693.37	2743.62
Range	1745.94	1296.98	1376.80	1296.98	1376.80	3651.90	4379.82	1995.36	1895.35	1995.36
GINI Coef.	0.2042	0.0981	0.1004	0.0797	0.0884	0.2354	0.3693	0.1128	0.1745	0.2343

Fig. 1. Negatively skewed distributions

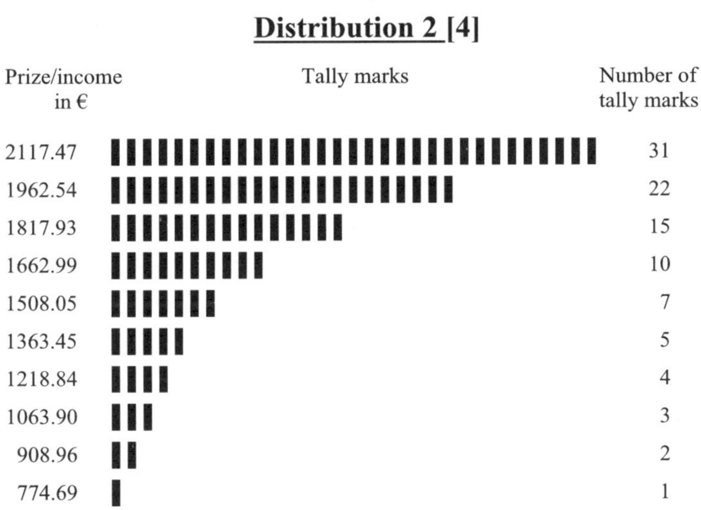

Fig. 2. Positively skewed distributions

In each session, the subjects were arranged in groups of about ten. At the beginning of the session, the subjects were asked to read carefully the instructions and the payment regulations. To make sure that they had properly understood the instructions,[6] we required subjects to pass a test before starting with the experiment. The test consisted of ten multiple-choice questions, which could be easily answered by any subject who had carefully read the instructions.[7] Subjects were informed that for each incorrectly answered question they had to face a 10% cut of their final payoff from the experiment.

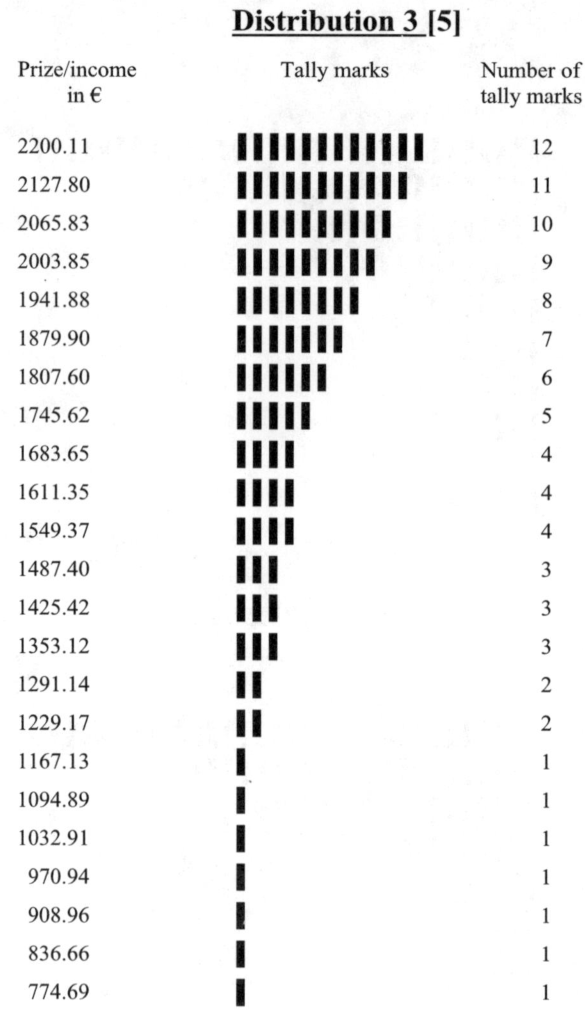

Fig. 3. Unimodal, rectangular and bimodal distributions

If they answered five or more questions incorrectly, they were excluded from any payoff.[8]

Recall that we applied two treatments: a lottery treatment and an income distribution treatment. Within each treatment, the subjects were given two booklets, both depicting either ten lotteries or ten income distributions, as shown in Figs. 1 to 3.

Let us first consider the lottery treatment. The lottery prizes were arranged in terms of 100 tally marks. Subjects were told that each tally mark, depicted in the lottery figures in the booklets, represented exactly one ticket equal in value to the amount listed on the left hand side of the lottery figure. For instance, in Lottery 1 there were 31 tickets bearing the prize "€ 2582.28",

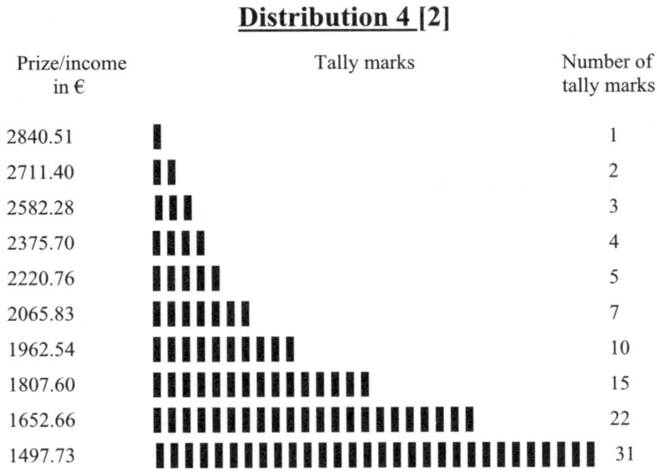

Fig. 4. Lorenz relations of stimulus distributions

22 tickets bearing the prize "€ 2065.83", etc. These prizes were paid in tokens. The subjects had an equal chance to draw one of the 100 tickets in a particular lottery. In the first booklet, subjects were asked to state on a 20-point rating scale their degree of happiness (1 means very unhappy, 20 means very happy) to play a particular lottery. In the second booklet, they were asked to state their *certainty equivalents* (CEs for short) of the ten lotteries as selling prices. The CEs were elicited by way of the Becker–DeGroot–Marschak (BDM) incentive scheme.[9]

The payment to subjects ran as follows: Concerning the first booklet, exactly two out of the ten lotteries were randomly selected for each subject, and the higher rated lottery[10] was played out and constituted one source of tokens. Concerning the second booklet, one out of the ten lotteries was randomly selected and constituted the second source of tokens stemming from the application of the BDM incentive scheme. A subject's total tokens were the sum of the two token sources. Thus, although a subject's total tokens came only, in effect, from two lotteries, each subject had an incentive to reveal his or her true preferences and CEs because each lottery had an equal chance of being selected and becoming the source of a subject's payoff.

The income distribution treatment differed only in minor points from the lottery treatment. The subjects were told that each income distribution represented a population of 100 million income earners, and that each tally mark in a distribution represented exactly 1 million income earners.[11] Each of these 1 million income earners had a monthly income as stated on the left side of the respective income distribution figure. The figures represented monthly disposable incomes because the subjects were more accustomed to monthly salaries in Italy and Spain. The subjects were asked to imagine that they had an equal chance to become one of the 100 million income earners in this population, but they would not know ex ante what their precise income will be in this population. All they would know was the distribution of monthly incomes. They were then asked to state on a 20-point rating scale their degree of happiness from becoming a member of a population characterized by a particular income distribution. The rating scale ranged from 1 (very unhappy) to 20 (very happy).

Thereafter, subjects were asked to imagine that they could alternatively become a member of a population in which all income earners had the *same* monthly income. This income has been termed the *equally distributed equal income* (EDE for short) by the profession. They were invited to indicate the level of income at which they would be *indifferent* between the respective income distribution and the alternative in which each income earner received the same income, viz. the EDE. The EDEs were also elicited by way of the BDM incentive scheme.

In contrast to the lottery treatment, the subjects were informed in the income distribution treatment that income distributions had to obtain for the group as a whole. Therefore, *one participant* in the group would be randomly selected, and, for this particular person, two income distributions would then be randomly selected. The higher rated income distribution would become the group's income distribution, and all the subjects in this group would be given tokens from independent draws according to this income distribution. Thereby, every subject had to assume responsibility for the income distribution of the whole group.[12] This constituted the first source of a subject's tokens. The second source of a subject's tokens stemmed from the application of the BDM incentive scheme to each subject's statement about the EDE for the selected income distribution. For this income distribution, a number was drawn from a uniform distribution defined on the support of the group's income distribution; if the number drawn was less than the stated EDE, then a draw of a new income level according to the group's income distribution was made; if the number drawn was greater than or equal to the stated EDE, then the subject was given tokens amounting to the number drawn. A subject's total tokens were then the sum of the two token sources. Notice that every subject had the same chance to become a random dictator. Thus, each subject had an incentive to reveal his or her true preferences and EDEs because he or she had a one-in-ten chance to decide for the whole group.

In both treatments, final payoffs (in lire or pesetas) were computed by dividing the total number of a subject's tokens by 500. The subjects received a mean payoff of about € 6.50.

3 Results

When screening the data, we noticed that subjects made different use of the 20-point rating scale. Some dwelled more on the lower end, some on the upper end, and some on the extremes. To avoid assigning different weights to the subjects, we calibrated the rating scales, assigning a 1 to the lowest rated lottery or income distribution, and a 10 to the highest rated lottery or income distribution according to the formula:

$$r_i = 1 + \left[R_i - \min_j\{R_j\}\right] \frac{9}{\max_j\{R_j\} - \min_j\{R_j\}},$$

where the R_i's denote the noncalibrated and the r_i's the calibrated ratings.

Recall that all our experimental lotteries and income distributions have the same mean. Then, for nonintersecting Lorenz curves, risk averse (inequality averse) subjects should prefer the lottery (income distribution) whose Lorenz curve is closer to the diagonal.[13] Risk loving (inequality loving) subjects, should prefer the lottery (income distribution) whose Lorenz curve is farther away from the diagonal (see Lopes (1984), p. 475).

What about intersecting Lorenz curves of two lotteries or income distributions with the same mean? Suppose that the Lorenz curve associated with x, $L(x)$, cuts the Lorenz curve associated with y, $L(y)$, from below. Then risk averse or inequality averse subjects, who want to avoid the risk of a relatively low prize or income level, should prefer the lottery or income distribution x, whose associated Lorenz curve is farther away from to the diagonal at the lower end, whereas risk loving or inequality loving subjects, who appreciate the chance of a relatively high prize or income level, should prefer the lottery or income distribution y, whose associated Lorenz curve is farther away from the diagonal at the upper end (see Lopes (1987), p. 270).

If response-mode effects were absent, then subjects should state their preferences according to their risk and inequality attitudes, irrespective of the mode used to elicit their preferences: ratings or valuations. This does not deny that subjects' responses may be affected by a framing effect, in that they exhibit different preferences for particular distributional shapes when they are framed one time as a lottery and the other time as an income distribution. For instance, a particular subject may, at the same time, be risk loving when dealing with lotteries and inequality averse when dealing with income distributions. Within a particular frame, however, subjects should state the same preferences, irrespective of the elicitation mode applied, if response-mode effects were absent.

Our experimental design allows us to study both sides of the medal: the framing effect and the response-mode effect. The former is related to systematic differences between the perception of lotteries and identically shaped income distributions. The later is related to the fact that the elicitation mode of subjects' preferences matters within a given frame. When response-mode effects matter, Lorenz-dominance or single-crossing Lorenz curves would be bad proxies for subjects' preferences because their articulation depends on the elicitation mode applied.[14]

Table 2 provides a summary statistics of subjects' responses. These data provide the basis for studying the mean conformity with the Lorenz relations.

Table 2. Average calibrated ratings and valuations

Distributions		Average calibrated ratings				Average valuations in €			
		Lotteries		Distributions		Lotteries (CE)		Distributions (EDE)	
		Italy	Spain	Italy	Spain	Italy	Spain	Italy	Spain
Negatively skewed	1	8.12	7.82	7.14	7.70	1589.99	1704.58	1532.86	1730.24
	2	6.53	7.11	6.54	7.30	1673.44	1645.35	1706.41	1741.24
	3	5.54	6.45	6.29	6.74	1682.54	1665.48	1801.64	1764.96
	Average	6.73	7.13	6.66	7.24	1648.66	1671.80	1680.30	1745.48
Positively skewed	4	4.99	4.70	6.23	5.61	1802.44	1757.28	1846.43	1841.86
	5	4.24	4.29	6.08	4.92	1808.59	1653.86	1853.54	1803.58
	6	4.37	4.42	3.65	3.32	1981.59	1836.57	2051.21	2039.45
	7	5.30	4.94	3.34	2.92	2096.53	2049.09	2129.63	2211.98
	Average	4.72	4.59	4.83	4.19	1922.29	1824.2	1970.20	1974.22
Unimodal	8	5.57	5.12	6.62	6.26	1769.40	1718.61	1853.71	1842.63
Rectangular	9	4.45	4.11	4.28	5.97	1850.54	1768.01	2014.40	1931.13
Bimodal	10	4.89	4.84	3.17	4.20	1759.23	1736.42	1913.46	1827.18

3.1 Mean Conformity with the Lorenz Relations

Based on the data shown in Table 2, Table 3 shows the conformity rates of subjects' mean responses with the Lorenz relations as shown in Fig. 4.

The entries in Table 3 represent the rates of conformity with the different Lorenz relations that result from the comparison of the ten lotteries or income distributions used as stimulus material in our experimental design as displayed in Fig. 4.

In Table 3, the rate of conformity is provided for the two types of Lorenz relations: Lorenz dominance and Lorenz cutting from below, and for the two elicitation modes used: ratings and valuations. The entries in Table 3 show the percentages of Lorenz relations confirmed according to Table 2. The number of Lorenz relations confirmed refers to the total number of Lorenz relations according to Fig. 4: 32 Lorenz dominance relations and 13 crossing Lorenz curves relations, which amounts to a total of 45 Lorenz relations. For instance, the entry 92.3% in the cell "Lotteries/Cutting Lorenz Curves/Ratings/Italy" means that 12 out of the 13 crossing Lorenz curves relations displayed in Fig. 4 are confirmed according to the mean responses in Table 2 for the Italian data on lottery ratings. The entries under "All cases" refer to the confirmation rate regarding all 45 Lorenz relations included in Fig. 4.

The inverse mirror-image of the first two and the second two columns in Table 3 constitutes a strong evidence of a response-mode effect regarding average responses. Note that, for the rating elicitation mode, subjects' stated preferences confirm the large majority of Lorenz relations, whereas, for the valuation elicitation mode, we find widespread violation of the Lorenz relations as displayed in Fig. 4.

Concerning lottery ratings, the conformity rates of the mean lottery ratings are higher in the case of crossing Lorenz curves than in the case of Lorenz dominance. This means that subjects prefer those lotteries in which the probability of the higher prizes is higher. For dominating Lorenz curves, a conformity rate of 62.5% and 68.7% for the Italian and Spanish data, respectively, again confirms risk aversion, but to a lower degree. Regarding the

Table 3. Conformity of mean responses with Lorenz relations in percentages

Mode		Ratings Italy	Ratings Spain	Valuations Italy	Valuations Spain
Lotteries	↑	92.3	100.0	0.0	23.1
	→	62.5	68.7	15.6	12.5
	All cases	71.1	77.8	11.1	15.6
Income	↑	76.9	100.0	7.7	15.4
Distributions	→	81.3	87.5	9.4	15.6
	All cases	80.0	91.1	8.9	15.6

income distribution ratings, the conformity rates for crossing Lorenz curves are again 100% for the Spanish data but only 76.9% for the Italian data, which means less mean inequality aversion of the Italian subjects. The availability of very high incomes seems to outweigh their small probability in about a quarter of cases for the Italian subjects. For dominating Lorenz curves, inequality aversion considerably exceeds risk aversion for the lottery domain.

Concerning lottery valuations, the conformity rates for lotteries (income distributions) are at rather low levels: 11.1% and 15.6% (8.9% and 15.6%) for the Italian and Spanish data, respectively. Inspecting Figs. 1 to 3 allows us to conclude that subjects are captured by the top prizes or income levels when valuating a particular lottery or income distribution. This shows that risk attitudes and inequality preferences are largely affected by response-mode effects: In the rating mode, subjects' preferences are more affected by risk and inequality aversion, whereas, in the valuation mode, subjects' preferences seem to be more affected by risk and inequality sympathy. This reflects a greater influence of the top prizes or income levels due to the compatibility hypothesis.[15] It predicts that subjects would pay more attention to the most spectacular (i.e., top) prizes or incomes in the valuation of lotteries or income distributions as compared to the rating mode, for which the probability is more compatible.

3.2 Individual Conformity with the Lorenz Relations

Conformity with the Lorenz relations can also be analyzed in terms of individual ratings and valuations. In this sub-section we look at each subject's 45 pairwise comparisons of lotteries and income distributions.

In Table 4 we present a summary statistics of the conformity rate with Lorenz relations. This conformity rate is computed, for each one of the 45 pairwise comparisons, as the mean percentage of subjects whose responses conform to the Lorenz relations. Although the results are less pronounced than for the mean ratings and valuations, the main results are confirmed. For more detailed data see the Appendix.

As far as response-mode effects are concerned, the majority of ratings conform to the Lorenz relations, whereas the majority of valuations violate them.[16] A Wilcoxon signed ranks test shows that the differences in the conformity rates between the rating and valuation modes are statistically significant.

Concerning ratings, Lorenz dominance is again more frequently confirmed for income distributions[17] than for lotteries. This demonstrates greater inequality aversion than risk aversion for Lorenz dominance. In contrast to that, the ratings of Lorenz curves, which cut others from below, conform less frequently to income distributions than to lotteries. This shows that, in this case, fewer subjects exhibit inequality aversion as compared to those who exhibit risk aversion.

Table 4. Conformity of individual responses with Lorenz relations in percentages

			Lotteries			Income distributions		
			↑	→	All cases	↑	→	All cases
Spain	Wilcoxon	*P*	0.002	0.000	0.000	0.001	0.000	0.000
	test	*Z*	−3.115	−3.526	−4.703	−3.184	−4.548	−5.459
	Valuation	S.D.	0.062	0.061	0.067	0.090	0.091	0.093
		Mean	48.4	42.0	43.9	41.1	36.1	37.6
	Rating	S.D.	0.093	0.149	0.147	0.065	0.160	0.139
		Mean	66.8	54.1	57.8	62.2	64.9	64.1
Italy	Wilcoxon	*P*	0.001	0.005	0.000	0.021	0.000	0.000
	test	*Z*	−3.181	−2.778	−4.191	−2.312	−4.038	−4.777
	Valuation	S.D.	0.048	0.077	0.071	0.124	0.113	0.116
		Mean	38.2	41.0	40.2	34.9	38.8	37.7
	Rating	S.D.	0.104	0.148	0.147	0.091	0.140	0.149
		Mean	63.6	51.3	54.9	48.9	65.7	60.8

Concerning valuations, the mirror image of the results for the ratings is also reflected in the individual data: the valuation rates of lotteries and income distributions which conform to the Lorenz relations are down by one fifth to one fourth of the conformity rates of the ratings.

To analyze the framing effect, we compare the conformity rates within a particular elicitation mode, but between frames, that is, lotteries versus income distributions. In Table 5 we present the results of a Mann–Whitney test. We find that, in the Italian data, the framing effect is only significant for the ratings. In the Spanish data, this effect is found significant in all cases except for the ratings of crossing Lorenz curves. However, this test is based on the differences between means when only the framing effect is considered. Therefore, this test does not allow differentiating between both effects. Later on, a more detailed joint analysis of the response-mode and framing effects will be provided.

Table 5. Mann–Whitney test: Lotteries vs. income distributions

		Italy		Spain	
		Ratings	Valuations	Ratings	Valuations
↑	*Z*	−3.473	−0.026	−1.517	−2.287
	p	0.001	0.979	0.129	0.022
→	*Z*	−3.796	−1.284	−2.889	−3.096
	p	0.000	0.199	0.004	0.002

Concerning the joint analysis of the response-mode and the framing effects, Table 6 shows the results of the estimation using a logit panel data model with random effects for the Italian and Spanish data. The dependent variable is the conformity with Lorenz relations that should assume, for a particular rating or valuation, the value 1 for perfect conformity with the Lorenz relation, and 0 for perfect nonconformity. The explanatory variables are two dummies. The first, denoted as *Mode*, refers to the response mode, and assumes the value 0 for valuation and 1 for rating. The second, denoted as Frame, refers to the framing used and assumes the value 0 for a lottery and 1 for an income distribution.

The results shown in this table[18] reinforce our previous findings. Regarding the response-mode effect, the coefficient for the explanatory variable *Mode* confirms that a strong response-mode effect exists. In fact, the sign of this coefficient indicates that the probability of conformity with the Lorenz relations increases as we use rating as an elicitation mode instead of valuation. Moreover, we find no differences between the Italian and the Spanish data for this effect, since the coefficients for both countries do not differ significantly.[19]

Regarding the framing effect, the coefficient for the explanatory variable Frame is nonsignificant for both countries. However, we know from Table 5 that framing effects can be more easily observed when we differentiate between crossing and dominant Lorenz curves. Hence, we apply logit panel regressions separately to crossing and dominating Lorenz curves. The results are shown in Tables 7 and 8.

These tables show that, although the response-mode effect does not vary between countries, it is higher[20] in the case of crossing Lorenz curves than for the Lorenz dominance cases. In any case, the probability of conformity of the Lorenz relations is higher for the ratings than for the evaluations. However, concerning the framing effect, differences do exist between the Italian and Spanish data. While this effect is not statistically significant both for crossing and dominant Lorenz curves for the Italian data, it is significant

Table 6. Logit panel data model with random effects for Italy and Spain: Response-mode and frame effects

Explanatory Variables	Italy Coefficient	$p-$value	Spain Coefficient	$p-$value
Constant	−0.4344	0.005	−0.4326	0.000
Mode	0.8840	0.000	0.8497	0.000
Frame	−0.0289	0.854	−0.0806	0.449
Observations	9090		9720	
Number of groups	101		108	
σ_u	0.5839		0.7003	
ρ	0.2542		0.3291	

Table 7. Logit panel data model with random effects for Italy and Spain: Crossing Lorenz curves

Explanatory Variables	Italy Coefficient	$p-$value	Spain Coefficient	$p-$value
Constant	−0.0970	0.014	−0.4455	0.011
Mode	1.003	0.000	1.056	0.000
Frame	−0.2321	0.284	−0.5092	0.049
Observations	2626		2808	
Number of Groups	101		108	
σ_u	1.1181		1.2831	
ρ	0.5556		0.6221	

Table 8. Logit panel data model with random effects for Italy and Spain: Dominant Lorenz curves

Explanatory Variables	Italy Coefficient	$p-$value	Spain Coefficient	$p-$value
Constant	−0.5607	0.000	−0.6254	0.000
Mode	0.8944	0.000	0.8484	0.000
Frame	0.1012	0.450	0.3496	0.005
Observations	6464		6912	
Number of Groups	101		108	
σ_u	0.5941		0.6881	
ρ	0.2609		0.3213	

for the Spanish data. Note, in this case, that the framing effect has an opposite sign for crossing and dominant Lorenz curves. Moreover, for crossing Lorenz curves, the probability of conformity with Lorenz relations is higher for lotteries than for income distributions. The contrary obtains for dominant Lorenz curves, that is, the probability of conformity is higher in the case of income distributions.

4 Conclusion

Although there is a close relationship between income distributions and lotteries, their joint analysis is much in its infancy. Moreover, multiple-outcome payoff distributions have hardly ever been employed systematically and material incentives were only rarely used.

In this paper we investigate experimentally the violation of Lorenz relations in the case dominant and single-crossing Lorenz curves using multiple-outcome payoffs distributions. We use as stimulus material different types of payoff distributions: three negatively skewed, four positively skewed, one rectangular, one unimodal, and one bimodal.

Our experimental design consists of two treatments. In the first treatment, the ten distributions of payoffs were presented to the subjects as lotteries, whereas in the second treatment, they were presented as income distributions. In each treatment, subjects were asked to judge the ten multiple-outcome lotteries or n-dimensional income distributions in terms of both ratings and valuations (in terms of their CEs or EDEs using a BDM incentive scheme).

The experiment was administered to more than 200 subjects in Italy and Spain. Subjects' comprehension of the experimental setting was examined before the experiment started. In each session only one treatment was applied and each subject was allowed to participate only in one experimental session.

If no response-mode effect existed, subjects should state their preferences according to their risk attitude and inequality preference, irrespective of whether their preferences are elicited through ratings or valuations. This does not deny that subjects' responses may be affected by a framing effect. In fact, they may exhibit different preferences for particular distributional shapes when they are framed one time as a lottery and the other time as an income distribution. For instance, a particular subject may, at the same time, be risk loving when dealing with lotteries and inequality averse when dealing with income distributions.

Our results constitute strong evidence of the existence of response-mode effects. For average responses subjects' stated preferences conform largely to Lorenz relations when elicited as ratings, but widely violate Lorenz relations when elicited as valuations. This shows that risk and inequality attitudes are largely affected the response-mode effects: In the rating mode, subjects' preferences are more affected by risk and inequality aversion, while, in the valuation mode, subjects' preferences seem to be more affected by risk and inequality sympathy.

Regarding individual data, the main results continue to hold, although the effects are less pronounced than with the mean ratings and valuations.

Concerning framing effects, a Mann–Whitney test shows that for the Italian data the framing effect is only significant for the ratings, independently of the type of Lorenz relation. In the Spanish data, this effect is found significant in all cases except for the ratings of crossing Lorenz curves.

Finally, a joint analysis of the response-mode and framing effects based the use of panel logit regressions reinforces our previous findings. Regarding the response-mode effect, we find that the probability of conformity with the Lorenz relations increases as we use rating as the elicitation mode instead of valuation. Moreover, we find no differences between the countries. Regarding the framing effect, it is only significant for the Spanish data: for crossing Lorenz curves, the probability of conformity with Lorenz relations is higher for lotteries than for income distributions. The contrary obtains for dominant Lorenz curves, that is, the probability of conformity is higher in the case of income distributions.

Acknowledgement. This research was supported by a grant from the European Commission under contract No. ERBFMRXCT 980248. We are indebted to Michael Birnbaum, Stefan Traub, Martin Missong, and to two anonymous referees for helpful comments. We also thank Andrea Morone for helping in conducting the experiments. The usual disclaimer applies.

Instructions, the multiple-choice questions of subjects' test, and the data of the results of this experiment are available from Eva Camacho-Cuena, e-mail: eva.camacho@uam.es, upon request.

References

Becker, G.M., DeGroot, M.H., Marschak, J. (1964), Measuring Utility by a Single-Response Sequential Method, *Behavioral Science* 9: 226–232.

Beckman, S.R., Cheng, D., Formby, J.P., Smith, W.J. (1994), Preferences for Income distributions and Redistributions: Evidence from Experiments with Real Income at Stake, Discussion Paper, University of Colorado-Denver.

Camacho–Cuena, E., Seidl, C., Morone, A. (2005), Comparing Preference Reversal for General Lotteries and Income Distributions, *Journal of Economic Psychology* 26: 682–710.

Fitts, P.M. and Seeger, C.M. (1953), S-R Compatibility: Spatial Characteristics of Stimulus and Response Codes, *Journal of Experimental Psychology* 46: 199–210.

Lopes, L.L. (1984), Risk and Distributional Inequality, *Journal of Experimental Psychology: Human Perception and Performance* 10: 465–485.

Lopes, L.L. (1987), Between Hope and Fear: The Psychology of Risk, *Advances in Experimental Social Psychology* 20: 255–295.

Schneider, S. and Lopes, L.L. (1986), Reflection in Preferences Under Risk: Who and When May Suggest Why, *Journal of Experimental Psychology: Human Perception and Performance* 12: 535–548.

Slovic, P. and MacPhillamy, D. (1974), Dimensional commensurability and cue utilization in comparative judgment, Organizational Behavior and Human Performance 11: 172–194.

Traub, S., Seidl, C., Schmidt, U. (2006), An Experimental Study on Individual Choice, Social Welfare, and Social Preferences, Economics Department, University of Kiel [submitted].

Tversky, A., Slovic, P., Kahneman, D. (1990), The Causes of Preference Reversal, *American Economic Review* 80: 204–217.

A Appendix: Individual Conformity with Lorenz Relations

In Tables 9 to 12 we present the fractions of subjects whose responses concerning ratings and valuations conform to each one of the Lorenz relations illustrated in Fig. 4. We organized our results for Italy and Spain and for the two frames here considered: lotteries and income distributions. These table provide more detailed information than Tables. 3 and 4.

We used the entries in these tables to compute the entries in Table 4 and run the nonparametric tests used in Sect. 3.2. Note that the shaded cells refer to the case in which the Lorenz curve of the lottery or income distribution in a row dominates the Lorenz curve of the lottery or income distribution in the corresponding column, to differentiate this case from the case where the Lorenz curve of the lottery or income distribution in a row cuts the Lorenz curve of the lottery or income distribution in the corresponding column from below.

B Notes

1. Note that, whereas the rating and the valuation methods can be applied to larger sets of distributions, the choice method requires the arrangement of the distributions in terms of pairs, which requires subjects to make $\frac{m(m-1)}{2}$ instead of m comparisons for m distributions. Thus, the choice method of preference elicitation is more appropriate for simple experiments, whereas the rating method is more appropriate for more complicated experimental designs. Note that both methods are equivalent. Having applied both methods, Tversky et al. (1990, p. 213) report: "The data reveal no discrepancy between choice and rating."
2. For an analysis of lotteries by means of Lorenz curves see Lopes (1984, 1987) and Schneider and Lopes (1986).
3. Due to such influences the average level of entries in terms of euros was some 3.4% lower in Spain than in Italy. The actual figures for the means were about € 1807 in Italy and € 1745 in Spain.
4. This approach was adopted to control for ordering effects of presentation. Had we presented the lotteries at random to the subjects, ordering effects would have evened out *if* they were present. The comparison of two orderings of presentation allows, however, ordering effects of presentation to be identified, or to be ruled out. As shown in our earlier paper (Camacho et al. (2005), Sect. 3.3), we can rule out ordering effects of presentation. Only for the distribution ratings did we observe cultural effects for the Italian and Spanish subjects.
5. Kurtosis is defined as the fourth central moment of the distribution less 3 (i.e., the value of the fourth central moment of a normal distribution with parameters $\mu = 0$ and $\sigma = 1$).

Table 9. Conformity of lotteries (Italy)

		1	2	3	4	5	6	7	8	9	10
1	Rating				0.750	0.731	0.731	0.615	0.788	0.788	0.788
	Valuation				0.385	0.385	0.346	0.346	0.346	0.385	0.462
2	Rating	0.135		0.596	0.558	0.654	0.692	0.558	0.654	0.654	0.692
	Valuation	0.519		0.423	0.481	0.423	0.365	0.385	0.423	0.346	0.423
3	Rating	0.192			0.519	0.558	0.558	0.519	0.481	0.654	0.538
	Valuation	0.500			0.451	0.308	0.327	0.327	0.365	0.288	0.404
4	Rating					0.558	0.500	0.423		0.519	0.442
	Valuation					0.577	0.442	0.385		0.519	0.500
5	Rating						0.404	0.423		0.442	0.346
	Valuation						0.346	0.365		0.442	0.442
6	Rating							0.365			
	Valuation							0.308			
7	Rating										
	Valuation										
8	Rating				0.596	0.654	0.615	0.558		0.654	0.577
	Valuation				0.423	0.385	0.577	0.365		0.385	0.442
9	Rating						0.481	0.423			
	Valuation						0.404	0.308			
10	Rating						0.538	0.462			
	Valuation						0.327	0.308			

6. More complicated experiments often suffer from the subjects' being insufficiently acquainted with the experimental design, the experimental procedure, and the incentive schemes. In this case, they become sources of data distortions which cannot easily be controlled.
7. The instructions and the test are available from Eva Camacho, email: eva.camacho@uam.es upon request.

Table 10. Conformity of income distributions (Italy)

		1	2	3	4	5	6	7	8	9	10
1	Rating				0.500	0.500	0.714	0.804	0.464	0.768	0.857
	Valuation				0.268	0.214	0.268	0.268	0.143	0.179	0.179
2	Rating	0.304		0.464	0.500	0.482	0.768	0.750	0.429	0.607	0.732
	Valuation	0.643		0.304	0.393	0.357	0.411	0.339	0.250	0.214	0.304
3	Rating	0.321			0.446	0.464	0.768	0.768	0.375	0.625	0.768
	Valuation	0.768			0.446	0.446	0.393	0.339	0.411	0.339	0.321
4	Rating					0.500	0.679	0.661		0.643	0.732
	Valuation					0.464	0.339	0.357		0.357	0.411
5	Rating						0.661	0.696		0.625	0.768
	Valuation						0.357	0.393		0.375	0.429
6	Rating							0.464			
	Valuation							0.464			
7	Rating										
	Valuation										
8	Rating				0.482	0.500	0.804	0.768		0.696	0.857
	Valuation				0.482	0.518	0.411	0.411		0.339	0.357
9	Rating						0.571	0.571			0.554
	Valuation						0.518	0.500			0.411
10	Rating						0.446	0.518			
	Valuation						0.429	0.446			

8. Note that this test only served the purpose of inducing subjects to acquaint themselves properly with the setup of the experiment. Indeed, this precaution worked well: Out of 110 subjects in Italy, only five answered only 3 or 4 questions incorrectly for the test in each treatment. In Spain only 11 out of 102 subjects answered 3 questions incorrectly in the test in each treatment. All others scored better. This meant that we could rely

Table 11. Conformity of lotteries (Spain)

		1	2	3	4	5	6	7	8	9	10
1	Rating				0.745	0.784	0.706	0.627	0.784	0.765	0.804
	Valuation				0.510	0.529	0.353	0.314	0.490	0.373	0.451
2	Rating	0.275		0.627	0.686	0.706	0.686	0.627	0.667	0.784	0.725
	Valuation	0.392		0.412	0.431	0.549	0.431	0.412	0.412	0.451	0.451
3	Rating	0.176			0.647	0.647	0.647	0.569	0.667	0.765	0.647
	Valuation	0.392			0.451	0.569	0.490	0.431	0.510	0.412	0.451
4	Rating					0.510	0.490	0.353		0.588	0.392
	Valuation					0.647	0.451	0.490		0.392	0.451
5	Rating						0.490	0.392		0.510	0.412
	Valuation						0.392	0.373		0.333	0.314
6	Rating							0.373			
	Valuation							0.392			
7	Rating										
	Valuation										
8	Rating				0.510	0.569	0.608	0.588		0.588	0.471
	Valuation				0.529	0.529	0.451	0.451		0.392	0.412
9	Rating						0.490	0.490			0.373
	Valuation						0.451	0.431			0.431
10	Rating						0.510	0.529			
	Valuation						0.412	0.353			

that the subjects were being sufficiently acquainted with the rules of the experiment.

9. This means that for any lottery a number was drawn from a uniform distribution defined on the support of this lottery. If the number drawn was less than the CE stated for this particular lottery, the respective lottery was played out and the subject was given tokens amounting to the value of the respective prize. If the number drawn was greater than

Table 12. Conformity of income distributions (Spain)

		1	2	3	4	5	6	7	8	9	10
1	Rating				0.680	0.720	0.800	0.780	0.700	0.680	0.780
	Valuation				0.380	0.400	0.360	0.260	0.380	0.280	0.360
2	Rating	0.260		0.660	0.580	0.640	0.800	0.820	0.580	0.680	0.800
	Valuation	0.440		0.360	0.340	0.440	0.320	0.240	0.340	0.340	0.320
3	Rating	0.260			0.480	0.580	0.760	0.820	0.600	0.620	0.760
	Valuation	0.480			0.360	0.520	0.300	0.260	0.440	0.380	0.500
4	Rating					0.560	0.680	0.680		0.360	0.580
	Valuation					0.500	0.380	0.220		0.380	0.520
5	Rating						0.560	0.640		0.360	0.520
	Valuation						0.280	0.220		0.360	0.380
6	Rating							0.540			
	Valuation							0.300			
7	Rating										
	Valuation										
8	Rating				0.580	0.620	0.804	0.820		0.580	0.680
	Valuation				0.520	0.600	0.411	0.260		0.420	0.440
9	Rating						0.740	0.800			0.640
	Valuation						0.420	0.300			0.580
10	Rating						0.640	0.660			
	Valuation						0.340	0.300			

or equal to the stated CE, then the subject was given tokens amounting to the number drawn. For a more detailed explanation see Becker et al. (1964).

10. Ties were resolved by flipping a coin.
11. This design was adopted to minimize computational errors. We tried to avoid using different dimensions such as having 10 million income earners,

and associated tally marks each of which represented 100,000 income earners. Indeed, no subject found this design unrealistic.

12. Beckman et al. (1994, p. 8) used a similar assumption to "create a group identity," but they employed majority voting instead of a random dictator. This was possible in their experimental setting because they had only two distributions to choose from for any decision.
13. For lotteries with the same mean, a dominating Lorenz curve is associated with a lottery derived from another lottery by way of a sequence of mean-preserving contractions. In the case of income distributions, these mean-preserving contractions are nothing else but progressive transfers.
14. Another cause may be due to order effects of stimulus presentation. Recall that they were ruled out for our experiment.
15. The compatibility hypothesis was originally developed by Fitts and Seeger (1953) and rediscovered by Slovic and MacPhillamy (1974). It states that attributes which are more compatible with the dimension of the response mode are assigned greater weight.
16. Except for the income distribution ratings of the Italian subjects in the case of cutting Lorenz curves, where the conformation rate is only 48.9%. However, it is still markedly higher than the conformity rate for valuations.
17. In a related paper, Traub et al. (2006) observed Lorenz dominance conformity rates for income distributions between 61.3% and 64.4% for three treatments. Their experimental design was based on asking subjects directly for their preference orderings of twelve income distributions. This method is equivalent to the rating method of eliciting preferences as used in the present paper.
18. Notice that conformity with the Lorenz relations is either 0 or 1. We model the probability of conformity as the logistic distribution:
    $P(\text{conformity} = 1) = \frac{e^x}{1+e^x}$;
    $P(\text{conformity} = 0) = \frac{1}{1+e^x}$.
    The value of z is estimated from:
    $z = \beta_0 + \beta_1 \text{Mode} + \beta_2 \text{Format}$,
    using the logit panel data method with random effects. Note that $\frac{\partial P(\text{Conformity}=1)}{\partial z} > 0$, so that $P\,(\text{Conformity} = 1)$ increases as z increases.
19. A χ^2-test shows that the null hypothesis that the values of the two coefficients do not differ significantly cannot be rejected ($\chi^2_1 = 0.68$ (p-value $= 0.409$)).
20. A χ^2-test shows that the null hypothesis that the values of the coefficients for cutting and dominant Lorenz curves within each country are not statistically different can be rejected. (The values of the statistic for Italy and Spain are $\chi^2_1 = 4.16$ (p-value $= 0.041$) and $\chi^2_1 = 15.95$ (p-value $= 0.000$), respectively.)

A Deeper Look at Hyperbolic Discounting

Barry Sopher[1] and Arnav Sheth[2]

[1] Department of Economics Rutgers University, NJ, USA
[2] Department of Finance Rutgers University, NJ, USA
e-mail: sopher@econ.rutgers.edu, shetha@andromeda.rutgers.edu

Abstract. We conduct an experiment to investigate the degree to which deviations from exponential discounting can be accounted for by the hypothesis of hyperbolic discounting. Subjects are asked to choose between an earlier or later payoff in series of forty choice questions. Each question consists of a pair of monetary amounts determined by compounding a given base amount at a constant rate per period. Two bases (8 and 20 dollars), three compounding rates (low, medium and high) and three delays (two, four, and six weeks) are each used. There are also two initial periods (today and two weeks) and there are two separate questionnaires, one with lower "realistic" compounding rates and the other with higher compounding rates, typical of those used in previous studies. We analyze the detailed patterns of choice in 6 groups of 6 related questions each (in which the base and rate is fixed but the initial period and delay varies), documenting the frequency of patterns consistent with exponential discounting and with hyperbolic discounting. We find that exponential discounting is the clear modal choice pattern in virtually all cases. Hyperbolic discounting is never the modal pattern (except in the sense that constant discounting is a special case of hyperbolic discounting). We also estimate a linear probability model that takes account of individual heterogeneity. The estimates show substantial increases in the probability of choosing the later option when the compounding rate increases, as one would expect. There are small, sometimes significant, increases in this probability when the delay is increased or the initial period is in the future. Such behavior is consistent with hyperbolic discounting, but can account for only a small proportion of choices. Overall, deviations from exponential discounting appear to be due to error, or to other effects not accounted for by hyperbolic discounting. Principal among these is an increase in later choices when the base is larger.

Keywords: intertemporal choice, hyperbolic discounting

1 Introduction

Since the late 1930s, when Samuelson introduced discounted utility, the concept of discounting has been used by economists in analyses of intertemporal choice. Although the descriptive accuracy of this model has been called into question by many, the analytic convenience and the normative logic of the model have kept it alive. In other words, it is easy to use, and it seems, for many purposes quite sensible. It is a workhorse in dynamic models in labor economics and macroeconomics, and it is, indeed, hard to imagine what model one would use in its place.

Loewenstein and Thaler (1989) summarize and review the literature illustrating many shortcomings in the model, and Loewenstein and Prelec (1992) have proposed an alternative formulation with a structure similar that proposed by Kahneman and Tversky's prospect theory for risky choice. Indeed, on closer consideration, it is often only functional form restrictions that are being questioned by these authors, rather than the fundamental notion that future utility is somehow discounted. For example, Loewenstein and Prelec (1993) pointed out that an individual might well value different sequences of restaurant meals differently, and thus they questioned the simple adding-up (with appropriate discount factors) of sequences of utility. The intuition that one might prefer to vary the cuisine of one's meals out rather than have the same thing week after week is attractive, of course, and perhaps a utility function that explicitly accounts for complementarities between adjacent periods would be more appropriate in this case.

The time period of analysis and the consumption basket that Samuelson envisioned, though, was something more like total consumption year by year. More to the point, in a dynamic model of, say, lifetime labor supply, considerations of the particulars of week by week consumption patterns is too fine a level of detail, and such models were never intended to capture such factors with perfect accuracy. The key element of the discounted utility model is, after all, not the utility function but the discounting function. We could assume risk-neutral income-maximizing behavior as the baseline behavioral model and not affect the predictions of the theory in any substantial way.

The specifics of the discounting function have, in fact, been the focus of many writers in recent years, and here there is perhaps a bit more room for improvement, even for the big-picture uses to which the discounted utility model has been put. The logical inconsistencies associated with non-constant discounting were first explored by Strotz (1955), and the tendency of some people (and rats, too) to discount in a non-constant manner have been documented in many experimental studies since then. Here, as in the criticisms of the utility function in discounted utility, a little intuition seems to go a long ways. The story, told by Thaler (1981), that I might prefer an apple today over two apples tomorrow, but that I would more likely prefer two apples in two weeks and a day to one apple in two weeks has, again, a certain appeal. Coupled with the observation that this would violate discounted utility, this is persuasion enough for some. But again, on closer examination, is it so persuasive? Will \$ 100 today be chosen over \$ 100 compounded at a constant daily rate tomorrow? (Surely \$ 100 compounded at some constant rate for 15 days will be chosen over \$ 100 compounded at the same rate for 14 days.) We suggest, provisionally, that some might violate stationarity in this way, but that most people would not.

We would suggest, further, that it is not enough that significantly fewer individuals choose the later option in the (today, tomorrow) case than in the (two weeks, two weeks and a day) case to prove that the apple story is descriptively accurate. Such a result would certainly falsify the predictions of the constant discounting assumption embodied in the discounted utility model, but it does not clearly support some other clear alternative model, such as hyperbolic discounting, as some studies seem to conclude, if only implicitly. Thaler (1981), Benzion et al. (1989), Mischel (1966, 1974) Mischel and Ebbenson (1970) and Ainslie and Haendel (1983) all fall in this class. It should be pointed out as well that all of these studies made use of hypothetical payoffs only.

More specifically, most studies have focused on the two main implications of the constant discounting. The first implication is *stationarity.* Stationarity means that in a choice between two consequences, it matters only how far apart in time the consequences are delivered, and not their absolute position in time. The apple story above is meant to show how stationarity will be routinely violated. The second implication we will refer to as *linearity.* This means that if one prefers $ 100 compounded at a constant rate for two weeks over $ 100 today, then one ought to also prefer $ 100 compounded at the same constant rate for 4 weeks over $ 100 today.

The studies cited above typically find that a significant proportion of subjects will violate stationarity, and a significant proportion will violate linearity as well. There is a tendency for immediate consequences to be chosen over delayed consequences, leading to the stationarity violation. There is also a tendency for later outcomes to be chosen more often, the more delayed they are. That is, later consequences seem to be discounted at a lower rate than early consequences.

Holcomb and Nelson (1992) have conducted one of the few studies that carefully tried to induce monetary incentives in the manner that experimental economists do. They found support for stationary but not linearity in their study. This analysis is a significant improvement, statistically, over previous studies, but not enough detail is provided in the analysis reported in the paper to answer the question posed above, whether the violations of constant discounting can be interpreted as, equally, support of hyperbolic discounting. In this paper, we use a design much like that of the Holcomb–Nelson experiment, but expand the design to include a larger variety of compounding rate. After making the obvious aggregate comparisons of choice frequencies between appropriate sets of questions, we then conduct two further types of analysis to more deeply probe into this question. First, we consider *patterns* of choices over large sets of questions and investigate to what degree the frequencies with which individuals choose different patterns help us to differentiate between alternative discounting schemes. Second, we treat the data as a panel of observations and conduct regression analysis, controlling for individual heterogeneity, to quantify more precisely the magnitudes of the various departures from constant discounting that we observe.

2 The Experiment

2.1 The Discounting Function

For purposes of generating testable predictions, we will focus on the special case of discounted utility in which the utility is linear. In other words, we will focus on the present value maximization as the baseline model. In this model, we suppose that individuals choose between alternative income streams in a way that maximizes present discounted value. For example, if one has a choice between $\$\,x$ today and $\$\,y = \$\,x(1+r)$ in two weeks, then one compares the present value of each alternative and chooses appropriately. That is, $x >$ or $< y\delta$, where δ is the appropriate 2-week discount factor. The *stationarity* property of discounted utility is seen as follows. If, without loss of generality, x today is preferred to y in 2 weeks, then this means that $x \geq y\delta$. Note that the choice between x in two weeks and y in four weeks is evaluated by comparing $x\delta$ to $y\delta^2$, and, since $x\delta \geq y\delta^2$, one continues to prefer the earlier to the later option in this case. Note also that we could compound both alternatives at the same rate over the additional 2 weeks with out changing the choice pattern. That is, $x(1+r)\delta \geq y(1+r)\delta^2 = x(1+r)^2\delta^2$. Similarly, the *linearity* property of discounted utility can be seen by asking how our individual would choose between x today and $z = x(1+r)^2$ in four weeks. Since we know now that $(1+r)\delta \leq 1$, we also can say that $x \geq x(1+r)^2\delta^2$ so, again, the earlier option continues to be chosen.

The hyperbolic discount function is an alternative to constant discounting that has been proposed to accommodate the types of violations of constant discounting commonly observed in experimental studies. Although it is usually presented as a generalized hyperbola, a certain limit of which is equivalent to constant (exponential) discounting, we follow Holcomb and Nelson (1992) and use a discretized version of this discounting function, as it has a more natural and intuitive look. The general form of the function is $1/(1+k_1)(1+k_2)(1+k_3)\ldots(1+k_t)$ for a payment to be received t periods into the future. Moreover, we have $k_1 \geq k_2 \geq k_3 \geq \ldots \geq k_t$. Obviously, this is equivalent to constant discounting if the k's are all equal, and $1/(1+k) = \delta$, the discount factor. With this hyperbolic discounting function, both linearity and stationarity can be violated. We return to investigate this function in more detail in Sect. 4, when we analyze in detail the individual choice patterns in the data.

2.2 Design

The experiment was a survey consisting of forty choice questions. Each question offered a choice between an earlier and a later monetary payment. The payments were arrived at by applying a constant compounding rate to a given base amount. Subjects chose either the earlier or the later payment for each question. It was explained that one of the forty questions would be chosen at random at the end of the experiment, and the subject's choice on that question would be his or her payment for the experiment. If the choice involved a delay, the subject was required to return to the same room where the experiment was held on the appointed day to collect the payment. Payments varied from \$ 8 to more than \$ 40. A total of 86 subjects participated in the experiment. The latest payment was 8 weeks from the date of the experiment, which always fell with the semester in which the experiment was conducted, so that student subjects were still on campus at the time the payment was due.

Four factors were varied between questions in the questionnaires: the base amount (8000 or 20,000 francs[1]), the compounding rate (low, medium or high), the initial time (today or in two weeks) and the time delay between choices (two, four or six weeks). There were two distinct questionnaires, and subjects were randomly assigned to answer one or the other, not both. The low-rate questionnaire used lower compounding rates of 0.1, 0.5 and 1 percent per week. These translate into implied annual rates of approximately 5.2, 26 and 52.26 percent, respectively. The high-rate questionnaire used higher compounding rates of 1, 5 and 10 percent per week. These translate into implied annual rates of approximately 52.26, 266.5 and 546 percent, respectively. A total of 44 subjects answered the lower-rates questionnaire and 42 subjects answered the higher-rates questionnaire. The questionnaires are contained in the Appendix.

This 2(base) $\times$ 2(initial period) $\times$ 3(delays) $\times$ 3(rates) $=$ 36 accounts for only 36 of the 40 questions. The other four questions on each questionnaire were designed to test for simple monotonicity by asking if the subject wanted the same amount of money earlier or later, with no compounding of the initial base amount. Finally, it should be noted that order of the questions was randomized for each subject, so that no subject saw the questions in the orderly fashion shown in the appendix, where all questions for a given base amount and compounding are shown in order.

The subjects were students at Rutgers University and at New York University. The experiment was conducted in computer labs at the two institutions in the fall of 2002 and the spring of 2003. Subjects were recruited through email notification and electronic sign-up procedures. The subjects were seated at individual computer and completed the online questionnaire individually. Subjects were paid at the end of the session or given a reminder notice telling them how much they were scheduled to receive at some specified future date, and where they should come to collect their payments. All but three subjects collected their payments on schedule. Of these, two collected their payments late, and one never collected a payment.

[1] The dollar to franc exchange rate was 1 to 1000. The large franc values were used so that all amounts to be accurately expressed as whole numbers.

3 Aggregate Results

For the basic monotonicity issue (questions 37 through 40 on both questionnaires) subjects were quite consistent. In all, 92 percent of the subjects (81 out of 86) chose to take the money earlier rather than later in all four questions. None of the subjects chose the later choice in all four questions. The worst violator of monotonicity chose to take the money later in three of the four questions.

We now focus on the 36 main questions that comprise the actual experiment. Table 1 shows the percentage of subjects choosing the delayed alternative for each question in which the initial time is today, while Table 2 shows the percentage choosing the delayed option when the initial period is two weeks from today. There is a clear increase in the number of people choosing the delayed option as the interest rate increases, as one would expect. There is also some evidence, though not so obvious and uniform, that more individuals choose the delayed option as the delay grows bigger. If this is a systematic result, then this is a violation of the linearity property of constant discounting.

The rates at which subjects choose the more delayed alternative in Table 2 are broadly consistent with those in Table 1, though in many cases the rates are higher in Table 2 than the corresponding entries in Table 1, suggesting that the stationarity property of constant discounting is often violated.

Figures 1 and 2 help us to visualize the data in Tables 1 and 2. The figures show essentially the same information as in the tables, except that the high and low-base questions are aggregated for each interest rate/initial pe-

Table 1. Percentage choosing more delayed alternative for initial time = today

	Lower-rates questionnaire		Higher-rates questionnaire	
	High base	Low base	High base	Low base
Today versus two weeks from today				
Low rate	9.09	11.36	21.43	16.67
Medium rate	43.18	20.45	40.48	33.33
High rate	50.00	45.45	52.38	57.14
Today versus four weeks from today				
Low rate	15.91	15.91	16.67	19.05
Medium rate	40.91	34.09	54.76	50.0
High rate	54.55	50.00	80.95	64.29
Today versus six weeks from today				
Low rate	18.18	18.18	21.43	16.67
Medium rate	38.64	40.91	59.52	54.76
High rate	56.82	56.82	88.10	66.67

Table 2. Percentage choosing more delayed alternative for initial time = 2 weeks from today

	Lower-rates questionnaire		Higher-rates questionnaire	
	High base	Low base	High base	Low base
2 weeks from today versus six weeks from today				
Low rate	25.00	29.55	23.81	28.57
Medium rate	43.18	43.18	45.24	50.00
High rate	54.55	54.55	66.67	64.29
2 weeks from today versus six weeks from today				
Low rate	36.36	22.73	26.19	11.90
Medium rate	47.73	52.27	42.86	47.62
High rate	65.91	50.00	90.48	71.43
Two weeks from today versus eight weeks from today				
Low rate	25.00	22.73	33.33	14.29
Medium rate	47.73	40.91	61.90	30.95
High rate	56.82	50.00	85.71	71.43

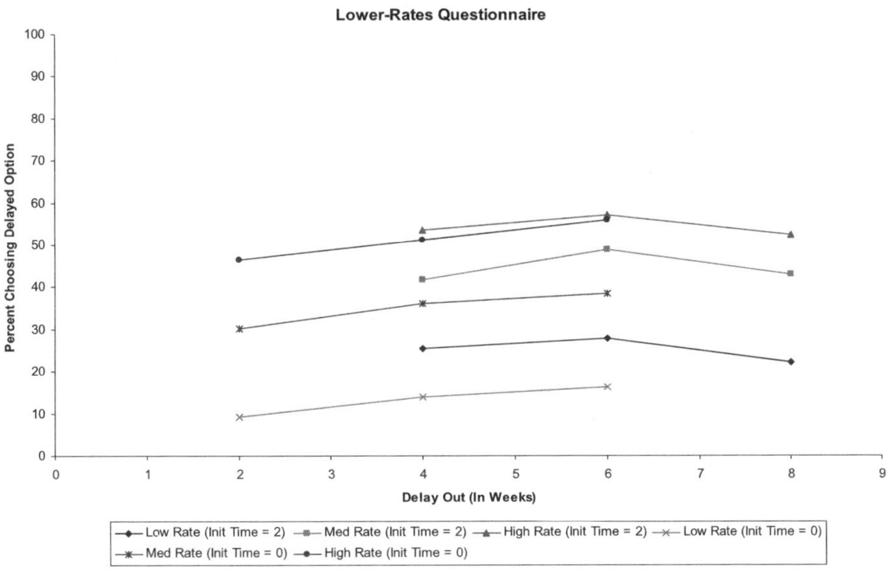

Fig. 1. Increasing the delay in the lower-rates questionnaire for the average of both the bases

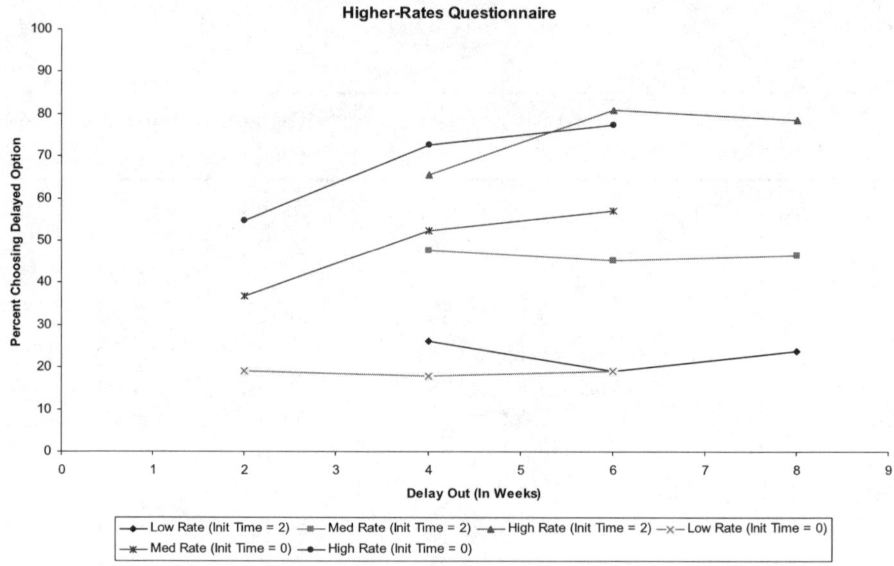

Fig. 2. Increasing the delay in the higher-rates questionnaire for the average of both the bases

riod/delay combination. The figures show the percentage choosing the later option on the vertical axis, and the delay from today of the later option in each pair. The three averages for a given interest rate and initial period are connected by straight line segments. As a point of reference, if the subjects used constant discounting unfailingly and without error, the two sets of line segments corresponding to each interest rate should coincide and form a perfectly horizontal line. Instead, there is a tendency for the line segments associated with the choice questions where the earliest period was also delayed tend to lie above the segments associated with the choice questions where the earliest period was today. This illustrates the extent to which stationarity is violated. Also, there is some tendency for the line segments to slope upwards, especially when the initial period is today. This illustrates the extent to which linearity is violated.

The conventional wisdom that both stationarity and linearity are violated receive some support from the aggregate averages, but it not clear if the departures from constant discounting are limited to a distinct set of individuals, or if it is a general phenomenon. We now turn to analysis of the data in which we take careful account of individual patterns of choice behavior.

4 Choice Patterns

Each questionnaire has thirty-six questions with varying bases and rates (the last four questions are to test for monotonicity and so are ignored in this analysis). There are three rates (high, medium and low) and two bases (high and low). We form six sets of six questions. Each set of six questions corresponds to a given base and compounding rate.

There are $2^6 = 64$ possible choice patterns for each set of six questions. Patterns 1 and 64 correspond to constant discounting and are, in fact, the only "legal" choice patterns one should see if individuals discount at a constant rate. We have enumerated all of the possible patterns[2] and computed the distribution of choice patterns for each of the six sets of six questions. The numbering of the patterns corresponds roughly to the number of late choices: the larger the number, the greater the number of late choices in the pattern.

Table 3 illustrates the eight choice patterns that are consistent with hyperbolic discounting. A choice pattern may be represented by six indicator variables, specifying whether the earlier (0) or later (1) choice was made on each question. The columns in the matrices have a common initial time: Each indicator in column 1 corresponds to a question where the initial time is today (zero weeks) and each member of column 2 has an initial time of two weeks. The rows in the matrices have a common delay: Each indicator in row 1 has a delay of two weeks, each indicator in row 2 has a delay of four weeks and each indicator in row 3 has a delay of six weeks. Thus, for example, Pattern 1 is the pattern in which all choices are early choices, and Pattern 64 is the pattern in which all choices are late choices. The other six patterns shown have various combinations of early and late choices, but each can be rationalized as a possible "strictly hyperbolic" pattern for some set of k_i's. Note that any pattern in which there is a 1 above a 0 in a column, or any row with a 1 to the left of a 0, cannot be a hyperbolic pattern. No other patterns can be rationalized as consistent with either constant discounting or strict hyperbolic discounting.[3]

Table 3. Hyperbolic choice pattern definitions

0 ≡ early and 1 ≡ late							
Pattern 1:	$\begin{bmatrix}0&0\\0&0\\0&0\end{bmatrix}$	Pattern 7:	$\begin{bmatrix}0&0\\0&0\\0&1\end{bmatrix}$	Pattern 22:	$\begin{bmatrix}0&0\\0&1\\0&1\end{bmatrix}$	Pattern 41:	$\begin{bmatrix}0&0\\0&1\\1&1\end{bmatrix}$
Pattern 42:	$\begin{bmatrix}0&1\\0&1\\0&1\end{bmatrix}$	Pattern 56:	$\begin{bmatrix}0&1\\0&1\\1&1\end{bmatrix}$	Pattern 63:	$\begin{bmatrix}0&1\\1&1\\1&1\end{bmatrix}$	Pattern 64:	$\begin{bmatrix}1&1\\1&1\\1&1\end{bmatrix}$

[2] See Appendix B for the detailed definitions of all 64 possible patterns.

[3] One might also think that Pattern 19: $\begin{bmatrix}0&0\\0&0\\1&1\end{bmatrix}$ and Pattern 55: $\begin{bmatrix}0&0\\1&1\\1&1\end{bmatrix}$ should also be included, as they seem intuitively to fit the criteria for hyperbolic patterns,

The logic of the experimental design was to observe the change in behavior of the subject as the delay or the initial time increases. In each of the patterns, as one goes down a particular column, the delay is increasing. Furthermore, as one switches from the left column to the right column in any row, the initial time period is being pushed out. An individual falling into Pattern 7, for example, chooses the delayed alternative in the sixth question – the one which has the longest delay (six weeks) and an initial time of two weeks. This individual shows characteristics we would expect in a hyperbolic type; that is, if the initial time and delay are pushed out enough, then the subject's effective discount rate (which is determined by the k_is) falls to a point below the compounding rate in the six questions. As a consequence, he or she will choose the later option. An individual of Pattern 22 would choose the later option in the fifth and sixth questions, for an initial time of two weeks for a delay of either four or six weeks.

In general, each hyperbolic pattern implies a particular configuration for the way that the k_is decline, and it is difficult (and perhaps pointless) to try to isolate or infer the precise pattern. The generalized hyperbola (a special case being the exponential) was shown by Loewenstein and Prelec (1992) to be the only possible form for the discounting function that will permit both violations of stationarity and linearity. Such a discounting function is "smooth" convex function in discount rate-time space. For the discretized version of the discounting function that we have adopted, we determined which patterns were consistent with hyperbolic discounting simply by showing that there is some non-increasing sequence of k_i, $i = 1, 2, 3, 4$, that allows one to rationalize the pattern. It is worth considering in more detail whether any such sequence is necessarily consistent with a generalized hyperbola.

In this way, as one progressively goes from Pattern 7 to Pattern 63, the k_i's are progressively getting smaller. For example, for pattern 7, we can infer that $k_1 > r$, that $k_2 > r$, that (redundantly, and roughly speaking) the average of k_1 and k_2 is greater than r, that the average of k_1, k_2 and k_3, and the average of k_2 and k_3 are each bigger than r, but that the average of k_1, k_2, k_3, and k_4 is less than r. In other words, k_4 is sufficiently small to bring the average discount factor down far enough so that the later choice is made for a sufficiently delayed payment. At the other extreme, in pattern 63, all but the first of the conditions have been reversed. That is, we still infer that $k_1 > r$, but k_2, k_3 and k_4 are sufficiently small, on average and independently, so that all but one choice is the later choice. For patterns in between, Table 7 maps out the inequalities that must hold.

i.e., there is no 1 above a 0 in a column, and no 1 the left of a 0 in a column. But these patterns can be rationalized only if $k_1 = k_2 = k_3 = k_4 = r$, the compounding rate, that is, only if the discount factor is constant and exactly equal to the compounding rate. Since any of the 64 patterns can be rationalized in this way, there is little point in singling these out for consideration either.

Table 4. Conditions on the discounting factors in the hyperbolic function implied by various choice patterns

Pattern	Conditions on k_1	Conditions on k_1, k_2	Conditions on k_1, k_2, k_3	Conditions on k_1, k_2, k_3, k_4
7	$(1+k_1) > (1+r)$	$\prod_{i=1}^{2}(1+k_i) > (1+r)^2$ and $(1+k_2) > (1+r)$	$\prod_{i=1}^{3}(1+k_i) > (1+r)^3$ and $\prod_{i=2}^{3}(1+k_i) > (1+r)^2$	$\prod_{i=1}^{4}(1+k_i) < (1+r)^3$
22	$(1+k_1) > (1+r)$	$\prod_{i=1}^{2}(1+k_i) > (1+r)^2$ and $(1+k_2) > (1+r)$	$\prod_{i=1}^{3}(1+k_i) > (1+r)^3$ and $\prod_{i=2}^{3}(1+k_i) < (1+r)^2$	$\prod_{i=1}^{4}(1+k_i) < (1+r)^3$
41	$(1+k_1) > (1+r)$	$\prod_{i=1}^{2}(1+k_i) > (1+r)^2$ and $(1+k_2) > (1+r)$	$\prod_{i=1}^{3}(1+k_i) < (1+r)^3$ and $\prod_{i=2}^{3}(1+k_i) < (1+r)^2$	$\prod_{i=1}^{4}(1+k_i) < (1+r)^3$
42	$(1+k_1) > (1+r)$	$\prod_{i=1}^{2}(1+k_i) > (1+r)^2$ and $(1+k_2) < (1+r)$	$\prod_{i=1}^{3}(1+k_i) > (1+r)^3$ and $\prod_{i=2}^{3}(1+k_i) < (1+r)^2$	$\prod_{i=1}^{4}(1+k_i) < (1+r)^3$
56	$(1+k_1) > (1+r)$	$\prod_{i=1}^{2}(1+k_i) > (1+r)^2$ and $(1+k_2) < (1+r)$	$\prod_{i=1}^{3}(1+k_i) < (1+r)^3$ and $\prod_{i=2}^{3}(1+k_i) < (1+r)^2$	$\prod_{i=1}^{4}(1+k_i) < (1+r)^3$
63	$(1+k_1) > (1+r)$	$\prod_{i=1}^{2}(1+k_i) < (1+r)^2$ and $(1+k_2) < (1+r)$	$\prod_{i=1}^{3}(1+k_i) < (1+r)^3$ and $\prod_{i=2}^{3}(1+k_i) < (1+r)^2$	$\prod_{i=1}^{4}(1+k_i) < (1+r)^3$

5 Distributions of Choice Patterns

How should one interpret the observed choice patterns? One possibility is to take the observed choice pattern as a genuine and perfectly accurate reflection of the subject's preferences. In this case, one naturally would want to look at what happens to the observed distribution of patterns as the compounding rate is increased. The hypothesis of constant discounting leads one to posit that there is some threshold of the compounding rate at which individuals switch from all early to all late choices. The hypothesis of hyperbolic discounting is a little tricky (since there is not a single threshold), but still relatively straightforward. As the compounding rate is increased one would expect to see plenty of "all early" patterns at low rates and plenty of "all late" patterns with, one would hope, some strict hyperbolic patterns in between. One might also find strict hyperbolic patterns at low rates (presumably followed by "all late" patterns at medium and higher rates), or "all early" patterns for low and medium rates, with strict hyperbolic patterns at higher rates.

More realistically, one has to allow for the possibilities that there is a stochastic element in the choices subjects make. Indeed, we have given the subjects plenty of rope to hang themselves, so to speak. There are only 2 patterns out of 64 consistent with constant discounting, and 6 others consistent with strict hyperbolic discounting (though, as noted above, a strictly hyperbolic type may choose constant discounting patterns for sufficiently high or low compounding rates). This leaves 56 of 64 patterns as clear mistakes or errors, even under the rather generous allowances that the hyperbolic discounting hypothesis provides. In order to begin to get a feel for the data, Tables 5 and 6 show the set of possible patterns that one might observe, and the distributions of the actual choices made, for each base level and compounding rate. We have combined the results from the low-rate and high-rate questionnaires for these tables, as the same point could be made for each questionnaire separately or together, even though the distributions are not identical. In later regression analysis we will treat these results separately.

As Table 5 shows, for each set of 6 related choice questions, there are $\binom{6}{n}$ patterns involving n late choices, adding up to 64 possible patterns in all. Table 5 also indicates which numbered patterns correspond to patterns with the various number of late choices. The most notable feature of Table 6, keeping in mind the information contained in Table 5, is the wide divergence between the expected frequency of patterns of each sort and the actual. Table 7 summarizes this information. What is striking is that the frequency (proportion) of patterns with 1, 2, 3, 4 or 5 late choices is approximately equal, at about 0.09, on average. A second thing to note in Table 6 is that the frequency of observed hyperbolic patterns for a given number of late choices (1 to 5) is generally higher than one would expect if the patterns were purely random error. This is not so unreasonable, as the hyperbolic patterns, unlike the other patterns, do in fact satisfy some basic dominance properties. This is also summarized in Table 7.

Table 5. Distribution of possible choice patterns

# Late choices	# of patterns	Pattern #s	Hyperbolic-discounting?
No late choices	1	Pattern 1	1
1 Late Choice	6	Patterns 2–7	7
2 Late Choices	15	Patterns 8–22	22
3 Late Choices	20	Patterns 23–42	41, 42
4 late Choices	15	Patterns 43–57	56
5 Late Choices	6	Patterns 58–63	63
6 Late Choices	1	Pattern 64	64

A provisional interpretation of all of this is that, while the absolute frequency of non-constant discounting patterns is non-trivial (44% of all choices), the proportion of these patterns that are hyperbolic is only about a third (34%), implying that only about 15% of all choices are hyperbolic patterns, which is not such strong evidence for hyperbolic discounting as an overall account for intertemporal choice behavior. On the other hand, the last column of Table 7 suggests that deviations from constant discounting are not purely random errors, and are somewhat inclined to be driven by the sorts of factors that hyperbolic discounting is meant to account for: stationarity violations and linearity violations.

Up to now the analysis has been descriptive and has dealt in aggregate behavior. We now proceed to conduct regression analysis that will allow us to formally account for individual-specific factors, and to formally quantify the effects of extending the time horizon and of shifting all payment periods.

6 Regression Analysis of the Probability of Choosing Later Options

In order separate the effects of differences in individual propensities to choose later choices from the effects of such things as the length of time between payments and the placement in time of the payment, we organize the data as a panel of observations and take specific account of individual heterogeneity. Individuals may have different propensities to choose later choices due to higher or lower average subjective[4] discount rates. A linear probability model (LPM) for a binary response y may be specified as

$$P(y=1|\mathbf{x}) = \beta_0 + \beta_1 \times_1 + \cdots + \beta_K \times_K \,, \tag{1}$$

[4] We use the phrase "average discount rate" to cover the possibility that individuals may have non-constant discount rates, as in hyperbolic discounting.

Table 6. Distributions of actual choice patterns

Low compound rate				Medium compound rate				High compound rate			
Low base		High base		Low base		High base		Low base		High base	
Pattern #/ Frequency		Pattern #/ Frequency		Pattern #/ Frequency		Pattern #/ Frequency		Pattern #/ Frequency		Pattern #/ Frequency	
1	56	**1**	45	**1**	29	**1**	22	**1**	17	**1**	16
4	1	2	1	4	1	2	1	2	1	8	1
5	6	3	1	5	3	3	1	3	1	13	1
7	1	4	3	6	2	4	2	4	1	***22***	1
8	1	5	3	***7***	1	5	1	5	2	32	1
10	1	6	2	8	1	***7***	3	6	2	33	1
13	1	***7***	3	9	2	8	1	***7***	1	38	1
17	2	8	1	13	2	9	1	8	1	39	2
20	1	15	1	17	2	10	1	9	1	***41***	4
21	1	20	2	18	1	12	1	10	1	***42***	1
22	2	***22***	5	20	3	13	2	17	1	44	1
23	1	26	1	***22***	1	15	1	20	1	46	1
24	2	28	1	32	1	19	2	21	1	49	1
35	1	40	1	33	1	20	2	***22***	2	55	5
42	3	***41***	1	34	2	21	1	23	1	***56***	1
48	1	***42***	4	***42***	2	***22***	2	34	1	57	2
53	1	53	1	44	1	24	1	***41***	2	58	1
58	1	***56***	1	46	1	35	2	***42***	1	59	1
59	1	**64**	9	47	1	***42***	3	43	2	60	2
62	1			53	2	43	1	46	1	62	1
63	2			55	1	44	2	48	1	***63***	7
64	3			***56***	2	45	1	55	3	**64**	34
				58	2	47	1	***56***	3		
				60	1	55	2	59	3		
				63	8	***56***	2	***63***	3		
				64	13	59	2	**64**	32		
						60	2				
						61	2				
						62	1				
						63	2				
						64	18				

Table 7. Summary information about actual proportions of patterns according to number of later choices

Number of late choices	Expected proportion	Actual proportion (average over all cases) (516 cases in all)	Expected proportion of hyperbolic patterns within category	Actual proportion of hyperbolic patterns in each category
0	1/64 = .02	185/516 = .36	1.0	185/185 = 1.0
1	6/64 = .09	44/516 = .09	.17	9/44 = .20
2	15/64 = .23	47/516 = .11	.07	13/47 = .28
3	20/64 = .31	42/516 = .08	.10	21/42 = .50
4	15/64 = .23	42/516 = .08	.07	9/42 = .21
5	6/64 = .09	43/516 = .08	.17	22/43 = .51
6	1/64 = .02	109/516 = .21	1.0	109/109 = 1.0

where $P(y = 1|\mathbf{x})$ is the probability that the later of the two choices in a question is chosen. That is, the event that the later choice is chosen is coded as $y = 1$, and otherwise $y = 0$. Assuming that x_i is not functionally related to the other explanatory variables, $\beta_i = \partial P(y = 1|\mathbf{x})/\partial x_i$. Therefore, β_i is the change in the probability of success given a one-unit increase in x_i. If x_i is a binary explanatory variable (as it always will be in our analysis), then β_i is just the difference in the probability of success when $x_i = 0$ and $x_i = 1$, holding the other x_j fixed. Since all of the regressors are 0-1 variables, our analysis is not vulnerable to one of the usual criticisms made of the linear probability model, that the fitted values for P may be larger than 1 or less than 0.

Another advantage of using the linear probability model, instead of some nonlinear transformation function, such as the logit or probit, is that it is straightforward to allow for individual heterogeneity in choice behavior. We estimate the model by specifying the error term as $u_{im} = e_{im}+c_i$. That is, the error is modeled as being the sum of an individual specific component c_i and an idiosyncratic component e_{im} that varies from observation to observation on an individual. We use a random effects specification in the estimation.

$$P(y = 1|\mathbf{x}, c_i) = \beta_0 + \beta_1 x_1 + \cdots + \beta_K x_K + c_i + e_{im} . \tag{2}$$

The regressors are:

ratem = 1 if compounding rate is the medium rate, 0 otherwise,
rateh = 1 if compounding rate is the high rate, 0 otherwise,
delay4 = 1 if the delay between the earlier and later payment is 4 weeks, 0 otherwise,
delay6 = 1 if the delay between the earlier and later payment is 6 weeks, 0 otherwise,

initial2 = 1 if the initial period (when the earlier payment is made) is 2 weeks, 0 otherwise,
baseh = 1 if the base amount is the higher level ($ 20), 0 otherwise (if it is $ 8).

To summarize, the estimated coefficients on these variables in the linear probability model indicate the marginal increase or decrease in the probability of choosing the later option relative to the probability of choosing earlier when the initial period is today, the compounding rate is low, the delay between payments is two weeks, and the base amount is $ 8. We estimate these effects separately for the lower-rate and the higher-rate questionnaires. We estimate the linear probability model using generalized least squares with random effects. The estimation results are reported in Tables 8 and 9.

The results in Table 8 for the lower-rates questionnaire may be interpreted as follows. The "baseline" probability of choosing the later option in the first choice question with the low compounding rate, the low base amount, the early payment today and the later payment in two weeks, is .09 (the constant term). The medium rate raises this probability by .20, the high rate by .33, as one would expect (i.e., the change should be positive and non-trivial). A delay of either four or six weeks (compared to two weeks) raises this probability by about the same amount, roughly .05. Shifting the initial payment time to two weeks raises this probability by .08. These last two effects are the effects that the hyperbolic hypothesis is meant to accommodate. Raising the base amount also has a significant effect, although it is quite small in magnitude (less than .005). The results in Table 9 for the higher-rates questionnaire are qualitatively similar.

Table 8. Random effects LPM analysis of the lower-rates questionnaire

R^2 within = .14		Wald $X^2(6) = 259.11$	Number of obs. = 1584
R^2 between = .00		Prob. > $X^2 = .00$	Number of groups = 44
Overall R-sq. = .09			Obs. Per group = 36
Variable	Coefficient	z-statistic	$P > \|z\|$
Ratem	.20	9.21	.00
Rateh	.33	14.98	.00
Delay4	.05	2.15	.03
Delay6	.04	1.64	.10
Initial2	.08	4.57	.00
Baseh	.00	2.18	.03
constant	.09	1.72	.09
$\sigma_c = .30$			
$\sigma_e = .36$		$\rho = .42$ (fraction of var. due to c_i)	

Table 9. Random effects LPM analysis of the higher-rates questionnaire

R^2 within = .27		Wald $X^2(6) = 552.31$	Number of obs. = 1512
R^2 between = .00		Prob. > X^2 = .00	Number of groups = 42
Overall R-sq. = .19			Obs. per group = 36
Variable	Coefficient	z-statistic	$P > \|z\|$
Ratem	.27	11.96	.00
Rateh	.51	22.68	.00
Delay4	.06	2.83	.01
Delay6	.09	3.90	.00
Initial2	.03	1.59	.11
Baseh	.01	4.34	.00
constant	.05	.96	.34
$\sigma_c = .28$			
$\sigma_e = .36$		$\rho = .38$ (fraction of var. due to c_i)	

The violations of stationarity and linearity implied by the coefficients on Initial2 and the Delay variables here are significant, but not huge. The linearity effect does not seem to be too strong after the initial increase from a two week delay to a four week delay, as shown by the similarity of the coefficients on the two delay variables (especially in Table 8). The stationarity effect ranges from .03 to .08, implying an increase in later choices of 3 to 8% when the initial payment period is shifted out in time. Overall, the results are consistent with what we have already observed: violations of constant discounting are widespread, but the degree to which they can be accounted for by the hypothesis of hyperbolic discounting is modest.

One further bit of regression analysis that we think helps to organize the data is motivated by the discussion in Sect. 5 about the frequency of choice patterns with various numbers of late choices. If we organize the data so that the unit of observation is a set of six related choice questions, and the "choice" observed is the number of later choices made in those six questions, rather than the simple binary choice of earlier or later on each question individually, then we can estimate the probability that the number of late choices per set of questions will occur. We divide the 36 choice questions into six sets of six questions each. The only variation over these sets is in the base amount and the compounding rate. That is, for each group of six questions the base amount and compounding rate is the same. The initial period and delays are subsumed within each set, so no effect of these variables can be estimated in this instance. This analysis, then, can be thought of as quantifying the pure effect of the compounding rate and the base amount, once individual-specific effects are accounted for.

There are seven possible "responses" for each such set of six questions. We simply record the number of late choices in each set, regardless of whether the patterns are hyperbolic or not, and without regard to which specific questions had an early or late response. Responses with n late choices, $n = 0, \ldots, 6$, are coded as n. We then estimate a random effects ordered probit model[5]. "Cut-points" (essentially separate constant terms for each instance of n) as well as coefficient estimates for the base amount and indicator variables for the different compounding rates are the output of the estimation. Random effects are assumed, meaning, as in the LPM estimation, that we are allowing for an individual-specific error component that is fixed over all six observations on an individual.

Tables 10 and 11 contain estimates from the ordered probit procedure, and Table 12 contains the fitted (predicted) values for the probabilities of each category (0 through 6 late choices in a six-choice set). The probabilities, for a given base value and compounding rate, are calculated as follows:

Probability of 0 Late Choices $= \Phi(c_0 - b - r)$.
Probability of 1 Late Choice $= \Phi(c_1 - b - r) - \Phi(c_0 - b - r)$.
Probability of 2 Late Choice $= \Phi(c_2 - b - r) - \Phi(c_1 - b - r)$.
Probability of 3 Late Choice $= \Phi(c_3 - b - r) - \Phi(c_2 - b - r)$.
Probability of 4 Late Choice $= \Phi(c_4 - b - r) - \Phi(c_3 - b - r)$.
Probability of 5 Late Choice $= \Phi(c_5 - b - r) - \Phi(c_4 - b - r)$.
Probability of 6 Late Choices $= 1 - \Phi(c_5 - b - r)$.

In these calculations, Φ is the standard normal cumulative distribution function, c_i $i = 0, \ldots, 5$ are the estimated cut-points, and b and r stand for the estimated coefficients on dummy variables for the specific base amount and compounding rate in question[6]. In the estimates reported in Tables 10 and 11, Ratem and Rateh are the dummies for the medium and high rates, respectively, and Baseh is the dummy for the high base amount.

The estimated coefficients in Table 10 and 11 are difficult to interpret; the fitted values in Table 12 provide a more intuitive picture of the experiment. Several things are notable. First, we see now (unlike in Table 7, where all of the treatments were pooled), how there is a movement towards patterns with a larger number of late choice as the base amount increases and as the compounding rate increases (both within a questionnaire, and between the two questionnaires). Second, there is a remarkable amount of inertia in the choices: large increases in the compounding rate lead to smaller shifts towards late choices than one would expect if people were doing constant discounting. We already know from our earlier analysis that most of these "in between"

[5] We used the REOPROB procedure, an "Ado" procedure in *Stata* authored by Guillaume Frechette.

[6] There are indicator (dummy) variables for the high base and for the medium and high compounding rates only. So, effectively, the coefficients are b=0 and r=0 for the low base and low compounding rate.

Table 10. Ordered probit estimates, lower-rates questionnaire

Likelihood. Ratio $X^2(3) = 96.80$	Prob $> X^2 = .00$	$N = 264$	Log Likelihood $= -335.39$
Variable	Coefficient	z-statistic	$P > \|z'$
Baseh	.28	1.75	.08
Ratem	1.35	6.34	.00
Rateh	2.13	9.13	.00
Cut1	.87	3.92	.00
Cut2	1.31	5.83	.00
Cut3	2.07	8.58	.00
Cut4	2.56	10.10	.00
Cut5	3.00	11.24	.00
Cut6	3.53	12.29	.00
Rho	.85	30.52	.00

Table 11. Ordered probit estimates, higher-rates questionnaire

Likelihood. Ratio $X^2(3) = 183.89$	Prob $> X^2 = .00$	$N = 252$	Log Likelihood $= -326.05$
Variable	Coefficient	z-statistic	$P > \|z'$
Baseh	.50	3.10	.00
Ratem	1.84	8.19	.00
Rateh	3.26	11.84	.00
Cut1	.94	4.23	.00
Cut2	1.67	7.24	.00
Cut3	2.21	9.13	.00
Cut4	2.70	10.28	.00
Cut5	3.36	11.25	.00
Cut6	4.11	12.17	.00
Rho	.81	22.84	.00

choices (neither all early or all late within a set) are not, in fact, hyperbolic choices.

These estimation results should not be taken too seriously (say, as a forecasting model), but they are suggestive. There is a strong assumption implicit in the ordered formulation that the categories coded as "larger" and "smaller" lie in some natural ordering. If the categories between 0 and 6 later choices are, in fact, truly errors, then the ordered formulation may well overstate the degree to which those categories are likely to be chosen, especially in

Table 12. Estimated probabilities of choosing n late choices out of six

Lower-rates questionnaire

N	Base = $ 8, Rate = .001	Base = $ 8, Rate = .005	Base = $ 8, Rate = .01	Base = $ 20, Rate = .001	Base = $ 20, Rate = .005	Base = $ 20, Rate = .01
0	.81	.32	.10	.72	.22	.06
1	.10	.17	.10	.13	.15	.07
2	.08	.28	.27	.11	.30	.23
3	.01	.12	.19	.03	.15	.19
4	.00	.06	.14	.01	.09	.16
5	.00	.04	.11	.00	.06	.15
6	.00	.01	.08	.00	.03	.13

Higher-rates questionnaire

N	Base = $ 8, Rate = .01	Base = $ 8, Rate = .05	Base = $ 8, Rate = .1	Base = $ 20, Rate = .01	Base = $ 20, Rate = .05	Base = $ 20, Rate = .1
0	.86	.18	.01	.67	.08	.00
1	.13	.25	.05	.21	.17	.02
2	.03	.23	.09	.08	.20	.04
3	.01	.16	.14	.03	.19	.08
4	.00	.13	.25	.01	.20	.20
5	.00	.05	.27	.00	.12	.30
6	.00	.01	.20	.00	.04	.36

making predictions via the fitted values of the choice probabilities. The fitted values tend to be "smeared" over more categories than, in fact, were seen to be chosen in the raw data. This is particularly evident in the high-rates questionnaire results for the high compounding rate.

Something we have not yet remarked upon but which is quite clear, even in the summary statistics in Tables 1 and 2, is that subjects make choices largely based upon the relative magnitudes of the compounding rates that they (implicitly) face. Although the highest compounding rate in the lower-rate questionnaire is the same as the lowest rate in the higher-rate questionnaire, the patterns of choice are only very slightly skewed towards more late choices in the higher rates questionnaire. It is not clear that this presents a major problem for the theory of intertemporal choice, though it surely does cast doubt upon the notion that the discount rate is a hard-wired part of an individual's preference structure. In particular, one cannot with confidence forecast choices for a given compounding rate when choice behavior is evidently so context-dependent. A better account may be that intertemporal preferences are constructed from the context in which one is choosing. In "real life," one is, however vaguely, aware of the options available, and tries to choose the best option, with a variety of constraints in place. In an artificial experimental setting, though the money is quite real, the options vary more widely than in the natural setting, and there is, perhaps, a tendency to try to establish what is "usual" or "normal" in the context of the experiment. Nonetheless, to the extent that subjects settle upon a notion of what is more and less preferred, even if it is context-dependent, the results may be perfectly reliable as an indicator of what people do in other naturally-occurring environments.

7 Conclusions

Our initial motivation for conducting this experiment was to try to quantify more precisely the degree to which violations of constant discounting, which we accept to be common and pervasive, can be accounted for by the hypothesis that individuals use hyperbolic discounting. We have approached this question in a number of ways: comparisons of the raw averages of choice frequencies, detailed examination of the choice patterns, a linear probability model, accounting for individual-specific effects, and an exploratory estimation of an ordered-probit formulation, also accounting for individual-specific effects. To summarize the results, without restating them, we can say that the absolute magnitude of the evidence supporting the hyperbolic discounting hypothesis is rather small. We suggest, provisionally, that a better account of the data may lie in thinking more generally of propensities to choose earlier or later that are stochastic, and that result in choice patterns that are nearer to constant discounting, the stronger are the factors that influence these propensities. Put differently, it may be better to try to come up with a plausible statistical account of the observed behavior than to enshrine what may be, after all, just a collection of biases, into a formal theoretical account of intertemporal choice behavior.

Acknowledgement. We acknowledge the Rutgers Research Council for financial support, the Center for Experimental Social Science at New York University for the use of laboratory facilities, and Joshua Greenberg for computer programming.

References

Ainslie, G., Haendel, V. (1983), "The Motives of Will," in *Etiologic Aspects of Alcohol and Drug Abuse*, ed. E Gotthell, Druley, K., Skoloda, T., Waxman, H., Springfield, IL: Thomas, Charles.

Benzion, U., Rapoport, A., Yagil, J. (1989) "Discount Rates Inferred From Decisions: An Experimental Study," *Management Science*, 35:270–284.

Holcomb, J., Nelson, P.S. (1992) "Another Experimental Look at Individual Time Preference," *Rationality and Society*, 4:2:199–220.

Loewenstein, G., Prelec, D. (1992) "Anomalies In Intertemporal Choice: Evidence and an Interpretation," *The Quarterly Journal of Economics*, 107:2:573–597.
Loewenstein, G., Prelec, D. (1993) "Preference for Sequences of Outcomes," *Psychological Review*, 100(1):91–108.
Loewenstein, G., Thaler, R. (1989) "Anomalies: Intertemporal Choice," *Journal of Economic Perspectives*, 3:181–193.
Mischel, W. (1966) "Theory and Research On The Antecedence of Self-Imposed Delay of Reward", *Progress In Experimental Personality Research*, 3:85–132.
Mischel, W. (1974) "Processes In Delay of Gratification," in *Advances in Experimental Social Psychology*, ed. L. Berkowitz, New York: Academic Press.
Mischel, W., Ebbenson, E. (1970) "Attention and Delay of Gratification," *Journal of Personality and Social Psychology*, 16:329–337.
Rachlin, H. (1974) "Self-control," *Behaviorism*, 2:94–107.
Strotz, R.H. (1955) "Myopia and Inconsistency in Dynamic Utility Maximization," *Review of Economic Studies*, 8:165–180.
Thaler, R. (1981) "Some Empirical Evidence on Dynamic Inconsistency," *Economics Letters*, 8:201–207.

A Appendix I: Questionnaires

A.1 Higher-Rates Questionnaire

1 Which do you prefer, 8000 francs in 0 weeks, or 8161 francs in 2 weeks?
2 Which do you prefer, 8000 francs in 0 weeks, or 8325 francs in 4 weeks?
3 Which do you prefer, 8000 francs in 0 weeks, or 8492 francs in 6 weeks?
4 Which do you prefer, 8161 francs in 2 weeks, or 8325 francs in 4 weeks?
5 Which do you prefer, 8161 francs in 2 weeks, or 8492 francs in 6 weeks?
6 Which do you prefer, 8161 francs in 2 weeks, or 8663 francs in 8 weeks?
7 Which do you prefer, 20000 francs in 0 weeks, or 20402 francs in 2 weeks?
8 Which do you prefer, 20000 francs in 0 weeks, or 20812 francs in 4 weeks?
9 Which do you prefer, 20000 francs in 0 weeks, or 21230 francs in 6 weeks?
10 Which do you prefer, 20402 francs in 2 weeks, or 20812 francs in 4 weeks?
11 Which do you prefer, 20402 francs in 2 weeks, or 21230 francs in 6 weeks?
12 Which do you prefer, 20402 francs in 2 weeks, or 21657 francs in 8 weeks?
13 Which do you prefer, 8000 francs in 0 weeks, or 8820 francs in 2 weeks?
14 Which do you prefer, 8000 francs in 0 weeks, or 9724 francs in 4 weeks?
15 Which do you prefer, 8000 francs in 0 weeks, or 10721 francs in 6 weeks?
16 Which do you prefer, 8820 francs in 2 weeks, or 9724 francs in 4 weeks?
17 Which do you prefer, 8820 francs in 2 weeks, or 10721 francs in 6 weeks?
18 Which do you prefer, 8820 francs in 2 weeks, or 11820 francs in 8 weeks?
19 Which do you prefer, 20000 francs in 0 weeks, or 22050 francs in 2 weeks?
20 Which do you prefer, 20000 francs in 0 weeks, or 24310 francs in 4 weeks?
21 Which do you prefer, 20000 francs in 0 weeks, or 26802 francs in 6 weeks?
22 Which do you prefer, 22050 francs in 2 weeks, or 24310 francs in 4 weeks?
23 Which do you prefer, 22050 francs in 2 weeks, or 26802 francs in 6 weeks?

24 Which do you prefer, 22050 francs in 2 weeks, or 29549 francs in 8 weeks?
25 Which do you prefer, 8000 francs in 0 weeks, or 9680 francs in 2 weeks?
26 Which do you prefer, 8000 francs in 0 weeks, or 11713 francs in 4 weeks?
27 Which do you prefer, 8000 francs in 0 weeks, or 14172 francs in 6 weeks?
28 Which do you prefer, 9680 francs in 2 weeks, or 11713 francs in 4 weeks?
29 Which do you prefer, 9680 francs in 2 weeks, or 14172 francs in 6 weeks?
30 Which do you prefer, 9680 francs in 2 weeks, or 17149 francs in 8 weeks?
31 Which do you prefer, 20000 francs in 0 weeks, or 24200 francs in 2 weeks?
32 Which do you prefer, 20000 francs in 0 weeks, or 29282 francs in 4 weeks?
33 Which do you prefer, 20000 francs in 0 weeks, or 35431 francs in 6 weeks?
34 Which do you prefer, 24200 francs in 2 weeks, or 29282 francs in 4 weeks?
35 Which do you prefer, 24200 francs in 2 weeks, or 35431 francs in 6 weeks?
36 Which do you prefer, 24200 francs in 2 weeks, or 42872 francs in 8 weeks?
37 Which do you prefer, 8000 francs in 0 weeks, or 8000 francs in 2 weeks?
38 Which do you prefer, 8000 francs in 0 weeks, or 8000 francs in 4 weeks?
39 Which do you prefer, 20000 francs in 0 weeks, or 20000 francs in 2 weeks?
40 Which do you prefer, 20000 francs in 0 weeks, or 20000 francs in 4 weeks?

A.2 Lower-Rates Questionnaire

1 Which do you prefer, 8000 francs in 0 weeks, or 8016 francs in 2 weeks?
2 Which do you prefer, 8000 francs in 0 weeks, or 8032 francs in 4 weeks?
3 Which do you prefer, 8000 francs in 0 weeks, or 8048 francs in 6 weeks?
4 Which do you prefer, 8016 francs in 2 weeks, or 8032 francs in 4 weeks?
5 Which do you prefer, 8016 francs in 2 weeks, or 8048 francs in 6 weeks?
6 Which do you prefer, 8016 francs in 2 weeks, or 8064 francs in 8 weeks?
7 Which do you prefer, 20000 francs in 0 weeks, or 20040 francs in 2 weeks?
8 Which do you prefer, 20000 francs in 0 weeks, or 20080 francs in 4 weeks?
9 Which do you prefer, 20000 francs in 0 weeks, or 20120 francs in 6 weeks?
10 Which do you prefer, 20040 francs in 2 weeks, or 20080 francs in 4 weeks?
11 Which do you prefer, 20040 francs in 2 weeks, or 20120 francs in 6 weeks?
12 Which do you prefer, 20040 francs in 2 weeks, or 20161 francs in 8 weeks?
13 Which do you prefer, 8000 francs in 0 weeks, or 8080 francs in 2 weeks?
14 Which do you prefer, 8000 francs in 0 weeks, or 8161 francs in 4 weeks?
15 Which do you prefer, 8000 francs in 0 weeks, or 8243 francs in 6 weeks?
16 Which do you prefer, 8080 francs in 2 weeks, or 8161 francs in 4 weeks?
17 Which do you prefer, 8080 francs in 2 weeks, or 8243 francs in 6 weeks?
18 Which do you prefer, 8080 francs in 2 weeks, or 8326 francs in 8 weeks?
19 Which do you prefer, 20000 francs in 0 weeks, or 20201 francs in 2 weeks?
20 Which do you prefer, 20000 francs in 0 weeks, or 20403 francs in 4 weeks?
21 Which do you prefer, 20000 francs in 0 weeks, or 20608 francs in 6 weeks?
22 Which do you prefer, 20201 francs in 2 weeks, or 20403 francs in 4 weeks?
23 Which do you prefer, 20201 francs in 2 weeks, or 20608 francs in 6 weeks?
24 Which do you prefer, 20201 francs in 2 weeks, or 20814 francs in 8 weeks?
25 Which do you prefer, 8000 francs in 0 weeks, or 8161 francs in 2 weeks?

26 Which do you prefer, 8000 francs in 0 weeks, or 8325 francs in 4 weeks?
27 Which do you prefer, 8000 francs in 0 weeks, or 8492 francs in 6 weeks?
28 Which do you prefer, 8161 francs in 2 weeks, or 8325 francs in 4 weeks?
29 Which do you prefer, 8161 francs in 2 weeks, or 8492 francs in 6 weeks?
30 Which do you prefer, 8161 francs in 2 weeks, or 8663 francs in 8 weeks?
31 Which do you prefer, 20000 francs in 0 weeks, or 20402 francs in 2 weeks?
32 Which do you prefer, 20000 francs in 0 weeks, or 20812 francs in 4 weeks?
33 Which do you prefer, 20000 francs in 0 weeks, or 21230 francs in 6 weeks?
34 Which do you prefer, 20402 francs in 2 weeks, or 20812 francs in 4 weeks?
35 Which do you prefer, 20402 francs in 2 weeks, or 21230 francs in 6 weeks?
36 Which do you prefer, 20402 francs in 2 weeks, or 21657 francs in 8 weeks?
37 Which do you prefer, 8000 francs in 0 weeks, or 8000 francs in 2 weeks?
38 Which do you prefer, 8000 francs in 0 weeks, or 8000 francs in 4 weeks?
39 Which do you prefer, 20000 francs in 0 weeks, or 20000 francs in 2 weeks?
40 Which do you prefer, 20000 francs in 0 weeks, or 20000 francs in 4 weeks?

B Appendix II: Definitions of Choice Patterns

Table 13. Definitions of choice patterns

Pattern #	X today vs $X(1+r)$ in 2 weeks	X today vs $X(1+r)^2$ in 4 weeks	X today vs $X(1+r)^3$ in 6 weeks	$X(1+r)$ in n 2 weeks vs $X(1+r)^2$ in 4 weeks	$X(1+r)$ in 2 weeks vs $X(1+r)^3$ in 6 weeks	$X(1+r)$ in 2 weeks vs $X(1+r)^4$ in 8 weeks
$0 \rightarrow$ earlier choice, $1 \rightarrow$ later choice						
1	0	0	0	0	0	0
2	1	0	0	0	0	0
3	0	1	0	0	0	0
4	0	0	1	0	0	0
5	0	0	0	1	0	0
6	0	0	0	0	1	0
7	0	0	0	0	0	1
8	1	1	0	0	0	0
9	1	0	1	0	0	0
10	1	0	0	1	0	0
11	1	0	0	0	1	0
12	1	0	0	0	0	1
13	0	1	1	0	0	0
14	0	1	0	1	0	0

Table 13. (continued)

Pattern #	X today vs $X(1+r)$ in 2 weeks	X today vs $X(1+r)^2$ in 4 weeks	X today vs $X(1+r)^3$ in 6 weeks	$X(1+r)$ in n 2 weeks vs $X(1+r)^2$ in 4 weeks	$X(1+r)$ in 2 weeks vs $X(1+r)^3$ in 6 weeks	$X(1+r)$ in 2 weeks vs $X(1+r)^4$ in 8 weeks
0 → earlier choice, 1 → later choice						
15	0	1	0	0	1	0
16	0	1	0	0	0	1
17	0	0	1	1	0	0
18	0	0	1	0	1	0
19	0	0	1	0	0	1
20	0	0	0	1	1	0
21	0	0	0	1	0	1
22	0	0	0	0	1	1
23	1	1	1	0	0	0
24	1	1	0	1	0	0
25	1	1	0	0	1	0
26	1	1	0	0	0	1
27	1	0	1	1	0	0
28	1	0	1	0	1	0
29	1	0	1	0	0	1
30	1	0	0	1	1	0
31	1	0	0	1	0	1
32	1	0	0	0	0	1
33	0	1	1	1	0	0
34	0	1	1	0	1	0
35	0	1	1	0	0	1
36	0	1	0	1	1	0
37	0	1	0	1	0	1
38	0	1	0	0	1	1
39	0	0	1	1	1	0
40	0	0	1	1	0	1
41	0	0	1	0	1	1
42	0	0	0	1	1	1
43	1	1	1	1	0	0
44	1	1	1	0	1	0
45	1	1	1	0	0	1
46	1	1	0	1	1	0

Table 13. (continued)

Pattern #	X today vs $X(1+r)$ in 2 weeks	X today vs $X(1+r)^2$ in 4 weeks	X today vs $X(1+r)^3$ in 6 weeks	$X(1+r)$ in n 2 weeks vs $X(1+r)^2$ in 4 weeks	$X(1+r)$ in 2 weeks vs $X(1+r)^3$ in 6 weeks	$X(1+r)$ in 2 weeks vs $X(1+r)^4$ in 8 weeks
0 → earlier choice, 1 → later choice						
47	1	1	0	1	0	1
48	1	1	0	0	1	1
49	1	0	1	1	1	0
50	1	0	1	1	0	1
51	1	0	1	0	1	1
52	1	0	0	1	1	1
53	0	1	1	1	1	0
54	0	1	1	1	0	1
55	0	1	1	0	1	1
56	0	0	1	1	1	1
57	0	1	0	1	1	1
58	1	1	1	1	1	0
59	1	1	1	1	0	1
60	1	1	1	0	1	1
61	1	1	0	1	1	1
62	1	0	1	1	1	1
63	0	1	1	1	1	1
64	1	1	1	1	1	1

Part III

The Assessment of Several Alternative Decision Rules

"Take-the-Best" and Other Simple Strategies: Why and When They Work "Well" with Binary Cues

Robin M. Hogarth[1] and Natalia Karelaia[2]

[1] ICREA & Universitat Pompeu Fabra, Barcelona, Spain
e-mail: robin.hogarth@upf.edu

[2] HEC Université de Lausanne, Lausanne, Switzerland
e-mail: natalia.karelaia@upf.edu

Abstract. The effectiveness of decision rules depends on characteristics of both rules and environments. A theoretical analysis of environments specifies the relative predictive accuracies of the "take-the-best" heuristic (TTB) and other simple strategies for choices between two outcomes based on *binary* cues. We identify three factors: how cues are weighted; characteristics of choice sets; and error. In the absence of error and for cases involving from three to five binary cues, TTB is effective across many environments. However, hybrids of equal weights (EW) and TTB models are more effective as environments become more compensatory. As error in the environment increases, the predictive ability of all models is systematically degraded. Indeed, using the datasets of Gigerenzer, Todd et al. (1999), TTB and similar models do not predict much better than a naïve model that exploits dominance. Finally, we emphasize that the results reported here are conditional on *binary* cues.

Keywords: decision making, bounded rationality, lexicographic rules, Take-the-Best, equal weighting

1 "Take-the-Best" and Other Simple Strategies: Why and When They Work "Well" with Binary Cues

Imagine that you are facing a binary choice. You must decide which of two alternatives, A or B, is "better" in the sense of having more of a specific criterion. Examples could include choosing between two job candidates, two stocks, two restaurants, which route to take on a trip, and so on. Imagine further that the information on which you make your judgment is limited to several (k) binary cues, i.e., cues that indicate presence or absence of an attribute relevant to the task. Thus, if the number of cues (k) is, say, three, option A can be characterized by the vector or cue profile

$$A = \{x_{a1}, x_{a2}, x_{a3}\} , \tag{1}$$

where the x_{aj} can only take the values of 0 or 1 ($j = 1, \ldots, 3$).

Similarly, option B can be characterized by the vector or cue profile

$$B = \{x_{b1}, x_{b2}, x_{b3}\} , \tag{2}$$

where the x_{bj} can only take the values of 0 or 1 ($j = 1, \ldots, 3$).

In many studies, simple lexicographic rules have demonstrated remarkably accurate performance for choices of this type when compared to statistical benchmarks (Gigerenzer and Goldstein, 1996; Gigerenzer et al., 1999). Of particular interest is the rule known as "take-the-best" (henceforth TTB) which works as follows. First, the model assumes knowledge of the differential ability of the cues to predict the criterion, i.e., the cue validities (in this case, assume that the order is $x_{.1}$, $x_{.2}$, $x_{.3}$). Second, choice between A and B is made if the first cue ($x_{.1}$) can discriminate between the options; if the first cue cannot, the second cue is used to make the choice; and so on. Finally, if none of the cues can discriminate, choice is made at random. From a cognitive viewpoint, this rule can be easily implemented. It does not require many mental operations nor, in many cases, examining much information (often just the first cue). However, it does require ordering the cues by their validities.

TTB has been presented as an example of a "fast and frugal" heuristic, an element of the "adaptive toolbox" of bounded rationality (Gigerenzer and Goldstein, 1996; Gigerenzer and Selten, 2001). In addition to demonstrations of its predictive ability, several experimental studies have addressed whether and when people actually use TTB-like mental strategies (see, e.g., Rieskamp and Hoffrage, 1999, 2002; Bröder, 2000, 2003; Bröder and Schiffer, 2003; Newell and Shanks, 2003; Newell et al., 2003). Overall, there is some evidence that people do use TTB-like processes but, not to the exclusion of other strategies.

In this paper, we emphasize that the performance of response strategies or decision rules depends on characteristics of *both* the rules and of the environments in which they operate (Brunswik, 1952; Simon, 1956). A complete theory needs to specify both. However, whereas investigators have had little difficulty in identifying rules, the specification of task environments has proven more problematic. Our goal is to illuminate this issue and our approach is theoretical. It involves specifying abstract characterizations of tasks and noting how different models would be expected to perform in these environments (cf., Payne et al., 1993).

In conceptualizing environments for binary choice, we emphasize three dimensions. One is the class of functions that determines which alternative is correct. The second is the type of distribution of cue profiles in the choice set. The third is the role of error. This can be located in the application of the model, the environment, or both.

The paper is organized as follows. We first define the different models we consider. Second, we examine their theoretical performance under both non-compensatory and compensatory linear weighting functions in environments characterized by lack of error. This is done separately for models involving three, four and five cues (details of the latter are presented in the Appendix)[1]. Third, we investigate one aspect of error in models: namely, inappropriate application, i.e., failure to respect the ecological ordering of cues in TTB. Fourth, we identify ways of characterizing populations of data according to distributions of cue profiles and weighting functions and illustrate these using the 20 datasets of Czerlinski et al. (1999). We also subject the models to predictive tests on these data. Fifth, small differences between the actual predictive abilities of the models highlight the importance of error in the environment. We therefore provide a means of characterizing levels of error and show how this degrades model performance. Finally, we discuss our results.

In brief, we show that at a theoretical level TTB does work "well" as a model of binary choice. But to understand how "well" requires specifying appropriate benchmarks. The normative standard involves models such as Bayesian networks, multiple regression or exemplar-based approaches (cf., Chater et al., 2003). Whereas such comparisons are interesting from a normative viewpoint, we do not believe they are the most illuminating from a descriptive perspective. There are two reasons.

First, the advantage of simple models in the tradition of the "adaptive toolbox" (Gigerenzer and Selten, 2001) is that they can be used in many situations where people lack the experience necessary to develop more sophisticated processes[2]. Second, when dealing with small samples, it is well-known that regression analysis (and other optimizing techniques) produces parameter estimates with large standard deviations such that predictions to further samples are subject to much error. In these cases, regression analysis and similar tools become "straw men" that lack meaning (see, e.g., Einhorn and Hogarth, 1975).

Instead, we compare TTB to other simple models, some of which incorporate features of TTB. For example, we explore one previously unidentified model that combines features of both TTB and equal weighting – called EW/TTB. Moreover, we show that this model improves the predictive abil-

---

[1] We limit our analysis to three, four, and five cues for two reasons. One is to reduce analytical complexity. The second is that three, four, and five cues seem sufficient to understand what people can actually do within limited information processing constraints.

[2] As argued by Chater et al. (2003), it is clear that many basic physical and psychological processes can be well modeled by what most would classify as complex normative models. However, note that most of these processes (as an example, consider perception) have evolved over many years of evolution thereby implying much data in their "development." Paradoxically, many "higher order" mental processes are used in situations involving scarce data which effectively preclude using complex normative models (see also Todd and Gigerenzer, 2000).

ity of TTB in certain environments and yet, when the number of variables is small, does not require much additional information processing.

To establish a yardstick for simple models, we propose that all reasonable models of binary choice should exploit dominance. This leads to the following benchmark. Choose according to dominance. If there is no dominance, choose at random. As we show, this strategy – that we call DOMRAN – actually predicts quite well in the kinds of environments studied by Gigerenzer and his colleagues. In particular, when data are "noisy" its performance does not fall far behind that of TTB. Thus in interpreting the performance of TTB and similar models, it is important to investigate how they predict in cases that cannot be decided by dominance. What is the marginal predictive significance compared to the standard set by DOMRAN?

2 The Different Models

In comparing the different models, it is of interest to note: (1) what knowledge they require about the variables (i.e., the cues) such as relative importance and, if so, how accurate this needs to be; (2) how many cues must be examined to make a decision; (3) whether explicit calculations are required; (4) the number of comparisons to be made; and (5) whether random choice is used to break ties. Table 1 provides an overview of such characteristics. As will be seen, some of the models are combinations of different models.

2.1 DOMRAN

This exploits dominance. If one alternative dominates another, it is chosen; if not, choice is made at random. The psychological inspiration is provided by the work of Montgomery (1983) who has documented how people seek to find and exploit dominance and may even distort information so that dominance can be "justified." In general, we suspect that screening for dominance occurs frequently and thus this model provides a useful lower bound in terms of a "reasonable" simple strategy. On the other hand, we have found no explicit mention of this model in the literature to date.

2.2 EW

In the equal weighting or "tallying" model (Gigerenzer et al., 1999), each variable is given the same weight and the alternative chosen has the larger weighted sum. In practice, since variables take the values of 0 or 1, this is equivalent to summing the variables of each alternative and choosing the larger sum. When the sums are equal, choice is made at random. This model involves examining all data, making two sums, and one comparison. It has proven useful in predicting many phenomena when people do not know the relative weights to give to variables (Dawes and Corrigan, 1974; Dawes, 1979; Einhorn and Hogarth, 1975).

Table 1. Prior information and cognitive operations required by different models for binary choice

	Prior information*	Amount of information to consult	Calculations	Number of Comparisons	Random choice if tie
DOMRAN	None	All	None	Equal to number of variables	Yes
EW	None	All	Yes, two sums	One	Yes
TTB	Rank-order of importance of variables	Variable	None	Variable – from one to number of variables	Yes
EW/TTB hybrids	Rank-order of importance of variables	All	Yes, two sums	One, if choice by EW. Otherwise, more.	Yes

*For all models, the decision maker is assumed to know the sign of the zero order correlation between cues and the criterion

2.3 TTB

See description above.

2.4 EW/TTB hybrids

It is not unreasonable to imagine that people might use combinations of models that depend on the characteristics of the choices they face. For example, in many cases the use of EW results in ties. However, rather than resolving such ties at random, an alternative mechanism to resolve ties is to switch to another model that allows differential weighting, e.g., TTB. We postulate that the hybrid models operate in two phases. In the first, EW is used – on all or a subset of the variables (to be specified). If EW favors one alternative, it is chosen. Otherwise, choice is made by TTB. To the best of our knowledge, this is the first formal investigation of this class of models with binary cues.

It is important to note that in all of the models, the signs of the correlations between cues and criterion are assumed to be known, i.e., the variables are scaled such that a cue value of "1" implies a greater value on the criterion than a cue value of "0." In addition, we do not deal with cases involving missing values of cues.

3 A Starting Point: Error-Free Linear Environments

To choose between any pair of alternatives, A and B, we assume that their accompanying cue profiles have been evaluated according to some function that allows one to determine which is "better." We start by assuming an additive function where the sum of all the weights β_j $(j = 1, \ldots, k)$ given to the k cues is equal to 1 and there is no error. We make four arguments to justify the use of this simple error-free linear environment.

First, as noted by Dawes and Corrigan (1974), many nonlinear functions can be well approximated by linear functions and particularly when the former are conditionally monotonic with respect to the criterion. Second, the fact that the weights are constrained to sum to one is not important so long as they can be measured (conceptually) on an interval scale. Third, we believe it is appropriate to start our investigation with a simple model that can serve as a baseline. Fourth, later in the paper we relax the assumption of "no error" and investigate how error affects results.

4 Non-Compensatory and Compensatory Functions

We first characterize these linear environments by the types of functions used to classify choices. In doing this, we follow the lead of Martignon and Hoffrage (1999, 2002) who have distinguished between non-compensatory and compensatory functions for binary cues. Specifically, Martignon and Hoffrage define by non-compensatory any weighting scheme or function that has the property that, when weights are ordered from largest to smallest, each weight is larger than the sum of all weights that are smaller than it, i.e.,

$$\beta_j > \sum_i \beta_i, \text{ for any } i > j,\ j = 1, \ldots, k-1. \tag{3}$$

Martignon and Hoffrage define all other functions as *compensatory.* Thus, for three cues, $\beta_1 > (\beta_2 + \beta_3)$ is the non-compensatory case whereas $\beta_1 \leq (\beta_2 + \beta_3)$ is the compensatory case (assuming that $\beta_1 > \beta_2 > \beta_3$). An important theoretical result proven by Martignon and Hoffrage (1999, 2002) is that, in these simple error-free linear environments, TTB is the optimal model for choice when weighting functions comply with their definition of *non-compensatory*[3]. Thus, the bulk of our attention will be focused on what happens when weighting functions are compensatory. For expository purposes, we provide separate analyses for the cases involving three and four cues. Details concerning the 5-cue case are provided in the Appendix.

[3] Another way of interpreting this result is to note that TTB makes identical predictions to a linear model with non-compensatory weights. Thus, if the latter is optimal in the environment, so is the former. (It is important to emphasize that this is a *theoretical* result, i.e., the β_j's are environmental parameters.)

5 Different Cue Environments

5.1 The 3-Cue Environment

In empirical tests of TTB the basic task involves seeing how models predict between all possible pairs of a set of choice alternatives. Thus, given n alternatives, each characterized by binary vectors of length k, predictions are made for the $n(n-1)/2$ possible pairs of alternatives. Thus, with 30 alternatives there are 435 pairs to predict, with 40 alternatives, 780 pairs, and so on. However, even though there may be many pairs to predict, it should be clear that the distinct cue profiles that characterize alternatives are limited by the number of k binary cues. Specifically, the number of distinct cue profiles is 2^k. This means that, in large samples of alternatives, many predictions must involve cases involving identical cue profiles (so-called "repeats") and that the distribution of cue profiles among the alternatives affects results. This, as we shall show below, is an important insight.

To illustrate the effects of different cue profiles, consider the case of three cues and the eight different profiles that can result from these cues. These are shown in the top left section of Table 2 where the distinct cue profiles are given the labels A, B, C, D, E, F, G, and H. (Profile A is (1, 1, 1); profile B is (1, 1, 0); and so on.) Furthermore, assume that the variables have been ordered in importance, that is $\beta_1 > \beta_2 > \beta_3$.

Now consider the vectors of the arithmetical differences between the two vectors representing the attributes of the alternatives. In the case of A and B, above, this is

$$A - B = \{x_{a1} - x_{b1}, x_{a2} - x_{b2}, x_{a3} - x_{b3}\} \tag{4}$$

where each element of the vector can take values of 1, 0, or -1 depending on the characteristics of the alternatives. These are shown to the right of the eight distinct profiles. Thus, the first set of difference vectors under the letter A shows the differences between A and the seven other profiles (B to H); those under B the differences between B and its successors (C to H); those under C the differences between C and its successors (D to H); and so on.

The difference vectors provide a simple way of assessing the predictions of different models for each combination of cue profile types. First, if all the elements of a difference vector are non-negative and at least one is positive, the first cue profile dominates the second. Thus, as can be seen, cue profile A dominates all the other profiles – B through H. Similarly, B dominates D but not E, and so on. Cases involving dominance are indicated by a "d" in the matrix on the right hand side of Table 2.

Second, consider cases where the difference vectors contain negative elements such as the B-C pairing that takes the values 0, 1, -1 and restrict attention to weighting functions that are strictly compensatory, i.e., where $\beta_1 > \beta_2 > \beta_3$ and $\beta_1 < (\beta_2 + \beta_3)$. Indeed, in this paper, we only consider strict inequalities. TTB chooses B over C based on the first difference that

Table 2. Basic analysis of the 3-cue case

Profiles	cues			**Difference vectors (column minus row)**								**Classification of difference vectors**							
Profile type	x1	x2	x3		A	B	C	D	E	F	G		A	B	C	D	E	F	G
A	1	1	1	B	001							B	d						
B	1	1	0																
C	1	0	1	C	010	01 − 1						C	d	**t, c**					
D	1	0	0																
E	0	1	1	D	011	010	001					D	d	d	d				
F	0	1	0																
G	0	0	1	E	100	10 − 1	1 − 10	1 − 1 − 1				E	d	**t, c**	**t, c**	**w**			
H	0	0	0																
				F	101	100	1 − 11	1 − 10	001			F	d	d	**c**	**t, c**	d		
				G	110	11 − 1	100	10 − 1	010	01 − 1		G	d	**c**	d	**t, c**	d	**t, c**	
				H	111	110	101	100	011	010	001	H	d	d	d	d	d	d	d

Legend:	No.of cases
d = dominance relation	19
c = TTB predicts correctly in compensatory case ($\beta 1 > \beta 2 > \beta 3$ & $\beta 1 < (\beta 2 + \beta 3)$)	8
w = TTB predicts incorrectly in compensatory case	1
t = EW predicts tie	28

appears here between values of the second variable. Moreover, this prediction is consistent with any model where $\beta_2 > \beta_3$. Thus, TTB also predicts this case correctly when the weighting function is strictly compensatory. Continuing to examine all cases that do not involve dominance, we use the fact that $\beta_1 > \beta_2 > \beta_3$ to determine consistency between the choices of TTB and any strictly compensatory weighting function. These consistent cases are marked by a "c" in the appropriate places on the right hand side of Table 2.

Third, by the same logic, it is clear that TTB does not predict the D-E pairing correctly in the compensatory case where $\beta_1 < (\beta_2 + \beta_3)$. This is indicated by marking a "w" in the appropriate cell on the right of Table 2.

Fourth, we also indicate the letter "t" in cases where the EW model predicts ties, as in B-E and C-E.

Finally, Table 2 summarizes TTB's predictions between all cue profile types for strictly compensatory weighting functions. There are 19 cases involving dominance. Thus, all models considered in this paper would make the same predictions for these cases. Of the remaining nine cases, TTB predicts eight correctly and makes one error[4].

The left hand side of Table 3 shows the accuracy of the various models for three cues when the population of cue profiles consists of just one of each type (i.e., A through H), for both non-compensatory and compensatory weighting functions. First, TTB is 100% accurate for non-compensatory functions – as proven by Martignon and Hoffrage (1999, 2002). (To simplify reading of tables, we adopt the practice of highlighting the largest figures in relevant comparisons in **bold**.)

Second, TTB makes one error for compensatory functions where it achieves an overall accuracy rate of 96%.

Third, EW makes one error for non-compensatory functions and no errors for compensatory functions. However, as can be seen from Table 2, there are six cases where EW predicts ties for compensatory functions where decisions have to be made at random – hence its expected predictive accuracy of 89%. Thus the trade-off between the performances of TTB and EW is one certain error versus six cases that are decided by chance.

Fourth, EW/TTB makes one error with non-compensatory functions but no errors with compensatory functions. More specifically, it makes the same error as EW with the non-compensatory function but all ties are correctly resolved by the TTB mechanism. For the compensatory functions, EW/TTB has perfect performance because, first, it correctly predicts the D-E case for which TTB makes an error and, second, all of the EW ties are again correctly resolved by the TTB mechanism.

[4] It is clear from examining the experimental literature (e.g., Newell & Shanks, 2003) that some investigators are aware of the cue profile pairings for which TTB does and does not make appropriate predictions. However, we believe our work is unique in drawing more complete implications of this analysis.

Table 3. Expected predictive accuracy (%'s) for 3- and 4-cue cases

Models	3-cue cases NonCF	CF	4- cue cases NonCF	CF1	CF2	CF3	CF4*	CF5*
DOMRAN	84	84	77	77	77	77	77	77
EW	86	89	81	83	82	84	85	87
*No of errors***	1	0	9	7	8	6	4	0
TTB	**100**	96	**100**	**98**	**99**	**98**	96	94
*No of errors***	0	1	0	2	1	3	3	5
EW/TTB	96	**100**	93	94	93	95	95	**98**
*No of errors***	1	0	9	7	8	6	4	0
EW-3/TTB	x	x	97	95	98	96	**99**	96
*No of errors***			4	6	3	5	0	2

Notes:
NonCF = non-compensatory functions
CF = compensatory functions
* Functions contain some ambiguous cases (3 for CF4 and 5 for CF5). Thus, even though a model may make no errors, its expected predicted accuracy is less than 100% due to the presence of these ambiguous cases
** Errors involve misclassifications by the models (from totals of 28 and 120 choices for the 3- and 4-cue cases, respectively)

To summarize, *in the 3-cue case TTB is optimal for non-compensatory functions and EW/TTB is optimal for compensatory functions.* Moreover, these optimality statements can be made about TTB and EW/TTB irrespective of distributions of cue profiles precisely because they never make mistakes.[5]

5.2 The 4-Cue Environment

The operational definitions of compensatory and non-compensatory functions are quite straightforward in the 3-cue case. However, defining compensatory functions by violations of the condition for non-compensatory functions leads to several distinct classes of the former in the 4-cue case. Specifically, if – for four cues – we define non-compensatory by the conditions that, first, $\beta_j > \sum_i \beta_i$, for any $i > j$, $j = 1, \ldots, k-1$, and second, that $\beta_1 > \beta_2 > \beta_3 > \beta_4$,

[5] The expected predicted performances of these optimal models will not necessarily always be 100%. This can occur when the distribution of cue profiles contains one or more "repeats" of the same profile. We consider this issue in more detail below.

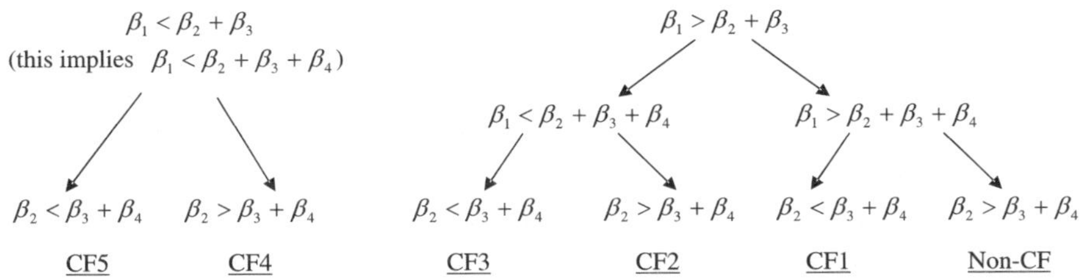

Fig. 1. Classification of compensatory (CF) and non-compensatory functions (non-CF) for the 4-cue case

there are several compensatory functions that violate the first condition to different extents.

For instance, if we specify that $\beta_1 < \beta_2 + \beta_3$ (which in turn implies that $\beta_1 < \beta_2 + \beta_3 + \beta_4$), this can be accompanied by either $\beta_2 < \beta_3 + \beta_4$ or $\beta_2 > \beta_3 + \beta_4$. In fact, there are five different classes of weighting functions that span the parameter space of compensatory environments – see Fig. 1. As in the 3-cue case, we only consider strict inequalities in dividing up the parameter space.

To illustrate differences between the weighting functions, Table 4 provides numerical examples. As can be seen, CF1, CF2, and CF3 are close to "non-compensatory" and CF5 can accommodate distributions of weights that vary from the first being much larger than the others to a set of almost equal weights.

With four cues, there are 120 distinctive profile pairings. The right hand side of Table 3 characterizes the performance of the different models. Compared to the 3-cue case, we have an additional hybrid model labeled EW-3/TTB. This is a modification of EW/TTB that works as follows. In the first stage, the decision maker uses an equal weighting model on the *three* most important variables (i.e., omitting the fourth). If this points to a decision, it is taken. If there is a tie, it is resolved by TTB[6].

Table 4. Exemplar weights of different weighting functions for 4-cue models

	β_1	β_2	β_3	β_4
NonCF	0.53	0.24	0.13	0.10
CF1	0.57	0.19	0.14	0.10
CF2	0.48	0.28	0.14	0.10
CF3	0.48	0.22	0.17	0.13
CF4	0.42	0.31	0.15	0.12
CF5a	0.40	0.25	0.20	0.15
CF5b	0.29	0.26	0.24	0.21

[6] Why, the reader may ask, do we not also define a EW-2/TTB model? The reason is that this latter model makes predictions that are identical to those of TTB.

In this 4-cue world, 54% of the distinctive pairings involve dominance such that the expected performance of DOMRAN is 77%, i.e., 54% + 0.5(100% − 54%). TTB makes, of course, no errors in the non-compensatory case and is unique in this respect. CF4 and CF5 are unable to provide unambiguous choices for three and five cue profile pairings respectively[7]. Operationally, these cases have been treated as ties which all models are assumed to predict correctly with probability of 0.5.

Overall, with populations of unique cue profile pairings, the pattern of results for the 4-cue case matches that of three cues. For non-compensatory (CF1) and close to non-compensatory functions (CF1, CF2, CF3), TTB makes the least numbers of errors. As the functions become more compensatory (CF4, CF5), it is the EW/TTB models that perform relatively better. In particular, under CF5, EW makes no errors such that the TTB contribution to EW/TTB is the correct allocation of all EW ties.

The 5-cue environment exhibits the same general trends. However, because it is more complex, we provide details in the Appendix.

5.3 Summary

As shown by Martignon and Hoffrage (1999, 2002), TTB is optimal when environments are non-compensatory (by their definition). In addition, for the 3-, 4-, and 5-cue cases, TTB is one of the best strategies when environments consist of unique cue profile pairings. Moreover, even in fairly compensatory environments, TTB does well. However, as the environments become more compensatory, hybrid strategies such as EW/TTB become more effective in a relative sense. In these strategies, TTB intervenes when EW predicts ties. The EW/TTB hybrid is 100% accurate with 3-cues and the strategy of preference for the most compensatory weighting functions in the 4- and 5-cue cases.

Finally, we note that although the DOMRAN strategy has the lowest expected performance in all cases, in absolute terms its expected performance is quite high, i.e., 84% in the 3-cue case, 77% in the 4 cue-case, and 71% in the 5-cue case. As we shall demonstrate below, the fact that DOMRAN provides such a high "lower benchmark" is important for understanding the relative success of simple models for binary choice.

6 Error in the Application of Models

All the simple models we investigate assume correct knowledge and use of the signs of the zero-order correlations between cues and criterion. In addition, TTB is assumed to know and use the relative sizes of the β-parameters

[7] For details, see Table A1 of the Appendix of Hogarth and Karelaia (2003).

associated with the cues[8]. What happens, therefore, when the cues used in the TTB process are not considered in the appropriate order, i.e., there is error in knowledge and/or application of the TTB model?

We begin by examining the 3-cue case where there are 3! (= 6) possible orderings of the cues. These are shown in Table 5 together with results of different models. (Once again, characters in **bold** indicate the best expected correct percentages within classes of parameters, non-compensatory and compensatory.) In the non-compensatory case, TTB remains the best strategy but only provided the most important variable is correctly identified as such. For all other orderings, EW/TTB has the best expected correct predictions. In the compensatory case, EW/TTB is better than the other strategies no matter the order in which variables enter the models. (However, note comments about EW below.)

At the foot of Table 5, we have also indicated the means of the different columns as well as the expected performance of models that are not affected by the order in which cues are examined. As noted previously, DOMRAN has expected performance of 84%. Indeed, this outperforms the last three cue orderings of all models with the exception of EW/TTB. EW achieves 86% and 89% for the non-compensatory and compensatory weighting functions, respectively. The means of the TTB columns equal the performance of what Gigerenzer et al. (1999) refer to as the MINIMALIST strategy. This is the performance that would be expected of a TTB model where the order of the variables entering the model is decided at random[9]. However, MINIMALIST fails to reach the expected performance level of EW which actually matches the mean of EW/TTB. Thus EW matches or exceeds EW/TTB in roughly half of the possible orderings (i.e., the lower orderings).

Parenthetically, one way to interpret the expected performance level of MINIMALIST is to consider situations where people cannot control the order in which cues are examined or, indeed, which cues will eventually become available. In these situations, decisions are made by using cues in the order in which they are accessed. Thus, if the environment essentially randomizes the ordering of cues, a TTB-like strategy will have the expected performance of MINIMALIST. However, if the more (less) important cues happen to be accessed first, the strategy will be more (less) effective than MINIMALIST.

As shown in Table 6, in the 4-cue case there are 24 different possible orderings of the variables. For clarity, we only present the results of three of the six possible weighting functions – NonCF, CF2, and CF5. Once again, we

---

[8] Recall, however, that it does not require precise knowledge of the sizes of the β-parameters and, in this sense, the prior knowledge requirements are not necessarily that onerous.

[9] In the MINIMALIST strategy, variables enter the model in a random order for each binary choice. The assumption being made here is that the expected performance level of MINIMALIST is equal to that of the mean of TTB across all possible orderings, i.e., the expected performance of MINIMALIST is equivalent to that obtained by sampling different cue orderings at random.

Table 5. Sensitivity to different cue orderings for 3-cue case (populations of distinctive pairings)

	Cue orderings			Expected correct - %			
				Non-compensatory		Compensatory	
	1st	2nd	3rd	TTB	EW/TTB	TTB	EW/TTB
1	x1	x2	x3	**100**	96	96	**100**
2	x1	x3	x2	**93**	89	89	**93**
3	x2	x1	x3	86	**89**	89	**93**
4	x2	x3	x1	79	**82**	82	**86**
5	x3	x1	x2	79	**82**	82	**86**
6	x3	x2	x1	71	**75**	75	**79**
		Means		85	86	86	89

Notes:
(1) **Bold** entries indicate best predictions within orderings
(2) Expected performance (%) of models not affected by ordering:

	NonCF	CF
DOMRAN	84	84
EW	86	89

emphasize the best predictions within an order in **bold** characters. Overall, results mirror the 3-cue case. When the functions are non-compensatory or least compensatory, TTB performs best provided the most important variable enters the model first. Otherwise, EW/TTB performs best and is best across the range of orderings as the parameters become more compensatory (see CF5). DOMRAN achieves 77% in this population and this is clearly a better score than achieved by different models that fail to identify the appropriate ordering of the variables. Interestingly, DOMRAN only exceeds EW/TTB in the most compensatory case (CF5) in the last (and most incorrect) ordering (77% vs. 76%). Finally, EW matches the mean of EW/TTB across orderings. (Analogous results for the 5-cue case are presented in the Appendix).

Overall, differences between the models across the different combinations of weighting functions are small. Moreover, provided the first two most important variables enter the models in the appropriate order, TTB and EW/TTB do quite well in an absolute sense, i.e., approximate expected success rates of 90% and above. However, the major result for populations of distinctive cue pairings is that TTB is best provided the most important cue or variable does enter the model first *and* the weighting functions are not the most compensatory. Otherwise, EW/TTB should be preferred.

It is perhaps surprising that all models seem to have quite high expected correct predictions even when cue orderings are inappropriate. However, once again the DOMRAN model provides a good naïve benchmark with which to

Table 6. Sensitivity to different cue orderings for some 4-cue models (populations of distinctive pairings)

	Cue orderings				Expected correct - %					
					Non-compensatory		Compensatory			
							CF2		CF5	
	1st	2nd	3rd	4th	TTB	EW/ TTB	TTB	EW/ TTB	TTB	EW/ TTB
1	x1	x2	x3	x4	**100**	93	**99**	93	94	**98**
2	x1	x2	x4	x3	**97**	89	**96**	90	90	**95**
3	x1	x3	x2	x4	**93**	89	**93**	90	90	**95**
4	x1	x3	x4	x2	**90**	86	**89**	87	87	**91**
5	x1	x4	x2	x3	**90**	86	**89**	87	87	**91**
6	x1	x4	x3	x2	**87**	83	**86**	83	84	**88**
7	x2	x1	x3	x4	87	**88**	**88**	**88**	88	**94**
8	x2	x1	x4	x3	83	**84**	84	**85**	85	**90**
9	x2	x3	x1	x4	80	**84**	81	**85**	83	**90**
10	x2	x3	x4	x1	77	**81**	78	**82**	80	**87**
11	x2	x4	x1	x3	77	**81**	78	**82**	80	**87**
12	x2	x4	x3	x1	73	**78**	74	**78**	76	**84**
13	x3	x1	x2	x4	80	**84**	81	**85**	85	**90**
14	x3	x1	x4	x2	77	**81**	78	**82**	81	**87**
15	x3	x2	x1	x4	73	**81**	74	**82**	80	**87**
16	x3	x2	x4	x1	70	**78**	71	**78**	76	**84**
17	x3	x4	x1	x2	70	**78**	71	**78**	76	**84**
18	x3	x4	x2	x1	67	**74**	68	**75**	73	**80**
19	x4	x1	x2	x3	77	**81**	78	**82**	79	**86**
20	x4	x1	x3	x2	73	**78**	74	**78**	75	**83**
21	x4	x2	x1	x3	70	**78**	71	**78**	75	**83**
22	x4	x2	x3	x1	67	**74**	68	**75**	72	**80**
23	x4	x3	x1	x2	67	**74**	68	**75**	72	**80**
24	x4	x3	x2	x1	63	**71**	64	**72**	69	**76**
			Means		79	81	79	82	81	87

Notes:
(1) **Bold** entries indicate best predictions within orderings
(2) Expected performance (%) of models not affected by ordering:

	NonCF	CF2	CF5
DOMRAN	77	77	77
EW	81	82	87

calibrate this impression. It is consistently superior to many results achieved with incorrect cue orderings.

7 Different Distributions of Cue Profiles

The above results are conditioned on populations consisting of unique cue profiles. However, characteristics of choice sets are an important dimension of environmental variability. In particular, we would expect both the general level of predictive ability as well as relative differences between models to depend on characteristics of the distributions of cue profiles in given populations. We consider three main factors that we illustrate by the three distributions shown in Table 7.

First, distributions can differ in the number of dominating cue profiles. In general, the greater the proportion of dominating cue profiles, the greater is the expected performance of all models. In Table 7, Distribution III has a lower proportion of dominating profiles than the other distributions. Hence, DOMRAN (as well as the other models) performs less well here than in the other distributions.

Second, when there are repeats of the same cue profile, all models would only be expected to discriminate correctly between such cases at a rate of 50%[10]. Thus, the general level of predictability between two populations depends, in part, on the number of repeated profiles in each. Specifically, repeats lower overall expected performance. In Table 7, Distribution III contains several repeats.

Third, the conflict implicit in the difference vectors has more impact on the relative success of some models than of others. For instance, when the weighting function is compensatory, TTB always makes mistakes for the D-E choice in the 3-cue case (see Table 2), but EW – and thus EW/TTB – does not. Hence, the presence or absence of D-E choices in a population can affect the relative success of these decision rules. As a case in point, D-E conflict is present in Distributions I and III but absent from Distribution II. Note, in particular, that TTB has predicted performance of 100% correct in Distribution II but 96% and 80% in Distributions I and III, respectively.

More generally, distributions or "choice environments" can be described as being "TTB-friendly" or "TTB-unfriendly" for compensatory functions depending on the absence or presence of cue profile pairings that TTB classifies

[10] In a world without error, repeat profiles must have identical values on the dependent variable. Thus, it could be argued, it does not matter which profile is selected because each must be "correct." In this work, however, we take a more conservative approach and assume that one of the two alternatives is indeed correct. Thus, models can only discriminate the correct alternative by chance, i.e., with probability of 0.50.

incorrectly (cf. Shanteau & Thomas, 2000)[11]. Thus, Distribution II in Table 7 can be described as TTB-friendly (there are no D-E pairings) whereas Distribution III is TTB-unfriendly.

Whether a distribution is TTB-friendly or TTB-unfriendly can be characterized by asking how it varies from a uniform distribution (e.g., Distribution I in Table 7) in terms of the number of errors made by TTB. Specifically, we describe choice environments with uniform distributions of cue profiles as "TTB-neutral." Thus, if the expected number of TTB errors in a distribu-

Table 7. Some distributions of cue profiles for the 3-cue case

Cue profile type	x1	x2	x3	Distributions (entries: number of profiles) I	II	III
A	1	1	1	1	1	0
B	1	1	0	1	1	2
C	1	0	1	1	1	0
D	1	0	0	1	1	2
E	0	1	1	1	0	6
F	0	1	0	1	1	0
G	0	0	1	1	1	3
H	0	0	0	1	1	3
Characteristics of distributions						
a) Total number of binary choices				28	21	120
b) Percentages of choices involving dominance				68	71	51
c) Presence of repeats?				No	No	Yes
d) TTB-error cases DE, as percentage of total				4	0	10
e) Overall characterization				TTB-neutral	TTB-friendly	TTB-unfriendly
Predicted correct (%'s) assuming compensatory weighting function:						
			TTB	96	**100**	80
			EW	89	91	83
			EW/TTB	**100**	**100**	**90**
			DOMRAN	84	86	75

Note: **Bold** entries indicate best predictions within distributions

[11] Environments can be classified as "friendly" or "unfriendly" to models as well as people. In this paper, we only consider the former. The latter depends on a decision maker's experience/knowledge.

tion is less (more) than that expected on the basis of a uniform distribution, it will be described as TTB-friendly (TTB-unfriendly). As an example, consider Distribution III in Table 7. This has 16 observations such that a uniform "equivalent" would have two observations of each cue profile type. This uniform distribution would have two D observations and two E observations and, consequently, make four TTB errors (i.e., there are 2×2 D-E pairings). In Distribution III, note that TTB makes 12 errors (i.e., there are 2×6 D-E pairings). Because 12 is greater than four, we describe Distribution III as TTB-unfriendly. Distribution II, on the other hand, is TTB-friendly because $0 < 1$. In short, when the number of predicted TTB errors is smaller (greater) than expected on the basis of a uniform distribution, we describe the distribution as TTB-friendly (TTB-unfriendly).

Parenthetically, we note that the predictive success of lexicographic models such as TTB has sometimes been attributed to correlation between the cues. However, this is not a complete explanation. As indicated above, TTB is quite successful in TTB-neutral environments in which the intercorrelations between cues are zero, i.e., uniform distributions of distinctive cue profiles. What is critical to the performance of TTB is the presence or absence of the specific cue pairings that it predicts incorrectly, i.e., whether the distributions are TTB-unfriendly or TTB-friendly.

7.1 Some Empirical Distributions

The data in Table 7 were constructed for illustrative purposes. What can be said about empirical data? To examine the characteristics of different distributions, we use 20 datasets created by Czerlinski et. al (1999)[12]. First, we ignore the empirical criterion variables and examine the characteristics of the datasets by cue profiles. What proportions of the choices in each dataset involve dominance and repeats? To what extent are these datasets TTB-friendly or TTB-unfriendly? Second, we test the different simple models by seeing how well they predict these data.

Table 8 reports characteristics of the 20 datasets that we have split into three groups according to numbers of cues (three, four, and five). The 5-cue set actually includes datasets that had more than five cues. However, in each of these we have only considered the five most important cues (determined by examining cue validities across all data). First, in addition to numbers of observations, we report the number of cases where all models make identical choices, i.e., for cases involving dominance and repeated cue profiles (in the second and third columns of the table). The total number of common choices (i.e., the sum of dominant pairs and repeats) is large, varying from 39% to 96% with an average of 67%. In other words, the predictions of models can

[12] We are very grateful to these researchers for making their data available on their website. See http://www-abc.mpib-berlin.mpg.de/sim/Heuristica/environments/

Table 8. Characterization of empirical datasets

Datasets	n	Choices (%) involving: Dominant Pairs	Repeats	Ratio of expected TTB errors: Uniform/actual		Overall characterized as	
3-cues:				**CF**			
Ozone	11	53	18	Infinite		F	
Attractiveness of women	30	51	17	Infinite		F	
Attractiveness of men	32	57	21	2.67		F	
Fish fertility	395	72	21	Infinite		F	
4-cues:				**CF2**	**CF5**	**CF2**	**CF5**
Oxidant	17	45	6	Infinite	2.50	F	F
Land rent	58	42	9	0.29	0.39	U	U
5-cues:				**CF9**	**CF23**	**CF9**	**CF23**
Homelessness	50	47	6	Infinite	1.24 to 4.94	F	F
Body fat	218	66	8	7.62	4.88	F	F
City populations	83	67	25	28/1 to 63/1	84/1 to 189/1	F	F
High school dropout rates	57	70	26	4.67	8.40	F	F
Cows' manure	14	13	33	Infinite	Infinite	F	F
Mortality	20	71	8	1.17	1.62	N	F
House prices	22	75	13	1.75	3.00	N	F
Car accidents	37	35	4	Infinite	1.11	F	N
Rainfall	24	49	4	1.75	1.24	N	N
Obesity at 18	46	72	10	0.41 to 1.65	0.55 to 2.21	N	N
Fuel consumption	48	47	6	0.13 to 0.52	0.38 to 1.50	U	N
Professors' salaries	51	48	9	0.47	0.89	U	N
Mammals' sleep	35	49	14	0.35	1.00	U	N
Biodiversity	26	42	7	0.35	0.68	U	U
Means		54	13				

Legend:
F stands for TTB-friendly; U stands for TTB-unfriendly; N stands for TTB-neutral

only be distinguished on about one-third of these data. In addition, given that overall dominant pairings account for some 54% of the data, we would expect DOMRAN to perform quite well. (Specifically, if there was no error, we would expect DOMRAN to have an overall predictive ability of 77%, i.e., $0.54 + 0.50(1 - 0.54)$).

Second, Table 8 illustrates the extent to which the datasets are TTB-friendly or TTB-unfriendly. That is, for each dataset we calculate the ratio of the number of TTB errors that would be expected in a uniform distribution of cue profiles (of size equal to the actual distribution) with the theoretical number of TTB errors implied by the actual distribution.

We begin by considering the 3-cue sets. The four 3-cue datasets are all TTB-friendly. Indeed, for three of the distributions, the ratios of expected errors are infinite because there are no expected TTB errors in the actual distributions. For these distributions, therefore, TTB would be expected to perform as well as EW/TTB for compensatory functions.

There are two 4-cue distributions. However, to assess whether a distribution is TTB-friendly or TTB-unfriendly, specific compensatory weighting function must be used. Here we use CF2 and CF5 to illustrate the range of cases. As can be seen, one distribution ("Oxidant") is TTB-friendly, whereas the other ("Land rent") is not.

The 5-cue datasets have a mix of TTB-friendly, TTB-unfriendly, and TTB-neutral distributions. Once again, classification of TTB-friendly or TTB-unfriendly depends on specifying particular weighting functions. In this case, we illustrate CF9 and CF23 (for details of weighting functions, see Appendix).

If the weighting functions applicable to each dataset were known, the characteristics of the datasets in Table 8 could be used to calculate the expected predictive performances of all the models in the absence of error. Indeed, using both non-compensatory and compensatory functions and averaging across datasets, the expected predictive abilities of the models in the absence of error was estimated to vary between 81% and 94% (Hogarth and Karelaia, 2003).

How well therefore do the various simple models actually predict the data? Figure 2 reports mean predictive accuracies of the models across all 20 datasets on holdout samples using 1,000 replications. Specifically, for each dataset we randomly sampled a proportion of the possible choices, fit parameters as appropriate (e.g., calculating cue validities in TTB), and then used these parameters to predict the remaining choices in the dataset (i.e., the holdout sample)[13]. We replicated this process 1,000 times and used different proportions of fitting and holdout samples – a 50/50 split and a 20/80 split.

[13] Once again, we limited the number of cues in any dataset to five. For the TTB and EW models, our cross-validated results for the 50/50 split are quite similar to those reported by Gigerenzer et al. (1999). Specifically, they reported 71%

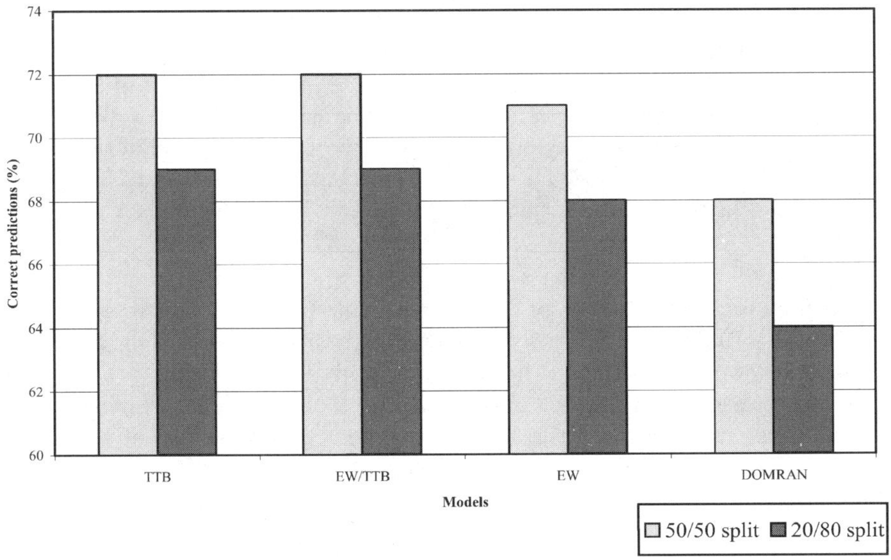

Fig. 2. Predictions in holdout samples with 1000 trials

Figure 2 reveals three major trends. First, the differences between TTB, EW/TTB and EW are small (this is also true of the results of all the datasets that have been averaged). Second, and as might be expected, predictability is somewhat greater in the 50/50 split than in the 20/80 split. Third, DOMRAN is the least successful of the models. However, the difference between DOMRAN and the other models is small (at most between 4% and 5% in predictive accuracy). Perhaps the surprising story of these data is not that TTB is the best of the simple models (an important finding), but that the naïve DOMRAN benchmark does so well.

8 Approximating the Effects of Error in the Data

Our theoretical analyses were conducted assuming no error in the data and, yet, when we look at model performance (Fig. 2), error clearly degrades overall performance. We now therefore consider the role of error in the data.

Recall, first, that – in the absence or error – TTB's expected performance in a specific dataset would, in percentage terms, be 100 *less* both (a) the percentage of incorrects among the pair-wise choices (conditional on the appropriate weighting function), and (b) 50% of the percentage of repeats. Denote this level of error-free performance $p*$. Second, imagine the other ex-

for TTB and 69% for EW. In addition, their results for multiple regression were 68% and 74% for Bayesian networks.

treme where the data are totally unreliable. In this case, TTB's expected performance would be 50.

These two estimates represent, respectively, expected maximum and minimum performance levels where variations (between p^* and 50) reflect error or inconsistency in the data. More formally, if we define the level of reliability or consistency by c $(0 \leq c \leq 1)$, expected performance – denoted $E(s)$ – can be expressed in percentage terms by

$$E(s) = 50 + c(p * -50)\,. \tag{5}$$

We now demonstrate the use of Eq. (5) with the four 3-cue datasets of Czerlinski et al. (1999)[14]. But first, we need to estimate c. Lacking independent assessments, we suggest using the R^2 that results from regressing the dependent variables of each dataset onto their corresponding cue-profiles on the grounds that this captures the maximum linear predictability of the data. Thus, replacing c in Eq. (5) by R^2, we estimate $E(s)$ by

$$s = 50 + R^2(p * -50) \tag{6}$$

As can be seen in Table 9, we have applied the reasoning behind Eq. (6) to make predictions for the EW/TTB, EW, and DOMRAN models as well as

Table 9. Predicted accuracy of models incorporating error for 3-cue models (% correct)

Predicted accuracy for datasets	Models			
	TTB	EW/TTB	EW	DOMRAN
Ozone	85	85	82	72
Fish fertility	71	71	70	70
Attractiveness of men	68	68	64	63
Attractiveness of women	67	67	63	60
Averages	73	73	70	66
Realizations for datasets (based on 50/50 splits, Fig. 2)				
Ozone	82	78	78	78
Fish fertility	73	73	70	70
Attractiveness of men	71	71	70	70
Attractiveness of women	67	67	65	64
Averages	73	72	71	71

[14] We limit our attention to these four datasets because, in order to estimate $p*$, one needs to know the appropriate weighting function. With 3-cue models there are only two alternatives and, as pointed out in Table 8, three of the four 3-cue models are entirely "TTB friendly" such that predictions for the non-compensatory and compensatory cases are identical.

TTB. Overall, correspondence between empirical realizations and predicted accuracy is quite good for TTB (the largest deviation is 3%). Error is somewhat larger for the other models (the DOMRAN predictions, in particular, are too modest). However, the qualitative ordering by datasets is almost perfectly respected.

9 Discussion

We have investigated why and when simple decision rules such as TTB are effective in binary choice when informational cues are binary in nature. Our discussion is organized as follows. We first summarize our results. Second, we consider how characteristics of models and environments interact in affecting performance. Finally, we discuss the use of TTB as a prescriptive model.

9.1 Principal Findings

Our main results can be summarized as follows. We first analyzed linear environments characterized by populations of distinctive pairings of alternatives, no error, and known relative sizes of cue validities. Whereas analytical work had already shown that TTB is optimal in this world when weighting functions are non-compensatory (Martignon and Hoffrage, 1999, 2002), we showed that TTB is also an effective strategy when weighting functions are compensatory. Moreover, we demonstrated that a previously unidentified hybrid strategy, EW/TTB, is optimal for the more compensatory functions. In particular, in the 3-cue case, EW/TTB is optimal for all compensatory functions.

Second, when errors are made in the relative sizes of cue validities, TTB typically remains the most effective strategy *provided* the most important cue is identified as such. When this is not the case, EW/TTB should be preferred even though TTB is still quite effective. This finding is important because it addresses the extent to which precise knowledge of cue validities is essential to the effectiveness of TTB (for an empirical example, see also Martignon and Hoffrage, 2002). That precise knowledge is not necessary suggests that the cognitive demands of calculating and storing cue validities may not be as daunting as suggested by critics of TTB (e.g., Juslin and Persson, 2002). However, to place these results in context, recall that both EW and DOMRAN frequently perform better than TTB when this uses incorrect cue orderings.

Third, both the absolute and relative expected performances of models are affected by characteristics of sets of choice alternatives. Specifically, the numbers of dominance pairs and repeats in a distribution of cue profiles affect overall levels of expected predictive accuracy – increasing in the former and decreasing in the latter. In addition, all simple models considered here make the same predictions for all dominance pairs and have the same expected performance with respect to repeats. Thus, differences between the models only occur in subsets of data. Within these subsets, performance of the models is differentially sensitive to the presence or absence of specific pairings of cue profiles (such as the D-E pairing in the 3-cue case). It is always possible to "engineer" environments that are more or less "friendly" to different models.

Fourth, in cross-validated predictive tests, TTB, EW/TTB, and EW all had similar performance and were only superior to DOMRAN by 3% or 4% (across datasets, DOMRAN averaged 68%). These analyses highlighted the importance of error in data that clearly degrades the performance of all models.

Fifth, we developed a method to approximate the effects of error in data which we demonstrated using some of Czerlinski et al.'s (1999) datasets.

9.2 Environments and Models

To illuminate how model performance varies according to environments, it is instructive to focus on the comparison of TTB with EW. Whereas TTB always makes choices by treating some variables as being more important than others, EW predicts ties between certain pairs of alternatives and is forced to choose between these pairs at random. However, to perform better than EW on these cases, TTB does not always need to be correct; its success rate only needs to exceed 50%. (A little knowledge is better than none.) On the other hand, on occasions when TTB is mistaken, EW sometimes makes correct decisions. In creating the EW/TTB composites, therefore, the advantages of both models can be achieved in the more compensatory environments.

TTB differs from the other models in two major respects: it imposes an order in which cues are examined, and it can exit the process before consulting all information (i.e., the models differ in their "stopping" rules). When the environmental weighting function is non-compensatory, stopping the process "early" is sensible precisely because subsequent cues cannot change the decision. However, as the environment becomes more compensatory, more information should be examined.

In a series of intriguing simulation analyses, Payne et al. (1993) showed the relative effectiveness of different heuristics for choosing between gambles – in an analysis based on continuous as opposed to binary variables. They too considered whether dominance was possible and, although they did not talk of non-compensatory and compensatory weighting functions *per se*, they varied the dispersion of distributions of weights. Their analyses showed how the environment affects the relative performance of different heuristics in a manner consistent with our results. For example, their lexicographic strategy outperformed equal weights in high dispersion environments (similar to our non-compensatory functions) and equal weights outperformed the lexicographic strategy in the low dispersion environments (similar to compensatory functions). Thus, as also noted above, environments can be created that are more or less "friendly" toward different models in terms of how they affect relative predictive performance (see also Shanteau and Thomas, 2000).

However, even if one holds weighting functions constant, an important feature of binary-cue environments is how the distribution of cue-profiles affects the relative "friendliness" of the environment to particular models. For example, an environment where EW faced many (few) ties would be "unfriendly" ("friendly") to EW. Similarly, we defined environments that were "TTB-friendly" or "TTB-unfriendly" by the extent to which they contained pairs leading to less or more errors made by TTB compared to the number of TTB errors that would be made in a uniform distribution of all possible pairings of alternatives with the same number of observations (cf. Table 8). An important question, therefore, is to understand the types of environments that people encounter in their decision making activities. For example, to what extent do the datasets compiled by Czerlinski et al. (1999) characterize the kinds of situations people face in their natural ecologies? We simply do not know.

A key characteristic of the environments considered in this work is that the cues are binary variables. In a related paper (Hogarth and Karelaia, 2005), we have analyzed the performance of simple models for binary choice based on continuous cues. As with binary cues, environmental conditions can be defined that are more or less friendly to the different models. However, care should be exercised in extrapolating conclusions from one case to the other. With continuous cues, for example, EW is not so disadvantaged compared to lexicographic models. DOMRAN, on the other hand, is far less effective. More research is clearly needed to examine generalizations of TTB-like models to more complex environments, e.g., involving choice from multiple alternatives, different types of non-linear weighting functions, continuous cues, the effects of error, and so on. Although it would be illuminating to explore the effects of many cues on relative model performance, given limitations on human information processing, we have limited the number of cues to five in the present work.

9.3 Prescriptive Considerations

In most of the environments examined in this work, TTB has been shown to be an effective, simple model of choice. To what extent, therefore, should it be prescribed as a way to choose?

Assume first that underlying assumptions are met, i.e., that the zero-order correlations between cues and criterion are known as well as the relative importance of the binary variables. In this case, key issues center on the extent to which the environmental weighting function is compensatory and characteristics of the cue profiles.

Given sufficient resources, e.g., time, we first recommend checking for dominance. Indeed, with few cues this may be a simpler strategy than accessing relative cue validities from memory. Moreover, exploiting dominance can imbue the decision maker with appropriate confidence. Failing this, our recommendation is to use TTB or EW/TTB (and certainly in the 3-cue case). Briefly, if the decision maker feels uncomfortable about relying on TTB alone (e.g., she senses that the environment is compensatory), then the choice should also be examined using EW/TTB. If there is uncertainty about which variable is most important, then EW/TTB is the model to follow.

At the outset of this paper, we noted that there is now a growing experimental literature that tests if and when people actually use TTB-like mental strategies (Bröder, 2000, 2003; Bröder and Schiffer, 2003; Newell and Shanks, 2003; Newell et al., 2003; Rieskamp and Hoffrage, 1999, 2002). The major descriptive violation is that people do not respect the TTB stopping rule and tend to use more information than needed. Indeed, participants in some experiments would have earned more money had they used strict TTB strategies (Newell and Shanks, 2003; Newell et al., 2003).

As a general point, however, it should be noted that all TTB errors occur because the decision process stops too soon. In the 4-cue case, for example, no errors ever occur if the process is decided by the third or fourth most important variable. Thus, the reluctance people express to base important decisions on one or two cues may be a consequence of having experienced errors when all available information was not consulted. Interestingly, an analysis by Karelaia (2005) shows that, in 3-cue environments, the performance of a TTB-like strategy that requires confirming evidence before deciding does not lag far behind that of TTB. The cost of seeking confirming evidence may not be that high.

For decisions taken under time pressure, people should exploit the fact that TTB has a high success rate. (In particular, in using this strategy they will automatically exploit dominance even though they may never know this.) Two further issues concern feedback and the relative importance of decisions. With good feedback, people can learn to make appropriate responses. Failing accurate feedback or being faced with important decisions under time pressure, however, they are not powerless. Specifically, they can "rehearse" similar decisions (e.g., through simulations) and then use this knowledge to know what to do in real situations, e.g., what would happen if TTB or another heuristic were used in similar circumstances?

Finally, we are puzzled by a definitional conundrum within the paradigm of "bounded rationality." Gigerenzer and his colleagues have argued against so-called "optimizing" models on the grounds that these require unrealistic computational powers (Gigerenzer et al., 1999). However, if the tools in the "adaptive toolbox" are so effective, would the superior rational powers of the "demons" not have already allowed them to determine this? If yes, then surely they would have never suggested using, for example, multiple regression instead of TTB in making predictions for the Czerlinski et al. (1999) datasets. Do the demons know about other "simple" models of which we mere mortals are ignorant?

Acknowledgement. Robin M. Hogarth is ICREA Research Professor at Universitat Pompeu Fabra, Barcelona, Spain, and Natalia Karelaia is Assistant Professor of École des Hautes Etudies Commerciales at Université de Lausanne, Switzerland. This research was financed partially by a grant from the Spanish Ministerio de Educación y Ciencia. We thank Manel Baucells, Nick Chater, Daniel Goldstein, Ulrich Hoffrage, Konstantinos Katsikopoulos, Laura Martignon, James Shanteau and Jack Soll for comments on earlier versions and Marc LeMenestrel for a most illuminating conversation. We also acknowledge comments by participants of workshops at Leeds University Business School (UK), the Center for Decision Research at the University of Chicago, and Universitat Pompeu Fabra.

References

Bröder, A. (2000), Assessing the empirical validity of the "Take-The-Best" heuristic as a model of human probabilistic inference. *Journal of Experimental Psychology: Learning, Memory, and Cognition*, 26(5):1332–1346.

Bröder, A. (2003), Decision making with the "adaptive toolbox": Influence of environmental structure, intelligence, and working memory load. *Journal of Experimental Psychology: Learning, Memory, and Cognition*, 29(4):611–625.

Bröder, A., Schiffer, S. (2003), Take the best versus simultaneous feature matching: Probabilistic inferences from memory and effects of representation format. *Journal of Experimental Psychology: General*, 132(2):277–293.

Brunswik, E. (1952), *The Conceptual Framework of Psychology.* Chicago: The University of Chicago Press.

Chater, N., Oaksford, M., Nakisa, R., Redington, M. (2003), Fast, frugal, and rational: How rational norms explain behavior. *Organizational Behavior and Human Decision Processes*, 90:63–86.

Czerlinski, J., Gigerenzer, G., Goldstein, D.G. (1999), How good are simple heuristics? In G. Gigerenzer, P.M. Todd and the ABC Research Group. *Simple heuristics that make us smart* (pp. 97–118). New York: Oxford University Press.

Dawes, R.M. (1979), The robust beauty of improper linear models. *American Psychologist*, 34:571–582.

Dawes, R.M., Corrigan, B. (1974), Linear models in decision making. *Psychological Bulletin*, 81:95–106.

Einhorn, H.J., Hogarth, R.M. (1975), Unit weighting schemes for decision making. *Organizational Behavior and Human Performance*, 13:171–192.

Gigerenzer, G., Goldstein, D. (1996), Reasoning the fast and frugal way: Models of bounded rationality. *Psychological Review*, 103:650–669.

Gigerenzer, G., Selten, R. (2001), Rethinking rationality. In G. Gigerenzer, R. Selten (Eds.), *Bounded Rationality: The Adaptive Toolbox* (pp. 1–13). Cambridge, MA: MIT Press.

Gigerenzer, G., Todd, P.M., the ABC Research Group. (1999), *Simple Heuristics That Make Us Smart.* New York: Oxford University Press.

Hogarth, R.M., Karelaia, N. (2003), *"Take-the-Best" and other simple strategies: Why and when they work "well" in binary choice.* Barcelona: Universitat Pompeu Fabra, DEE working paper no. 709.

Hogarth, R.M., Karelaia, N. (2005), Ignoring information in binary choice with continuous variables: When is less "more"? *Journal of Mathematical Psychology*, 49:115–124.

Juslin, P., Persson, M. (2002), PROBabilities from EXemplars (PROBEX): a "lazy" algorithm for probabilistic inference from generic knowledge. *Cognitive Science*, 26:563–607.

Karelaia, N. (2005), *Consistency or accuracy: how good is a decision-making strategy that looks for confirming evidence?* Barcelona: Universitat Pompeu Fabra, PhD Dissertation.

Martignon, L., Hoffrage, U. (1999), Why does one-reason decision making work? A case study in ecological rationality. In G. Gigerenzer, P.M. Todd and the ABC Research Group. *Simple Heuristics That Make Us Smart* (pp. 119–140). New York: Oxford University Press.

Martignon, L., Hoffrage, U. (2002), Fast, frugal, and fit: Simple heuristics for paired comparison. *Theory and Decision*, 52:29–71.

Montgomery, H. (1983), Decision rules and the search for a dominance structure: Towards a process model of decision making. In P. Humphreys, O. Svenson & A. Vari (Eds.), *Analysing and Aiding Decision Processes* (pp. 343–369). Amsterdam: North Holland.

Newell, B.R., Shanks, D.R. (2003), Take the best or look at the rest? Factors influencing "one-reason" decision making. *Journal of Experimental Psychology: Learning, Memory, and Cognition*, 29(1):53–65.

Newell, B.R., Weston, N.J., Shanks, D.R. (2003), Empirical tests of a fast-and-frugal heuristic: Not everyone "takes-the-best." *Organizational Behavior and Human Decision Processes*, 91:82–96.

Payne, J.W., Bettman, J.R., Johnson, E.J. (1993), *The Adaptive Decision Maker.* New York: Cambridge University Press.

Rieskamp, J., Hoffrage, U. (1999), When do people use simple heuristics, and how can we tell? In G. Gigerenzer, P.M. Todd and the ABC Research Group. *Simple Heuristics That Make Us Smart* (pp. 141–167). New York: Oxford University Press.

Rieskamp, J., Hoffrage, U. (2002), *The use of simple heuristics: Inferences and preferences under time pressure.* Manuscript, Max Planck Institute for Human Development, Berlin.

Shanteau, J., Thomas, R.P. (2000), Fast and frugal heuristics: What about unfriendly environments? *Behavioral and Brain Sciences*, 23:762–763.

Simon, H.A. (1956), Rational choice and the structure of environments. *Psychological Review*, 63:129–138.

Todd, P.M., Gigerenzer, G. (2000), Précis of *Simple Heuristics That Make Us Smart. Behavioral and Brain Sciences*, 23:727–780.

Appendix – The 5-Cue Case

The 5-cue case involves 496 distinctive profile pairings; 23 different types of compensatory functions (defined in the same manner as the 4-cue case); and many more cases where functions imply ambiguous predictions. (For full details, see the Appendix of Hogarth and Karelaia, 2003). Functions CF1 through CF17 are close to "non-compensatory" and, even with CF23 it is possible to have the weight of the first variable much larger than the others (cf. the 4-cue case).

Table A1. Expected predictive accuracy (%'s) for the 5-cue case

Models	NonCF	CF2	CF4	CF6	CF8	CF10	CF12	CF14	CF16	CF18	CF20	CF22	CF23
DOMRAN	71	71	71	71	71	71	71	71	71	71	71	71	71
EW	78	78	79	78	78	80	79	79	80	81	81	83	85
*No of errors**	55	53	45	54	52	44	49	47	39	34	30	14	2
TTB	**100**	**100**	**98**	**100**	**99**	**98**	**99**	**99**	**97**	96	96	93	91
*No of errors**	0	2	6	1	3	7	3	5	9	9	11	15	21
EW/TTB	89	89	90	89	90	91	90	90	91	91	92	93	**95**
*No of errors**	55	53	45	54	52	44	49	47	39	34	30	14	2
EW-4/TTB	93	93	94	93	93	94	94	94	95	95	96	**96**	**95**
*No of errors**	36	34	28	35	33	27	30	28	22	16	12	1	5
EW-3/TTB	97	96	95	97	97	95	98	97	96	**98**	**98**	95	93
*No of errors**	16	18	22	15	17	21	10	12	16	0	2	6	13
Ambiguous cases**	0	0	6	0	0	6	3	3	9	19	21	38	43

Notes:
NonCF = non-compensatory functions
CF = compensatory functions
* Errors involve misclassifications by the models (from total of 496 choices)
** Functions contain some ambiguous cases. All models are assumed to be 50% correct on these cases

Table A2. Sensitivity to different cue orderings for some 5-cue models (populations of distinctive pairings)

	Cue orderings					Expected correct - %									
						Non-compensatory		Compensatory							
								CF1		CF9		CF17		CF23	
	1st	2nd	3rd	4th	5th	TTB	EW/TTB	TTB	EW/TTB	TTB	EW/TTB	TTB	EW/TTB	TTB	EW/TTB
1	x1	x2	x3	x4	x5	**100**	89	**99**	90	**99**	90	**96**	93	91	**95**
2	x1	x2	x3	x5	x4	98	87	98	88	97	89	94	91	90	94
3	x1	x2	x4	x3	x5	97	87	98	88	97	89	94	91	90	94
4	x1	x2	x4	x5	x3	**95**	86	**96**	87	**95**	87	**92**	90	88	**92**
5	x1	x2	x5	x3	x4	**95**	86	**96**	87	**95**	87	**92**	90	88	**92**
⋮	⋮	⋮	⋮	⋮	⋮	⋮	⋮	⋮	⋮	⋮	⋮	⋮	⋮	⋮	⋮
11	x1	x3	x5	x2	x4	**89**	83	**88**	84	**88**	85	**89**	88	85	**90**
12	x1	x3	x5	x4	x2	**87**	82	**86**	83	**87**	83	**87**	86	83	**88**
13	x1	x4	x2	x3	x5	**90**	85	**91**	86	**91**	86	**91**	89	87	**92**
14	x1	x4	x2	x5	x3	**89**	83	**90**	84	**90**	85	**90**	88	85	**90**
15	x1	x4	x3	x2	x5	**87**	83	**88**	84	**88**	85	**89**	88	85	**90**
⋮	⋮	⋮	⋮	⋮	⋮	⋮	⋮	⋮	⋮	⋮	⋮	⋮	⋮	⋮	⋮
21	x1	x5	x3	x2	x4	**86**	82	**86**	83	**87**	83	**87**	86	83	**88**
22	x1	x5	x3	x4	x2	**84**	80	**85**	81	**85**	82	**85**	84	81	**86**
23	x1	x5	x4	x2	x3	**84**	80	**85**	81	**85**	82	**85**	84	81	**86**
24	x1	x5	x4	x3	x2	**82**	78	**83**	79	**83**	80	**84**	82	79	**85**
25	x2	x1	x3	x4	x5	**87**	85	**86**	**86**	**86**	**86**	85	**89**	86	**92**
⋮	⋮	⋮	⋮	⋮	⋮	⋮	⋮	⋮	⋮	⋮	⋮	⋮	⋮	⋮	⋮

Table A2. (continued)

	Cue orderings					Expected correct - %									
						Non-compensatory		Compensatory							
								CF1		CF9		CF17		CF23	
	1st	2nd	3rd	4th	5th	TTB	EW/TTB	TTB	EW/TTB	TTB	EW/TTB	TTB	EW/TTB	TTB	EW/TTB
31	x2	x3	x1	x4	x5	81	**83**	80	**83**	80	**84**	79	**86**	82	**91**
32	x2	x3	x1	x5	x4	79	**81**	78	**82**	78	**82**	77	**85**	81	**89**
33	x2	x3	x4	x1	x5	77	**81**	77	**82**	76	**82**	76	**85**	80	**89**
34	x2	x3	x4	x5	x1	76	**79**	75	**80**	75	**81**	74	**83**	78	**87**
35	x2	x3	x5	x1	x4	76	**79**	75	**80**	75	**81**	74	**83**	78	**87**
⋮	⋮	⋮	⋮	⋮	⋮	⋮	⋮	⋮	⋮	⋮	⋮	⋮	⋮	⋮	⋮
41	x2	x4	x5	x1	x3	73	**78**	73	**78**	73	**79**	73	**82**	77	**86**
42	x2	x4	x5	x3	x1	71	**76**	72	**77**	72	**77**	71	**80**	75	**84**
43	x2	x5	x1	x3	x4	76	**79**	77	**80**	76	**81**	76	**83**	79	**87**
44	x2	x5	x1	x4	x3	74	**78**	75	**78**	75	**79**	74	**82**	77	**85**
45	x2	x5	x3	x1	x4	73	**78**	73	**78**	73	**79**	73	**82**	76	**85**
⋮	⋮	⋮	⋮	⋮	⋮	⋮	⋮	⋮	⋮	⋮	⋮	⋮	⋮	⋮	⋮
61	x3	x4	x1	x2	x5	71	**78**	70	**79**	71	**80**	74	**83**	77	**87**
62	x3	x4	x1	x5	x2	69	**77**	69	**78**	69	**78**	72	**81**	75	**85**
63	x3	x4	x2	x1	x5	68	**77**	67	**78**	68	**78**	71	**81**	75	**85**
64	x3	x4	x2	x5	x1	66	**75**	65	**76**	66	**77**	69	**80**	73	**84**
65	x3	x4	x5	x1	x2	66	**75**	65	**76**	66	**77**	69	**80**	73	**84**
⋮	⋮	⋮	⋮	⋮	⋮	⋮	⋮	⋮	⋮	⋮	⋮	⋮	⋮	⋮	⋮

Table A2. (continued)

						Expected correct - %									
	Cue orderings					Non-compensatory		Compensatory							
								CF1		CF9		CF17		CF23	
	1st	2nd	3rd	4th	5th	TTB	EW/TTB	TTB	EW/TTB	TTB	EW/TTB	TTB	EW/TTB	TTB	EW/TTB
81	x4	x2	x3	x1	x5	68	**77**	69	**78**	69	**78**	71	**81**	75	**85**
82	x4	x2	x3	x5	x1	66	**75**	67	**76**	68	**77**	70	**80**	73	**83**
83	x4	x2	x5	x1	x3	66	**75**	67	**76**	68	**77**	70	**80**	73	**83**
84	x4	x2	x5	x3	x1	65	**74**	65	**74**	66	**75**	68	**78**	72	**82**
85	x4	x3	x1	x2	x5	68	**77**	69	**78**	69	**78**	72	**81**	75	**85**
⋮	⋮	⋮	⋮	⋮	⋮	⋮	⋮	⋮	⋮	⋮	⋮	⋮	⋮	⋮	⋮
116	x5	x4	x1	x3	x2	63	**72**	64	**73**	64	**73**	67	**76**	67	**78**
117	x5	x4	x2	x1	x3	61	**72**	62	**73**	63	**73**	65	**76**	67	**78**
118	x5	x4	x2	x3	x1	60	**70**	61	**71**	61	**72**	63	**74**	66	**77**
119	x5	x4	x3	x1	x2	60	**70**	61	**71**	61	**72**	63	**74**	66	**77**
120	x5	x4	x3	x2	x1	58	**69**	59	**70**	60	**70**	62	**73**	64	**75**
					Means	74	78	74	79	74	79	76	82	77	85

Notes:
(1) Bold entries predict the best predictions within orderings.
(2) Expected performance (%) of models not affected by ordering:

	NonCF	CF1	CF9	CF17	CF23
DOMRAN	71	71	71	71	71
EW	78	79	79	82	85

Table A1 presents the expected predictive accuracies of the different models for the 5-cue case for non-compensatory functions and 12 of the 23 compensatory functions. The models are the same as in the 4-cue case except that the EW/TTB hybrids include a version based on the first four most important cues. At the foot of Table A1 we also indicate the number of ambiguous choices for each set of weighting functions. These become much larger as the parameters indicate more compensatory environments.

TTB is 100% correct with the non-compensatory function (as must be the case) but its performance drops off in relative terms as the functions become more compensatory. It makes the smallest number of errors, as defined above, through CF16. EW-3/TTB has the best performance for CF18, CF19, and CF20, and the EW/TTB hybrids perform relatively well for the most compensatory functions: see EW-4/TTB for CF22 and EW/TTB for CF23.

Table A2 shows what happens when cues do not enter the TTB model in the appropriate order. Given that there are 120 different such ways and 24 different weighting functions, we neither show the results of all cue orderings nor of all weighting functions. Instead, we illustrate the trends by showing significant subsets of combinations of functions and cue orderings. Overall, these are similar to the results of the 3- and 4-cue cases. First, for non-compensatory functions (NonCF) as well as lower levels of compensatory functions (CF1 to CF17), TTB performs best provided the most important cue enters the model first. When this does not occur, EW/TTB performs better. Second, EW/TTB is dominant across all cue orderings for the most compensatory set of weighting functions (CF23). Third, DOMRAN is superior to many of the combinations of cue orders and models where the cue orderings are inappropriate (with the exception of EW/TTB). And fourth, MINIMALIST (the mean of TTB across cue orderings) is inferior to EW which is equal to the mean of EW/TTB.

On Generalized Secretary Problems

J. Neil Bearden and Ryan O. Murphy

University of Arizona, Department of Management and Organizations,
405 McClelland Hall, Tucson, AZ 85721, USA
e-mail: jneilb@gmail.com

Abstract. This paper is composed of two related parts. In the first, we present a dynamic programming procedure for finding optimal policies for a class of sequential search problems that includes the well-known "secretary problem". In the second, we propose a stochastic model of choice behavior for this class of problems and test the model with two extant data sets. We conclude that the previously reported bias for decision makers to terminate their search too early can, in part, be accounted for by a stochastic component of their search policies.

Keywords: sequential search, secretary problem, optimization

1 Introduction and Overview

The secretary problem has received considerable attention by applied mathematicians and statisticians (e.g., Ferguson, 1989; Freeman, 1983). Their work has been primarily concerned with methods for determining optimal search policies, the properties and implications of those policies, and the effects of introducing constraints on the search process (e.g., by adding interview costs). More recently, psychologists and experimental economists have studied how actual decision makers (DMs) perform in these sorts of sequential search tasks (e.g., Bearden, Rapoport and Murphy, 2004; Corbin, et al. 1975; Seale and Rapoport, 1997, 2000; Zwick, et al. 2003).

The current paper is composed of two main parts. First, we present a procedure for computing optimal policies for a large class of sequential search problems that includes the secretary problem. It is hoped that the accessibility of this procedure will encourage additional experimental work with this class of search problems. Second, we present a descriptive model of choice for the search problems, describe some of its properties, and test the model with two extant data sets. We conclude with a cautionary note on the difficulties researchers may face in drawing theoretical conclusions about the cognitive processes underlying search behavior in sequential search tasks.

2 Secretary Problems

2.1 The Problems

The *Classical Secretary Problem* (CSP) can be stated as follows:

1. There is a fixed and known number n of applicants for a single position who can be ranked in terms of quality with no ties.
2. The applicants are interviewed sequentially in a random order (with all $n!$ orderings occurring with equal probability).
3. For each applicant j the DM can only ascertain the *relative rank* of the applicant, that is, how valuable the applicant is relative to the $j-1$ previously viewed applicants.
4. Once rejected, an applicant cannot be recalled. If reached, the nth applicant must be accepted.
5. The DM earns a payoff of 1 for selecting the applicant with *absolute rank* 1 (i.e., the overall best applicant in the population of n applicants) and 0 otherwise.

The payoff maximizing strategy for the CSP, which simply maximizes the probability of selecting the best applicant, is to interview and reject the first $t-1$ applicants and then accept the first applicant thereafter with a relative rank of 1 (Gilbert and Mosteller, 1966). Further, they proved that t converges to n/e as n goes to infinity. In the limit, as $n \to \infty$, the optimal policy selects the best applicant with probability $1/e$. The value of t and the selection probability converge from above.

Consider a variant of the secretary problem in which the DM earns a positive payoff $\pi(a)$ for selecting an applicant with absolute rank a, and assume that $\pi(1) \geq \ldots \geq \pi(n)$. Mucci (1973) proved that the optimal search policy for this problem has the same threshold form as that of the CSP. Specifically, the DM should interview and reject the first $t_1 - 1$ applicants, then between applicant t_1 and applicant $t_2 - 1$ she should only accept applicants with relative rank 1; between applicant t_2 and applicant $t_3 - 1$ she should accept applicants with relative ranks 1 or 2; and so on. As she gets deeper into the applicant pool her standards relax and she is more likely to accept applicants of lower quality.

We obtain what we call the *Generalized Secretary Problem* (GSP) by replacing **5** in the CSP, which is quite restrictive, with the more general objective function:

5'. The DM earns a payoff of $\pi(a)$ for selecting an applicant with absolute rank a where $\pi(1) \geq \ldots \geq \pi(n)$.

Clearly, the CSP is a special case of the GSP in which $\pi(1) = 1$ and $\pi(a) = 0$ for all $a > 1$. Results for other special cases of the GSP have appeared in the literature. For example, Moriguti (1993) examined a problem in which a DM's objective is to minimize the expected rank of the selected applicant. This problem is equivalent to maximizing earnings in a GSP in which $\pi(a)$ increases linearly as $(n-a)$ increases.

2.2 Finding Optimal Policies for the GSP

We will begin by introducing some notation. The orderings of the n applicants' *absolute ranks* is represented by a vector $\mathbf{a} = \left(a^1, \ldots, a^n\right)$, which is just a random permutation of the integers $1, \ldots, n$. The *relative rank* of the jth applicant, denoted r^j, is the number of applicants from $1, \ldots, j$ whose absolute rank is smaller than or equal to a^j. A *policy* is a vector $\mathbf{s} = \left(s^1, \ldots, s^n\right)$ of nonnegative integers in which $s^j \leq s^{j+1}$ for all $1 \leq j < n$. The policy dictates that the DM stop on the first applicant for which $r^j \leq s^j$. Therefore, the probability that the DM stops on the jth applicant, conditional on reaching this applicant, is s^j/j; we will denote this probability by Q^j. A DM's *cutoff* for selecting an applicant with a relative rank of r, denoted t_r, is the smallest value j for which $r \leq s^j$. Hence, a policy $\mathbf{s}$ can also be represented by a vector $\mathbf{t} = (t_1, \ldots, t_n)$. Sometimes, the cutoff representation will be more convenient. Again, a DM's *payoff* for selecting an applicant with absolute rank a is given by $\pi(a)$.

Given the constraint on the nature of the optimal policy for the GSP proved by Mucci (1973), optimal thresholds can be computed straightforwardly by combining numerical search methods with those of dynamic programming. We will describe below a procedure for doing so. A similar method was outlined in Lindley (1961) and briefly described by Yeo and Yeo (1994).

The probability that the jth applicant out of n whose relative rank is r^j has an absolute (overall) rank of a is given by (Lindely, 1961):

$$Pr\left(A = a | R = r^j\right) = \frac{\binom{a-1}{r-1}\binom{n-a}{j-r}}{\binom{n}{j}}, \tag{1}$$

when $r^j \leq a \leq r^j + (n-j)$; otherwise $Pr(A = a|R = r^j) = 0$. Thus, the expected payoff for selecting an applicant with relative rank r^j is:

$$E\left(\pi^j | r^j\right) = \sum_{a=r^j}^{n} Pr\left(A = a | R = r^j\right) \pi(a). \tag{2}$$

The expected payoff for making a selection at stage j for some stage j policy $s^j > 0$ is:

$$E\left(\pi^j | s^j\right) = \left(s^j\right)^{-1} \sum_{i=1}^{s^j} E\left(\pi^j | r^j = i\right); \tag{3}$$

otherwise, when $s^j = 0$, $E\left(\pi^j | s^j\right) = 0$. Now, denoting the expected payoff for starting at stage $j+1$ and then following a fixed threshold policy $(s^{j+1}, \ldots, s^n)$ thereafter by v^{j+1}, the value of v^j for any $s^j \leq j$ is simply:

$$v^j = Q^j E\left(\pi^j | s^j\right) + \left(1 - Q^j\right) v^{j+1}. \tag{4}$$

Since the expected earnings of the optimal policy at stage n are $v^n = n^{-1}\sum_{a=1}^{n}\pi(a)$, we can easily find an s^j for each j ($j = n-1,\ldots,1$) that maximizes v^j by searching through the feasible s^j; the expected earnings of the optimal threshold s^{j*} we denote by v^{j*}. These computations can be performed rapidly, and the complexity of the problem is just linear in n[1]. From the monotonicity constraint on the s^j, the search can be limited to $0 \leq s^j \leq s^{j+1}$. Thus, given v^{n*}, starting at stage $n-1$ and working backward, the dynamic programming procedure for finding optimal policies for the GSP can be summarized by:

$$s^{j*} = \arg \max_{s\in\{0,\ldots,s^{j+1*}\}} v^j \,. \tag{5}$$

The expected payoff for following a policy $\mathbf{s}$, then, is:

$$E\left(\pi|\mathbf{s}\right) = \sum_{j=1}^{n}\left[\prod_{i=0}^{j-1}\left(1-Q^i\right)\right]Q^j E\left(\pi^j|s^j\right) = v^1 , \tag{6}$$

where $Q^0 = 0$. The optimal policy $\mathbf{s}^*$ is the policy $\mathbf{s}$ that maximizes Eq. 6. Denoting the applicant position at which the search is terminated by m, the probability that the DM stops on the $(j < n)$th applicant is:

$$Pr\left(m = j\right) = \left[\prod_{i=0}^{j-1}\left(1-Q^i\right)\right]Q^j , \tag{7}$$

and the expected stopping position is (Moriguti, 1993):

$$E\left(m\right) = 1 + \sum_{j=1}^{n-1}\left[\prod_{i=1}^{j}\left(1-Q^i\right)\right] . \tag{8}$$

Optimal cutoffs for several GSPs are presented in Table 1. In the first column, we provide a shorthand for referring to these problems. The first one, GSP1, corresponds to the CSP with $n = 40$. The optimal policy dictates that the DM should search through the first 15 applicants without accepting any and then accept the first one thereafter with a relative rank of 1. GSP2 corresponds to another CSP with $n = 80$. In both, the DM should search through roughly the first 37% and then take the first encountered applicant with a relative rank of 1. These two special cases of the CSP have been studied experimentally by Seale and Rapoport (1997). GSPs 3 and 4 were discussed in Gilbert and Mosteller (1966), who presented numerical solutions for a number of problems in which the DM earns a payoff of 1 for selecting either the best or second best applicant and nothing otherwise. GSPs 5 and 6 correspond to those studied by Bearden, Papoport and Murphy (2004) in Experiments 1 and 2,

[1] More elegant solutions can be used for special cases of the GSP. The method described here can be easily implemented for all special cases of the GSP.

respectively. In the first, the DM searches through the first 13 applicants without accepting any; then between 14 and 28 she stops on applicants with relative rank of 1; between 29 and 36, she takes applicants with relative rank 1 or 2; etc. Finally, GSP7 corresponds to the rank-minimization problem studied by Moriguti (1993). The results of our method are in agreement with all of those derived by other methods.

When inexperienced and financially motivated decision makers are asked to play the GSP in the laboratory, they have no notion of how to compute the optimal policy. Why then should one attempt to test the descriptive power of the optimal policy? One major reason is that tests of the optimal policies for different variants of the GSP (e.g. Bearden, Rapoport and Murphy, 2004; Seale and Rapoport, 1997, 2000; Zwick, et al. 2003) may provide information on the question of whether DMs search too little, just enough, or too much. This question has motivated most of the research in sequential search in economics (e.g., Hey, 1981, 1982, 1987) and marketing (e.g., Ratchford and Srinivasan, 1993; Zwick, et al.). However, tests of the optimal policy do not tell us what alternative decision policies subjects may be using in the GSP. And because they prescribe the same fixed threshold values for all subjects, they cannot account for within-subject variability across iterations of the sequential search task or between-subject variability in the stopping behavior.

Seale and Rapoport (1997, 2000) have proposed and tested three alternative decision policies in their study of two variants of the CSP. These decision policies (descriptive models) are not generalizable in their present form to the GSP. Moreover, because all of them are deterministic, they cannot account for within subject variability in stopping times across trials. Rather than attempting to construct more complicated deterministic choice models for the GSP, with a considerable increase in the number of free parameters, we propose an alternative stochastic model of choice for the GSP. Next, we describe the model and discusses its main properties. Then we summarize empirical results from some previous studies of the GSP and use them to test the model. Finally, we conclude by discussing some problems that arise in drawing theoretical conclusions about choice behavior in the GSP and related sequential search tasks.

3 A Stochastic Model of Choice in Secretary Problems

3.1 Background

Stochastic models have a long history in psychological theories. As early as 1927, L.L. Thurstone posited that observed responses are a function of an underlying (unobservable) component together with random error (Thurstone, 1927a, 1927b). For reviews of the consequences of Thurstone's ideas, see Bock and Jones (1968) and Luce (1977, 1994).

Table 1. Optimal policies for several GSPs

GSP	n	$\pi = (\pi(1), \ldots, \pi(n))$	$\mathbf{t}^* = (t_1^*, \ldots, t_n^*)$	$E(\pi\|\mathbf{s}^*)$	$E(m)$
1	40	$(1, 0, \ldots, 0)$	$(16, 40, \ldots, 40)$	.38	30.03
2	80	$(1, 0, \ldots, 0)$	$(30, 80, \ldots, 80)$	.37	58.75
3	20	$(1, 1, 0, \ldots, 0)$	$(8, 14, 20, \ldots, 20)$	.69	14.15
4	100	$(1, 1, 0, \ldots, 0)$	$(35, 67, 100, \ldots, 100)$	.58	68.47
5	40	$(15, 7, 2, 0, \ldots, 0)$	$(14, 29, 37, 40, \ldots, 40)$	6.11	27.21
6	60	$(25, 13, 6, 3, 2, 1, 0, \ldots, 0)$	$(21, 43, 53, 57, 58, 59, 60, \ldots, 60)$	12.73	41.04
7	25	$(25, 24, 23, \ldots, 1)$	$(8, 14, 17, 19, 21, 22, 23, 23, 24, 24, 24, 25, \ldots, 25)$	22.88	14.46

More recently, theorists have shown that unbiased random error in judgment processes can produce seemingly biased judgments. For example, Erev, et al. (1994) have shown that symmetrically distributed random error can produce confidence judgments consistent with overconfidence even when the underlying (unperturbed) judgments are well-calibrated (see also, Juslin, et al. 1997; Pfeifer, 1994; Soll, 1996).

In related work, Bearden, Wallsten and Fox (2004) have shown that unbiased random error in the judgment process is sufficient to produce subadditive judgments. Suppose we have an event X that can be partitioned into k mutually exclusive and exhaustive subevents $X = \bigcup_{i=1}^{k} X_i$. Denote a judge's underlying (or true) probability estimate for X by $C(X)$ and her overt expression of the probability of X by $R(X)$. Bearden et al. assumed that $R(X) = f(C(X), e)$, where e is a random error component that is just as likely to be above as below $C(X)$. They proved that under a range of conditions $R(X)$ is regressive, i.e., it will be closer than $C(X)$ to .50. As a result, the overt judgment for X can be smaller than the sum of the judgments for the X_i, even when $C(X) = \sum_i C(X_i)$. Put differently, the overt judgments can be subadditive even when the underlying judgments are themselves additive. A considerable body of research has focused on finding high-level explanations such as availability for subadditive judgments (e.g., Rottenstreich and Tversky, 1997; Tversky and Koehler, 1994). Bearden et al. simply demonstrated that unbiased random error in the response process is sufficient to account for the seemingly biased observed judgments. One need not posit higher-level explanations. We follow this line of research and look at the effects of random error in the GSP.

Empirical research on the GSP has consistently shown that DMs exhibit a bias to terminate their search too soon (Bearden, Rapoport and Murphy, 2004; Seale and Rapoport, 1997, 2000). At the level of description, this observation is undeniable. However, researchers have gone beyond this observation by offering psychological explanations to account for the bias. In a paper on the CSP, Seale and Rapoport (1997) suggested that the bias results from an endogenous search cost: Because search is inherently costly (see, Stigler, 1961), the DM's payoff increases in the payoff she receives for selecting the best applicant but decreases in the amount of time spent searching. Therefore, early stopping may be the result of a (net) payoff maximizing strategy. Bearden, Papoport and Murphy (2004) offered a different explanation. They had DMs estimate the probability of obtaining various payoffs for selecting applicants of different relative ranks in different applicant positions. Based on their findings, they argued that the bias to terminate the search too soon in a GSP results from DMs overestimating the payoffs that would result from doing so.

In Sect. 3.2 we present a simple stochastic model of search in the secretary problem and show that it produces early stopping behavior even when DMs use decision thresholds that are *symmetrically* distributed about the optimal thresholds.

3.2 The Model

Recall that under the optimal policy for the GSP, the DM stops on some applicant j if and only if the applicant's relative rank does not exceed the DM's threshold for that stage (i.e., when $r^j \leq s^{j*}$). Experimental results, however, conclusively show that DMs do not strictly adhere to a deterministic policy of this sort. Rather, we posit that DMs' thresholds can be modelled as random variables. Each time the DM experiences an applicant with a relative rank r, she is assumed to sample a threshold from her distribution of thresholds for applicants with relative rank r; then, using the sampled threshold, she makes a stopping decision[2]. Denoting the sampled threshold σ_r, she stops on an applicant with relative rank r^j if and only if $r^j \leq \sigma_r$. (Note that at each stage j, the DM samples from a distribution that depends on the relative rank of the applicant observed at that stage. The distribution *is not* conditional only on the stage; it is only conditional on the relative rank of the observed applicant at that stage.) We assume that the probability density function for the sampled threshold is given by:

$$f\left(\sigma_r\right) = \frac{e^{-(\sigma_r-\mu_r)/\beta_r}}{\beta_r\left[1+e^{-(\sigma_r-\mu_r)/\beta_r}\right]^2} . \tag{9}$$

Consequently, conditional on being reached, the probability that an applicant with relative rank r^j is selected is:

$$Pr\left(r^j \leq \sigma_r\right) = \frac{1}{1+e^{-(j-\mu_r)/\beta_r}} . \tag{10}$$

We assume that $\mu_1 \leq \ldots \leq \mu_n$ and $\beta_1 \geq \ldots \geq \beta_n$. This is based on the constraint of the GSP that payoffs are nonincreasing in the absolute rank of the selected applicant. Hence, it seems reasonable to assume that $Pr\left(r^j \leq \sigma_r\right) \geq Pr\left(r'^j \leq \sigma_{r'}\right)$ whenever $r \leq r'$. That is, the DM should be more likely to stop on any given j whenever the relative rank of the observed applicant decreases. The constraints on the ordering of μ and β do not guarantee this property but do encourage it[3].

Note that the model approaches a deterministic model as $\beta_r \to 0$ for each r. Further, the optimal policy for an instance of a GSP obtains when β_r is small (near 0) and $t_r^* - 1 < \mu_r < t_r^*$ for each r.

Examples of the distributions of thresholds and resulting stopping probabilities for a possible DM are exhibited in Fig. 1 for the GSP2 (i.e., for

[2] The thresholds are, of course, unobservable. The model specified here as an as-if one: We are merely suggesting that the DM's observed behavior is in accord with her acting as if she is randomly sampling thresholds subject to the constraints of the model we propose.

[3] Adding the strong constraint that $Pr\left(r^j \leq \sigma_r\right) \geq Pr\left(r'^j \leq \sigma_{r'}\right)$ for all $r \leq r'$ makes dealing with the model too difficult. The numerical procedures used below to derive maximum likelihood estimates of the model's parameters from data would be infeasible under the strong constraint.

Secretary Problem

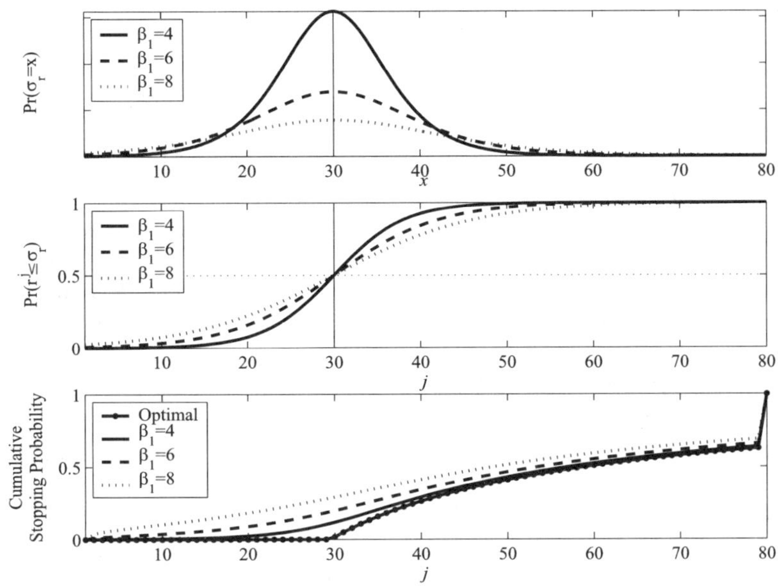

Fig. 1. Hypothetical threshold distributions and resulting stopping probabilities (conditional and cumulative) for one of the GSPs (GSP2) studied by Seale and Rapoport (1997) for various values of β. These results are based on $\mu_1 = t_1^*$. The *cumulative stopping probabilities* under the optimal policy are also shown in the bottom panel

a CSP with $n = 80$). In all cases shown in the figure, $\mu_1 = t_1^*$; that is, all of the threshold distributions are centered at the optimal cutoff point for the problem. The top panel shows the pdf of the threshold distribution. The center panel shows that for $j < \mu_1$ the probability of selecting a candidate (i.e., an applicant with a relative rank of 1) increases as β increases; however, for $j > \mu_1$, the trend is reversed. The bottom panel shows the probability of stopping on applicant j or sooner for the model and also for the optimal policy. Most importantly, in this example we find that the propensity to stop too early increases as the variance of the threshold distribution (β) increases, and in none of the model instances do we observe late stopping.

Under the model, the probability that the DM stops on the $(j < n)$th applicant, given that she has reached him, is:

$$\hat{Q}^j = \sum_{r^j=1}^{j} \frac{1}{j} Pr\left(r^j \leq \sigma_r\right) . \tag{11}$$

Replacing Q^j in Eq. 8 with $\hat{Q}^j$, we can easily compute the *model expected stopping position.* Some examples of model expected stopping positions for various values of μ_1 and β_1 for the GSP2 are presented in Table 2. Several features of the $E(m)$ are important. First, whenever $\mu_1 < t_1^*$, the expected stopping position under the model is smaller than the expectation under the optimal policy. Second, even when $\mu_1 \geq t_1^*$ and β_1 is non-negligible, we find that the model tends to stop sooner than the optimal policy. Also, when $t_1^* - 1 < \mu_1 < t_1^*$ (that is, when the mean of the model threshold distribution is just below the optimal cutoff), the expected stopping position under the model is *always* less than under the optimal. Finally, as β increases, the expected stopping position decreases. In other words, as the variance of the threshold distribution increases, the model predicts that stopping position move toward earlier applicants. This general pattern of results obtains for the other GSPs as well.

The optimal policies for the GSP are represented by integers, but we are proposing a model in which the thresholds are real valued (and can even be negative); hence, some justification is in order. Using Eq. 10 to model choice probabilities has a number of desirable features. First, we can allow for shifts in both the underlying thresholds (or the means of the threshold distributions) by varying μ_r, and we can control the steepness of the response function about a given μ_r by β_r. As stated above, this can (in the limit) allow us to model both deterministic policies and noisy policies. The logistic distribution was chosen for its computational convenience (its CDF can be written in closed form); we have tried other symmetric distributions (e.g., the normal) and reached roughly the same conclusions that we report here for the logistic. (Actually, the tails of the normal distribution tend to be insufficiently fat to well-account for the empirical data.) Again, we desire a distribution with a symmetric PDF to model the thresholds in order for the thresholds to be *unbiased*. Empirical data show that DMs in secretary search tasks tend to terminate their search too early. We wish to demonstrate that this may result from an essentially unbiased stochastic process.

Table 2. Expected stopping times under the model for the GSP2 for different values of β_1 and μ_1. Keep in mind that $E(m) = 58.75$ under the optimal policy and $t_1^* = 30$. The average value of m for this problem in Seale and Rapoport (1997) is 43.61

β_1	$E(m\|\mu_1 = 25)$	$E(m\|\mu_1 = 29.5)$	$E(m\|\mu_1 = 30)$	$E(m\|\mu_1 = 35)$
.01	53.83	58.74	59.24	63.80
1	53.72	58.65	59.15	63.73
2	53.29	58.32	58.83	63.48
4	51.08	56.72	57.29	62.33
8	41.46	48.54	49.27	55.94
10	36.63	43.55	44.29	51.27
12	32.62	39.03	39.74	46.59
16	26.86	32.04	32.63	38.57

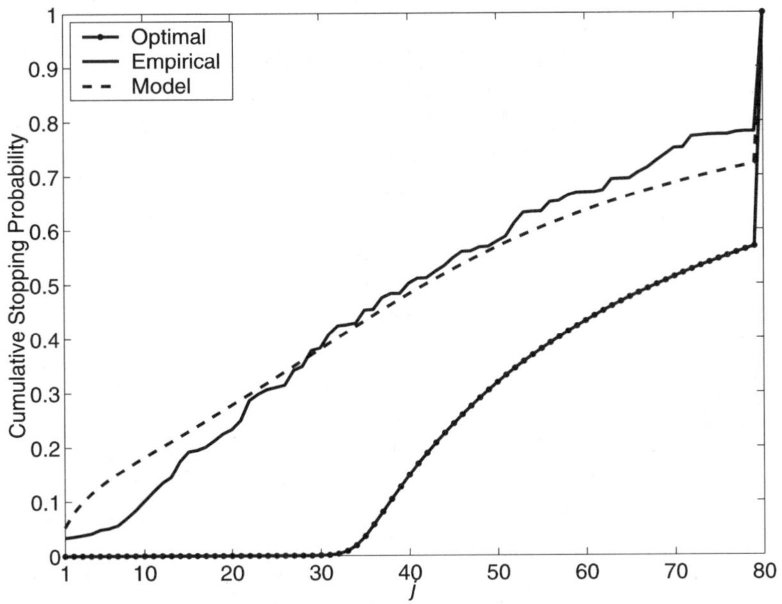

Fig. 2. Cumulative stopping probabilities for the GSP2 for the optimal and stochastic model policies and also for the empirical data reported by Seale and Rapoport (1997). The model probabilities are based on $\mu_1 = t_1^* = 30$ and $\beta_1 = 10$

Fig. 2 portrays optimal, empirical, and model cumulative stopping probabilities for the instance of the GSP that was studied empirically by Seale and Rapoport (1997). First, note that the empirical curve is shifted to the left of the optimal one. This indicates that DMs tended to stop earlier than dictated by the optimal policy. The model stopping probabilities are based on $\mu_1 = t_1^* = 30$, that is, the mean of the distribution from which the thresholds were sampled is set equal to the value of the optimal threshold. However, the model stopping probabilities are also shifted in the direction of stopping early. This is an important observation: In this example, the stochastic thresholds are distributed symmetrically about the optimal threshold and stopping behavior is biased toward early stopping. For example, it is just as likely that a DM's threshold will be 4 units above as below the optimal threshold, corresponding to too early and a too late thresholds, respectively; yet stopping behavior is biased toward early stopping.

The reason for early stopping under the stochastic model can be stated quite simply. First, there is a nonzero probability that a DM will stop sometime before it is optimal to do so; as a consequence, she will not have the opportunity to stop on time or stop too late. Secondly, though the threshold distribution itself is symmetric, the unconditional stopping probabilities are not. The probability of observing a given relative rank $r^j \leq j$ decreases in j. Consider $r^j = 1$. When $j = 1$, the probability of observing a relative rank of 1 is 1; when $j = 2$, the probability is 1/2; and in general it is $1/j$. Thus, for a given σ_r, the probability of stopping on applicant j is strictly decreasing in j. Therefore, properties of the problem itself can entail early stopping under the model. Researchers should, therefore, be cautious in attributing early stopping to general psychological biases.

Thus far we have only discussed the theoretical consequences of the stochastic model. Next, we evaluate the model using some of the empirical data reported in Seale and Rapoport (1997) and in Bearden, Papoport and Murphy (2004). We ask: Can the observed early stopping in these experiments be explained by unbiased stochastic thresholds?

3.3 Parameter Estimation

We estimated the model parameters for the stochastic choice model for individual subjects from two previous empirical studies of the GSP. Seale and Rapoport (1997) had 25 subjects play the GSP2 for 100 trials under incentive-compatible payoffs. They reported that their subjects exhibited a tendency to terminate their searches too early, and explained this by a deterministic cutoff rule of the same form as the optimal policy but whose cutoff was shifted to the left of the optimal cutoff. They evaluated alternative deterministic decision policies and concluded that the alternatively parameterized cutoff rule best accounted for the data. To determine a subject's cutoff – t_1, in our notation – they found the value of $1 \leq t_1 \leq 80$ that maximized the number of selection decisions compatible with the cutoff. For the GSP2, $t_1^* = 30$; Seale and Rapoport estimated that the modal cutoff for their subjects was 21.

Bearden, Rapoport and Murphy (2004) had 61 subjects perform the GSP6 for 60 trials under incentive-compatible payoffs. They, too, concluded that their subjects terminated search too early, and that the stopping behavior was most compatible with a threshold stopping rule. For the GSP6, $t_1^* = 21$, $t_2^* = 43$, $t_3^* = 53$, $t_4^* = 57$, $t_5^* = 58$, and $t_6^* = 59$; for their subjects, they estimated that the mean thresholds were $t_1 = 12$, $t_2 = 22$, $t_3 = 28$, $t_4 = 35$, $t_5 = 40$, and $t_6 = 44$. In both Seale and Rapoport and Bearden et al., the authors (implicitly) assumed that the subjects used deterministic or fixed thresholds. Hence, for a given subject, they could not account for stopping decisions inconsistent with that subject's estimated threshold.

In the current paper, we assume that the subjects' thresholds are random variables (whose pdf is given by Eq. 9) and use maximum likelihood procedures to estimate the parameters of the distribution from which the thresholds are sampled. For each set of data that we examine, the researchers reported learning across early trials of play, but in both, the choice behavior seems to have stabilized by the 20th trial. Hence, for the tests below, we shall eliminate the first 20 trials from each data set from the analyses, and we will assume that the choice probabilities are i.i.d.

For a given trial of a GSP problem, the DM observes a sequence of applicants and their relative ranks, and for each applicant she decides to either accept or continue searching. Denoting a decision function for applicant j by $\delta\left(r^j\right)$, we let $\delta\left(r^j\right)=0$ if the DM *does not stop* on applicant j and $\delta\left(r^j\right)=1$ if she *does stop*. Hence, decisions for a particular trial k can be represented by a vector $\Delta^k=\left(\delta\left(r^1\right),\ldots,\delta\left(r^m\right)\right)=(0,0,\ldots,1)$, where m denotes the position of the selected applicant. Under the stochastic model, if $m<n$, the likelihood of Δ^k can be written as:

$$L\left(\Delta^k|\mu,\beta\right)=\left[\prod_{i=1}^{m-1} Pr\left(r^i>\sigma_r\right)\right] Pr\left(r^m\leq\sigma_r\right). \tag{12}$$

When $m=n$ (i.e., when the DM reaches the last applicant, which she *must* accept), we simply omit the final term in Eq. 12 since the DM's choice is determined. Assuming independence, the likelihood of a DM's choice responses over K trials of the GSP is just:

$$L\left[\left(\Delta^1,\ldots,\Delta^K\right)|\mu,\beta\right]=\prod_{k=1}^{K} L\left(\Delta^k|\mu,\beta\right). \tag{13}$$

Due to the small numbers involved, it is convenient to work with the log of the likelihood, rather than the likelihood itself. Taking the log of Eq. 13, we get:

$$\ell\left[\left(\Delta^1,\ldots,\Delta^K\right)|\mu,\beta\right]=\sum_{k=1}^{K}\ln\left[L\left(\Delta^k|\mu,\beta\right)\right]. \tag{14}$$

For each subject we computed the parameters μ and β that maximized Eq. 14 under different constraints. We only estimated the parameters for relative ranks that can entail positive payoffs. For the GSP2, we restrict estimates to $r=1$, and for the GSP6 to $1\leq r\leq 6$. Therefore, we omit from the analyses trials on which the DM chose to stop on the applicants whose relative rank could not entail a positive payoff. Very likely these were errors. Fewer than 2% of the trials were omitted.

We are primarily interested in testing the following:

Optimal but stochastic threshold hypothesis: $\mu_r=t_r^*$ for all r.

If this hypothesis is supported, the bias toward early stopping behavior could be the result of the stochastic nature of the thresholds. We evaluate the optimal but stochastic threshold hypothesis (OBSTH) using standard likelihood ratio tests. Under the *constrained model*, we impose that $\mu_r=t_r^*$ for all r and allow the β_r to freely vary; under the *unconstrained model* we allow both the μ_r and β_r to freely vary. Denoting the maximum log-likelihood of the constrained model ℓ_c (based on Eq. 15) and of the unconstrained model ℓ_u, the likelihood ratio is:

$$LR=\left(\ell_c-\ell_u\right). \tag{15}$$

The statistic $-2LR$ is χ^2 distributed with degrees of freedom (*df*) equal to the number of additional free parameters in the unconstrained model. Hence, for the Seale and Rapoport (1997), $df = 1$; and for Bearden, Rapoport and Murphy (2004), $df = 6$.

A few words about estimating the model parameters are in order. To estimate the model parameters we used a constrained optimization procedure (fmincon) in Matlab. We imposed the constraint that $\mu_r \leq \mu_{r'}$ whenever $r \leq r'$, and imposed the corresponding constraint on the β parameters. For each subject, we used a large number of initial starting values. We are confident that the estimated parameters provide globally optimal results for each subject.

Seale and Rapoport Data

Based on the likelihood ratio test with $df = 1$, the OBSTH could not be rejected for 12 of the 25 experimental subjects at the $\alpha = .01$ level. Seale and Rapoport concluded that 21 of their 25 subjects had thresholds below the optimal cutoff. Our analyses suggest that they overestimated the number of subjects with biased thresholds. Fig. 3 shows a distribution of thresholds

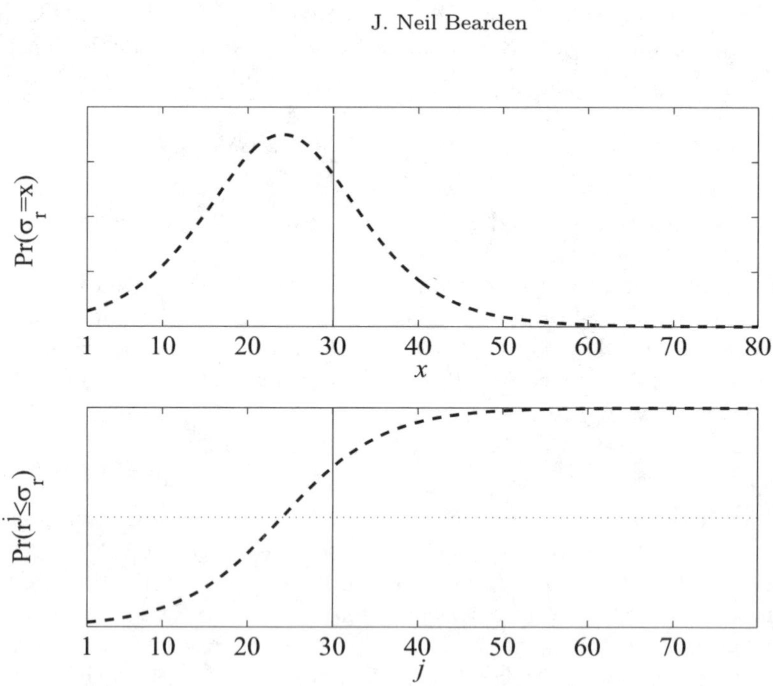

Fig. 3. Estimated threshold distribution and resulting stopping probabilities for the $n = 80$ CSP studied by Seale and Rapoport (1997) based on median estimated μ_1 and σ_1. The horizontal line is located at the optimal cutoff point ($t_1^* = 30$. The vertical line in the bottom panel corresponds to a probability of .50

(σ_1) that is based on the median estimated values of μ_1 and β_1 from the 25 experimental subjects. We find that the distribution of thresholds (based on the aggregate data) is, indeed, shifted to the left of the optimal cutoff, consistent with the observed early stopping behavior. Further, we find that the variance of the threshold distribution is considerably greater than 0. Thus, early stopping in Seale and Rapoport may be due both to thresholds that tend to be biased toward early stopping and also to stochastic variability in placement of the thresholds. Summary statistics from the MLE procedures are displayed in Table 3.

Bearden, Rapoport, and Murphy Data

The corresponding thresholds from Bearden, Rapoport and Murphy (2004) are displayed in Fig. 4. For these data, the OBSTH could not be rejected for 23 of the 61 subjects (i.e., for 37%). We find that the distribution of thresholds for $r = 1$ tends to be centered rather close to the optimal cutoff. Likewise for the $r = 6$ threshold. For $r = 2, \ldots, 5$, the thresholds tend to be shifted toward early stopping. The variances of the threshold distributions tend to decrease quite rapidly in r, but are all away from 0. Thus, as with the Seale and Rapoport (1997) data, the early stopping in the GSP6 seems to be driven by biased thresholds as well as the stochastic nature of those thresholds. Summary results are presented in Table 3.

Table 3. Summary of MLE results for Seale and Rapoport (n=80) condition and Bearden, Rapoport, and Murphy Experiment 1 data. Note: OBSTH compatible tests are based on $\alpha = .01$

Seale & Rapoport (1997) Data	
Number of subjects	25
Median μ	(24.08)
Median β	(5.97)
Median LR	4.19
Test df	1
OBSTH compatible	48%
Bearden, Rapoport, & Murphy (2004) Data	
Number of subjects	62
Median μ	(23.16, 34.71, 43.96, 48.70, 54.49, 58.53)
Median β	(4.13, 3.56, 2.49, 1.24, 0.69, 0.43)
Median LR	18.38
Test df	6
OBSTH compatible	37%

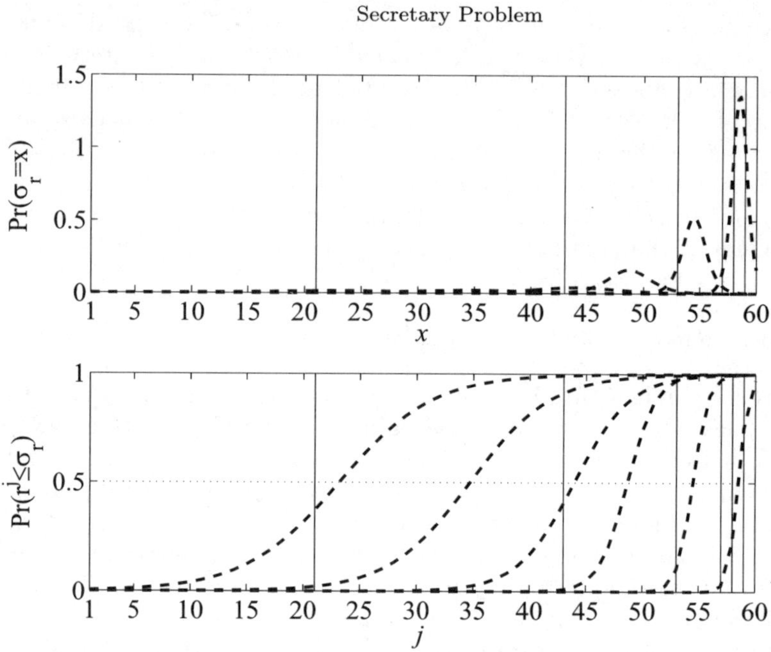

Fig. 4. Estimated threshold distribution and resulting stopping probabilities for the GSP6 studied by Bearden, Rapoport, and Murphy (2004). The curves are based on median estimated μ_r and σ_r ($r = 1, \ldots, 6$), and are ordered from left ($r = 1$) to right ($r = 6$). Note that the variances of the $Pr\,(\sigma_r = x)$ distributions for $r = 1, 2, 3$ relative to the variance of the $r = 6$ distribution are quite small, making the resulting distributions rather flat and difficult to see

The estimation results suggest that researchers should be cautious in drawing conclusions about the underlying causes of early stopping in GSPs without taking random error into account. A straightforward question must be addressed before any claims are made: What does it mean for subjects to be biased to stop early? Is the statement merely an empirical one that describes that observed stopping behavior or does it have some theoretical import? Does the "bias" refer to a property of the choice process? Seale and Rapoport (1997) suggested that the subjects in their task seemed to follow cutoff policies that were of the same form as the optimal policy but were parameterized differently. Specifically, the cutoffs for the experimental subjects tended to be positioned earlier than the optimal cutoff. They suggested that the shift might be a compensation for endogenous search costs. Our results suggest, however, that the threshold may not have been biased toward early stopping for nearly 50% of the subjects in their $n = 80$ condition. For these subjects, stochastic thresholds centered at the optimal cutoff can account for the early stopping. Likewise, for roughly 37% of the subjects in Experiment 1 of Bearden, Rapoport and Murphy (2004), we can account for early stopping by the OBSTH.

We do not argue that early stopping is not driven by some genuine choice or judgment bias (e.g., by overestimating the probability of obtaining good payoffs for selecting early applicants). Rather, we simply wish to demonstrate that the effects of random error should be taken into consideration before drawing sharp conclusions about the magnitude of the effects of these potential biases on the stopping behavior.

4 Conclusions

We began this paper by presenting a simple dynamic programming procedure for computing optimal policies for a large class of sequential search problems with rank-dependent payoffs. The generality of the permissible payoff schemes allows a number of realistic (especially in contrast to the CSP, which has an only-the-best payoff scheme) search problems to be modelled.

Next, we described a simple stochastic model of choice behavior for the GSP and described some previous experimental results. The empirical results show that DMs tend to terminate their search too early relative to the stopping positions dictated by the optimal policy. Previous explanations for this finding have invoked endogenous search costs (Seale and Rapoport, 1997) and probability overestimation (Bearden, Rapoport and Murphy, 2004) as explanations. Our results suggest that at least part of the observed early stopping *can* be explained by unbiased stochastic variability in stopping thresholds.

Future research should contrast the endogenous search cost and probability overestimation explanations of early stopping in generalized secretary problems. Importantly, in such tests, researchers should be cautious of the contribution of random error to the apparently biased search behavior.

Acknowledgement. We thank Amnon Rapoport for his comments and Darryl Seale for providing us with his data. We gratefully acknowledge financial support by a contract F49620-03-1-0377 from the AFOSR/MURI to the Department of Systems and Industrial Engineering and the Department of Management and Organizations at the University of Arizona.

References

Bearden, J.N., Rapoport, A., Murphy, R.O. (2004), Sequential oberservation and selection with rank-dependent payoffs: An experimental test. Unpublished manuscript.

Bearden, J.N., Wallsten, T.S., Fox, C.R. (2004), A stochastic model of subadditivity. Unpublished manuscript.

Bock, R.D., Jones, L.V. (1968), *The Measurement and Prediction of Judgment and Choice.* San Francisco: Holden-Day.

Corbin, R.M., Olson, C.R., Abbondanza, M. (1975), Context effects in optimal stopping rules. *Behavior and Human Performance*, 14, 207–216.

Erev, I., Wallsten, T.S., Budescu, D.V. (1994), Simulataneous over- and underconfidence: The role of error in judgment processes. *Psychological Review*, 101, 519–528.

Ferguson, T.S. (1989), Who solved the secretary problem? *Statistical Science*, 4, 282–296.

Freeman, P.R. (1983), The secretary problem and its extensions: A review. *International Statistical Review*, 51, 189–206.

Gilbert, J., Mosteller, F. (1966), Recognizing the maximum of a sequence. *Journal of the American Statistical Association*, 61, 35–73.

Hey, J.D. (1981), Are optimal search rules reasonable? And vice versa? *Journal of Economic Behavior and Organization*, 2, 47–70.

Hey, J.D. (1982), Search for rules of search. *Journal of Economic Behavior and Organization*, 3, 65–81.

Hey, J.D. (1987), Still searching. *Journal of Economic Behavior and Organization*, 8, 137–144.

Juslin, P., Olsson, H., Björkman, M. (1997), Brunswikian and Thurstonian origins of bias in probability assessment: On the interpretation of stochastic components of judgment. *Journal of Behavioral Decision Making*, 10, 189–209.

Lindley, D.V. (1961), Dynamic programming and decision theory. *Applied Statistics*, 10, 39–51.

Luce, R.D. (1977), Thurstone's discriminable process fifty years later. *Psychometrika*, 42, 461–489.

Luce, R.D. (1994), Thurstone and sensory scaling: Then and now. *Psychological Review*, 101, 271–277.

Moriguti, S. (1993), Basic theory of selection by relative rank with cost. *Journal of the Operations Research Society of Japan*, 36, 46–61.

Mucci, A.G. (1973), Differential equations and optimal choice problems. *Annals of Statistics*, 1, 104–113.

Pfeifer, P.E. (1994), Are we overconfident in the belief that probability forecasters are overconfident. *Organizational Behavior and Human Decision Processes*, 58, 203–213.

Ratchford, B.T., Srinivasan, N. (1993), An empirical investigation of return to search. *Marketing Science*, 12, 73–87.

Rottenstreich, Y., Tversky, A. (1997), Unpacking, repacking, and anchoring: Advances in support theory. *Psychological Review*, 104, 406–415.

Seale, D.A., Rapoport, A. (1997), Sequential decision making with relative ranks: An experimental investigation of the secretary problem. *Organizational Behavior and Human Decision Processes*, 69, 221–236.

Seale, D.A., Rapoport, A. (2000), Optimal stopping behavior with relative ranks: The secretary problem with unknown population size. *Journal of Behavioral Decision Making*, 13, 391–411.

Stein, W.E., Seale, D.A., Rapoport, A. (2003), Analysis of heuristic solutions to the best choice problem. *European Journal of Operational Research*, 51, 140–152.

Soll, J.B. (1996), Determinants of overconfidence and miscalibration: The roles of random error and ecological structure. *Organizational Behavior and Human Decision Processes*, 65, 117–137.

Stigler, G.L. (1961), The economics of information. *Journal of Political Ecnonomy*, 69, 213–225.

Thurstone, L.L. (1927a), A law of comparative judgment. *Psychological Review*, 34, 273–286.

Thurstone, L.L. (1927b), Psychophysical analysis. *American Journal of Psychology*, 38, 368–389.

Tversky, A., Koehler, D. (1994), Support theory: A nonextensional representation of subjective probability. *Psychological Review*, 101, 547–567.

Yeo, A.J., Yeo, G.F. (1994), Selecting satisfactory secretaries. *Australian Journal of Statistics*, 36, 185–198.

Zwick, R., Rapoport, A., Lo, A.K.C., Muthukrishnan, A.V. (2003), Consumer sequential search: Not enough or too much? *Marketing Science*, 22, 503–519.

Voting on Alternatives or on Criteria? The Meta-Decision on Group Decision Procedures

Niels P. Thomas

Fachgebiet Politische Ökonomie, Institut für Volkswirtschaftslehre, Darmstadt University of Technology, Residentially, 64283 Darmstadt, Germany
e-mail: thomas@vwl.tu-darmstadt.de

Abstract. For many decision problems the literature provides several types of solutions. By this, however, a preliminary meta-decision problem is created on the choice of the suitable decision procedure. This paper explores the meta-decision in the context of multicriterion group decisions, considering two alternative procedures: the group members may build individual preference orderings by solving the multicriterion problem on their own and then vote on the alternatives, or they may vote on the relative relevance of each criterion and then compose a group preference ordering with the results. Within a Monte Carlo simulation both procedures are confronted with an appropriate group welfare measure introduced by John Rawls. This paper demonstrates that voting on the alternatives in general results in higher group welfare, but voting on criteria may be superior, if the group faces fundamental value conflicts. The results are applied on the meta-decision of industrial and political decision committees.

Keywords: mcda, group decisions, meta-decisions, negotiations, simulation

1 Introduction

To an increasing degree, theoretical concepts of multicriterion decision making are being applied to complex real-life problems. A rich variety of different methods and procedures has emerged during the last decades, so that the decision maker or facilitator can typically choose between several types of solutions for each decision problem. Though, generally speaking, this methodical diversity is to be appreciated, an additional preliminary decision problem is created on the choice of the suitable decision procedure.

The issue of comparing different decision procedures is approximately as old as the scientific treatment of decision theory itself (de Borda, 1784, Condorcet, 1785), whereas comparisons between multicriterion decision methodologies can be found increasingly since the mid-1980s. Three different approaches have proven advantageous for this task:

- A predominantly descriptive comparison of different *decision procedures*, which puts emphasis on the relative strengths and weaknesses of them. "This will enable a decision analyst to perceive more clearly the choice she would be making between the two approaches and the implications of such a choice" (Belton, 1986: 8).
- A more theoretical and axiomatic comparison of the *'input' of the decision problem*: the nature of available data and preferences. This approach examines the relation between the decision method and scaling of the data, degree of uncertainty, measurability etc. and addresses their theoretical problems (e.g. Arrow and Raynaud, 1986).
- A comparison of the *'output' of the decision problem* by confronting the results of decision procedures with each other. These results are either attained from simulated decision settings (for single-criterion decisions, see Lindstädt, 1998) or from real cases. Typically, one of the decision procedures is declared the reference standard, and the others are described in terms of deviations from it (e.g. Karni et al., 1990, Zanakis et al., 1998).

In order to not only compare decision procedures, but also to make a substantial recommendation about a particular procedure, the third of these approaches can be enhanced in the following way[1]. The entire decision problem is taken as a sequence of two partial decision problems, first the choice of the decision procedure (hereinafter referred to as the *meta-decision*) and then the subsequent original decision (herein after called the *principal decision*). In this model, the meta-decision problem can be solved by means of backward induction: Those decision procedures are to be chosen, which provide the highest welfare to the decision maker(s), when applied to the principle decision problem. The structure of the entire decision problem is shown in Fig. 1.

This paper explores the above specified approach to the solution of the meta-decision problem within the context of multicriterion group decisions. Two potential decision procedures as alternatives of the meta-decision are being considered:

Typically, a small group confronted with a multicriterion decision problem proceeds a two stage process: first the group members build individual preference orderings by solving the multicriterion problem on their own, then they *vote on the alternatives.* It is known that different voting schemes may lead to different results (Nurmi, 1983) and yet no existing voting scheme guarantees collective rational and democratic results (Arrow, 1951). Nevertheless, the described decision procedure is widely used and accepted, and presumably leads to acceptable results in most cases.

However, the group members may as well *vote on the relative relevance of each criterion* and then compose a group preference ordering with help of the

[1] Parts of this paper are outlined in German in Thomas (2005).

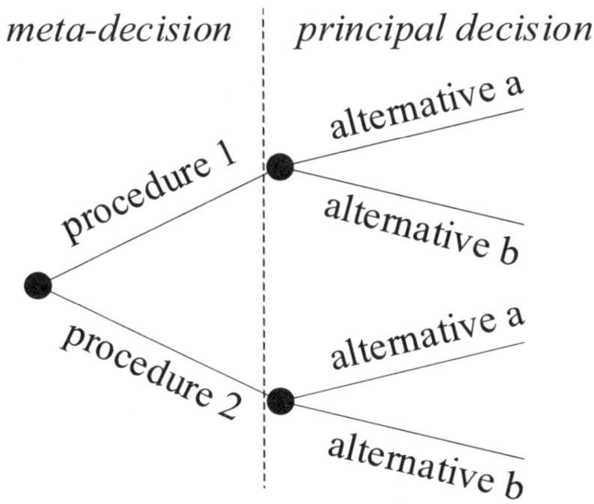

Fig. 1. Exemplary relationship between meta-decision and principal decision

results[2]. This decision procedure would replace one ballot on the alternatives by several ballots depending on the number of criteria which are considered by the group members. It is characterized by a much higher complexity in terms of transaction costs as well as cognitive effort.

In order to identify if *voting on alternatives* or *voting on criteria* is more beneficial to the decision committee in a certain situation, the following considerations have to be made: In the second section a simple multi-attribute value model is set up, from which is derived the result of *voting on alternatives* and *voting on criteria* subject to given alternatives and individual preferences. Within a Monte Carlo simulation of randomly generated decision situations, the election results of both decision procedures will be confronted with an appropriate group welfare measure, which will be introduced in Sect. 3. The results from the simulation will be presented in Sect. 4, and finally in Sect. 5 some conclusions are drawn in respect to the meta-decision problem.

2 The Principal Decision Setting

2.1 A simple multi-attribute value model

Let us consider a group of at least three persons

$$N = \{1, 2, \ldots, n\} \text{ with } n = |N| \geq 3\,, \tag{1}$$

---

[2] There are certainly several more possibilities to approach this decision problem, which will not to be examined here. See, for example, Keeney and Raiffa (1976) and Baucells and Sarin (2003).

voting on the set A of m alternatives:[3]

$$A = \{a_1, a_2, \ldots, a_m\} \text{ with } m = |A| \geq 2\,. \qquad (2)$$

For assessment of the alternatives, $c \geq 2$ criteria or attributes are considered. We will associate a level of attribute achievement q as an evaluator of each of these attributes for each of the alternatives. This permits us to characterize every alternative a_j $(1 \leq j \leq m)$ by the attribute vector

$$a_j = \begin{pmatrix} q_{j1} \\ q_{j2} \\ \ldots \\ q_{jc} \end{pmatrix} \text{ with } 0 \leq q_{jk} \leq 1\,. \qquad (3)$$

These levels of attribute achievement q are normalized, so that the worst possible achievement of the k-th attribute results in $q_{jk} = 0$, and the best possible achievement results in $q_{jk} = 1$. We assume that there exists a consensus about the attribute achievements, and that the attribute vectors are valid and accepted for every individual. Below we will only account for alternatives at the efficient frontier, so only non-dominated alternatives are considered.

An alternative a_j dominates $a_{j'}$, if $a_{j'}$ does not perform better in any attribute, but performs worse in at least one of them:

$$\forall a_j, a_{j'} \in A : a_j D a_{j'} \leftrightarrow [\forall k : q_{jk} \geq q_{j'k} \;\&\; \exists k : q_{jk} > q_{j'k}]\,. \qquad (4)$$

Dominated alternatives are excluded, because they are strictly Pareto inferior and therefore will not be evaluated as the best alternative by any individual regardless of her individual preferences.

Every individual i may then be specified by her criteria weights, which indicate the subjective relative appreciation or relevance of the k-th attribute. The row vector of all criteria weights is called the profile e_i of an individual:

$$e_i = (p_{i1}\, p_{i2} \ldots p_{ik}) \text{ with } 0 \leq p_{ik} \leq 1 \text{ and } \sum_{k=1}^{c} p_{ik} = 1\,. \qquad (5)$$

Further on we presume that an additive value function corresponds well with the preferences of each individual. To simplify matters, we presume as well a linear single value function for every attribute k, normalized on the interval $[0; 1]$. The complete value function v_{ij} then adds up to the scalar product of the attribute vector and the profile vector, the latter serving as scaling constants:

$$v_{ij} = e_i \times a_j = \sum_{k=1}^{c} p_{ik} \cdot q_{jk}\,. \qquad (6)$$

[3] On the mathematical notation see Tarski (1946) and Sen (1970).

By means of this value function every individual is able to generate her own complete, irreflexive and transitive preference ordering by sorting the set of alternatives by the corresponding value v_{ij}. Again, for reasons of simplicity, we do not take account of indifference to alternatives, but consider only those alternatives that can be put into strict preference. Of two alternatives a_j and $a_{j'}$ the one with the higher corresponding value is preferred:

$$\forall i \in N, a_j, a_{j'} \in A : a_j P_i a_{j'} \leftrightarrow [v_{ij} > v_{ij'}] . \tag{7}$$

The alternative with the highest achievable value for an individual is denoted alternative j_i^*. According to individual preferences, j_i^* may vary between individuals.

2.2 Voting on Alternatives

To solve the principal decision problem by *voting on alternatives*, the group members first build individual preference orderings according to the above presented considerations, then these are either aggregated to a complete group preference ordering, or at least the collectively most preferred alternative is extracted. Numerous voting systems or aggregation rules for this task are known and widely discussed. All of them have some strengths and weaknesses, while none satisfies all demands (Nurmi, 1983). We will focus here on the *simple majority rule*, which qualifies as the standard aggregation rule because of its wide recognition and intuitive accuracy.

The simple majority rule requires all alternatives to undergo a successive pairwise comparison. The first two alternatives a_1 and a_2 compete with each other, and of these two the alternative which is preferred by the majority of individuals, is compared to the next alternative a_3. After $m - 1$ comparisons the winner of the last comparison is declared the overall winner.

We assume that the individuals do not vote strategically, but according to their actual preferences. Further on, no individual may have the possibility to obtain an advantage by setting the voting agenda – e.g., by changing the sequence of comparisons.

2.3 Voting on Criteria

To solve the principal decision problem by *voting on criteria*, one ballot is necessary for each criterion, which adds up to a total of c elections. Within these elections, the set of alternatives under consideration consists of possible criteria weights of a collective preference profile. According to this, the number of potential alternatives is unlimited, because arbitrarily narrow differences between the alternative criteria weights may be shortlisted. Yet it is plausible to assume that, in a discussion about collectively determined criteria weights, only some exclusive focal points will be under consideration. Group members will probably express suggestions like "These criteria should determine the decision by one half each" or "The decision should depend on this attribute by ten percent only".

The assumption of a small number of d potential collective criteria weights, which are uniformly distributed, allows us to model a set of alternatives $\dot{A}$, all identical for each of the c criteria votings:

$$\dot{A} = \left\{ \dot{a}_1 = \frac{1}{2d}, \dot{a}_2 = \frac{3}{2d}, \dot{a}_3 = \frac{5}{2d}, \ldots, \dot{a}_d = \frac{2d-1}{2d} \right\} \text{ with } d \geq 2\,. \tag{8}$$

With $d = 5$ potential values the individuals may chose between criteria weights of 10%, 30%, 50%, 70% and 90%. Arrow calls this kind of synthetic alternatives 'equivalence classes', of which in general five to seven will be sufficient to adequately describe the individual's preferences (Arrow and Raynaud, 1986:10). The parameter d may be referred to as a measure of density or accuracy of the alternatives. The more alternatives are being considered, the better the individual preferences may be represented. However, each individual's effort increases too, when it is necessary to compare many similar alternative criteria weights.

For individual i regarding the k-th criteria election, the individual criteria weight p_{ik} is her ideal alternative, which may not always be contained in the set of alternatives $\dot{A}$. If not, her first preference will be the one that comes next to her ideal. The preference relation $\dot{P}_{ik}$ shows in general how an individual may chose between any two alternatives:

$$\forall i \in N, \dot{a}_x, \dot{a}_{x'} \in \dot{A} : \dot{a}_x \dot{P}_{ik} \dot{a}_{x'} \leftrightarrow [|\dot{a}_x - p_{ik}| < |\dot{a}_{x'} - p_{ik}|]\,. \tag{9}$$

Similar to the considerations in the previous chapters, the individuals may now construct a complete preference ordering on $\dot{A}$. Further on, simple majority voting results in c different collective criteria weights, which will be called $\dot{A}_k$. Once the collective criteria weights are figured out, the collective value $\dot{v}_j$ may be determined for every alternative a_j from the set of principle alternatives:

$$\dot{v}_j = \sum_{k=1}^{c} \dot{A}_k \cdot q_{jk}\,. \tag{10}$$

Though we have no reason to expect $\dot{v}_j$ being normalized to the interval $[0; 1]$, we can regard that alternative a_j, which maximizes $\dot{v}_j$, as the collective preferred alternative by *voting on criteria.*

3 A Group Welfare Measure

One of the first works in which different voting schemes are compared was published 1785 by Marquis de Condorcet. He established the idea of an "objectively best alternative" as a welfare measure, which is not chosen by every individual due to their limited power of judgement or incomplete information (Young, 1995). However, in a decision situation of complete information and rationality, as it is described above, every group member has an *individually correct* perception of the best alternative. From this point of view, no "objectively best alternative" exists, and all alternatives have to be regarded as equivalent. A different approach to obtain an appropriate welfare measure has to be explored. One possibility is to focus on the voting process as a method to make a *decision by consensus*, as Arrow (1951) points out explicitly. From that we can conclude that a good decision must always be a good consensus, which can be identified by a common consent and approval of all group members. Let us therefore derive the individual level of approval from the previous considerations.

As stated above, individual i prefers the alternative j_i^* most. The maximum achievable value v_{ij} and therefore complete satisfaction for individual i is thus obtained, if the group decides in favor of this alternative. Considering other alternatives than the individual optimum, it is plausible to define a relative level of approval z_{ij} as the ratio of the value of a certain alternative and the maximum achievable value:

$$z_{ij} = \frac{v_{ij}}{v_{ij_i^*}} \,. \tag{11}$$

This approval level, which can take a value between 0 and 1, may be considered as an indicator for individual satisfaction by an alternative. A group member facing a low approval level feels great regret about an alternative and is confronted with high opportunity costs, because changing to a different alternative may result in a much higher value of her individual value function. z is only known from an omniscient observer's perspective within the model. It is self-evident that an individual may not state the exact value of her z_{ij} in reality. Though it represents her actual approval about a certain alternative and may therefore contribute to a decision evaluation.

Let us therefore consider the average approval level $\bar{z}j$ of an alternative a_j as the group welfare measure:

$$\bar{z}_j = \frac{1}{n} \cdot \sum_{i=1}^{n} z_{ij} \,. \tag{12}$$

The average approval level may well be an indicator for the aggregated approval to a certain alternative, whereas it does not reflect the allocation of individual welfare within the group. To include this allocation, the minimum approval level $z_{j\,\min}$ of an alternative a_j may be considered. Obviously, the higher $z_{j\,\min}$, the lesser is the greatest opposition to the group decision, which can be seen as an indicator for decisions by consensus as well:

$$z_{j\,\min} = \min\left\{z_{ij} | i \in N\right\} \,. \tag{13}$$

Both indicators are reasonable, but on its own insufficient. We will therefore combine both to a group welfare measure first proposed by Rawls (1971), which is based upon the benefit of the least well-off individual. Rawls commences from a status quo and argues that from the viewpoint of social welfare an increase of individual well-being is only acceptable, if the situation of the least well-off individual is thereby improved as well. Transferring this idea to collective decision making, the choice of a_j instead of $a_{j'}$ improves group welfare, if and only if the average approval level as well as the minimum approval level increases. An increase in the average approval level is always connected with the increase of at least one individual's increase of value, whereas an increase in the minimum approval level can be interpreted as an improvement of the previous worst-off. Considering this, we can now define the following group welfare measure:

Definition. *The alternative a_{j^*} is Rawls optimal if and only if:*

$$\neg\left[\exists j : \bar{z}_j > \bar{z}_{j^*} \vee \exists j' : \ z_{j'\,\min} > z_{j^*\,\min}\right] . \tag{14}$$

According to this definition, an alternative maximizes group welfare if it combines the maximum of the average approval level as well as the maximum of the minimum approval level. If these maxima are not associated with the same alternative, Rawls optimality does not state which of these alternatives is to be preferred.

4 Analysis of Simulation Results

To solve the meta decision problem by backward induction, both voting procedures have to be applied to the principal decision situation. If *voting on alternatives* leads to higher group welfare than *voting on criteria*, then the first procedure is the better choice for the meta-decision – and vice versa. It is advisable to exercise this comparison on many principal decision situations, in order to identify their crucial characteristics. This is accomplished by a Monte Carlo simulation, where decision situations are generated randomly. A decision situation consists of alternatives (characterized by its attribute vector) and the group members' preferences (characterized by their criteria weight judgements). Five parameters can be chosen for the simulation: the group size $n \geq 3$, the number of alternatives $m \geq 2$, the number of criteria $c \geq 2$, the number of potential criteria weights $d \geq 2$ for *voting on criteria* and the number of iterations of the simulation. For reasons of clarity, only n, m and c will be varied, whereas d will be held constant at $d = 5$. All test runs described below consist of 1000 iterations.

A typical committee decision setting may consist of $1 = 5$ individuals, voting on $1 = 5$ alternatives, considering $1 = 5$ different criteria. The computer simulation leads to some interesting preliminary results. 88.6% of the decision situations feature a distinct welfare maximizing alternative in respect to the group welfare measure stated in formula 14. In 65.3% of all simulated decisions, both voting procedures obtain this best alternative. In 13% only *voting on alternatives*, and in 4.9% only *voting on criteria* is accurate. In all remaining decision situations (5.4%), both voting procedures chose an alternative that was inferior to the optimum. One could jump to the conclusion, that in most cases (82.1%), the choice of decision procedure is irrelevant, because either we do not know which alternative is best (11.4%), or both procedures lead to identical results (70.7%). However, this conclusion has to be doubted, since we have no reason to assume that alternatives or even preferences are distributed randomly in real-life decision situations. We should rather focus on the changes of the performance of voting procedures subject to changes in simulation parameters, instead of focusing on exact numerical values.

To illustrate this, let us define the *superiority rate* S of a decision procedure as the percentage of decision situations in which this decision procedure outperforms the other. For instance, a superiority rate $S_{\text{alt}} = 13\%$ states that in 13% of all simulated decision situations *voting on alternatives* is the *only* decision procedure choosing the optimal alternative. In Fig. 2, 3 and 4, the superiority rates S_{alt} and S_{crit} are shown under variation of group size n, number of alternatives m, and number of criteria c respectively.

With appropriate caution, three conclusions can be drawn from the figures above. First, the superiority rate of *voting on alternatives* S_{alt} always exceeds the superiority rate of *voting on criteria* S_{crit}. This may be interpreted as an indicator, that voting on alternatives typically leads to better results. Second, the superiority rate of *voting on criteria* is nevertheless always different from zero. We may therefore assume that real decision situations exist, in which *voting on criteria* leads to better results. Third, considering the shape of the graphs, the likelihood that *voting on criteria* outperforms *voting on alternatives* seems to increase if group size decreases and the number of alternatives increases. From Fig. 4 no correlation between the number of criteria and the performance of *voting on criteria* can be derived.

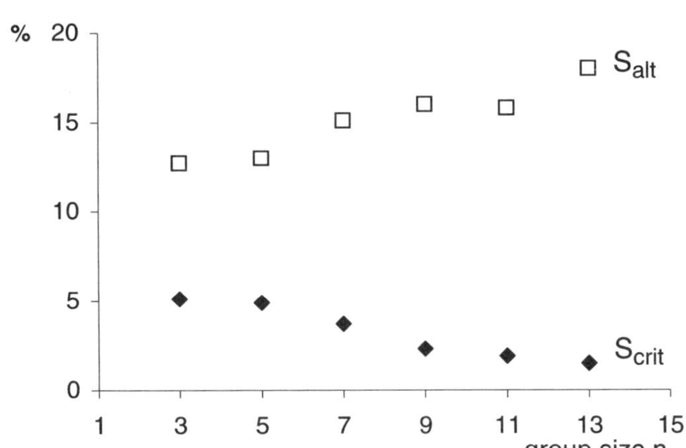

Fig. 2. Superiority rate of decision procedures – subject to group size

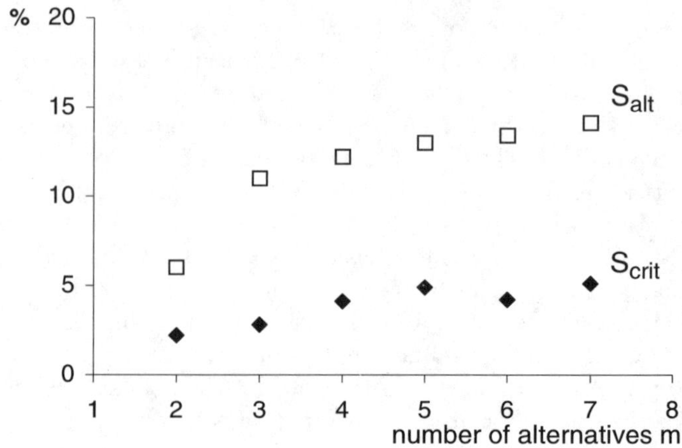

Fig. 3. Superiority rate of decision procedures – subject to the number of alternatives

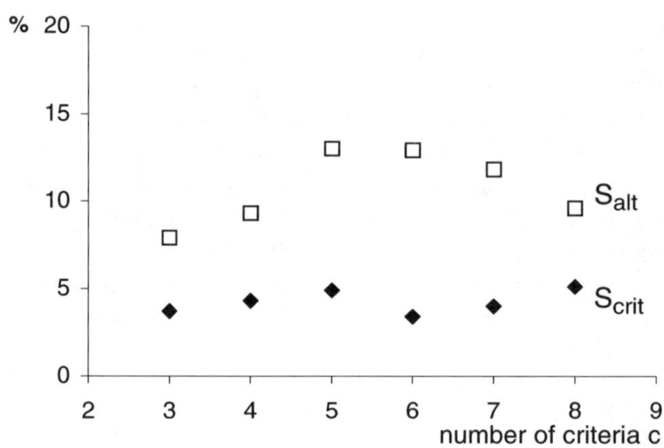

Fig. 4. Superiority rate of decision procedures – subject to the number of criteria

The previous results state that both decision procedures are essential, even though not in the same degree. It is then necessary to know in what types of decision situations one procedure outperforms the other, in order to chose the group welfare maximizing decision procedure in advance. One way of identifying the crucial similarities is to survey the degree of homogeneity of group member's preferences. The standard deviation of individual criteria weights may serve as an indicator for this unity. Similar criteria weights always result in standard deviations close to zero, while wide differences naturally result in higher standard deviations. Indeed, consideration of the criteria weights' standard deviation in the simulation leads to an interesting additional conclusion. With the exemplary simulation parameters $1 = 3$, $1 = 3$ and $1 = 2$, in decision situations in which voting on alternatives is superior to voting on criteria, the mean standard deviation of criteria weights is significantly lower as in situations in which voting on criteria is superior ($SD_{\text{alt>crit}} = 0.145$ compared to $SD_{\text{crit>alt}} = 0.195$). The highest deviation is registered in decision situations, in which neither procedure chooses the welfare maximizing alternative ($SD = 0.209$), and a middle deviation when both procedures choose this theoretical optimum ($SD = 0.164$). Though this coherence has to be further investigated, it can be assessed as an indication, that voting on criteria is rather superior to voting on alternatives in cases where the group faces highly unequal criteria weights.

5 Conclusion and Recommendations

The simulation demonstrated that different decision situations require different decision procedures. Both considered procedures apparently feature strengths and weaknesses. In principle, *voting on alternatives* is characterized by lower transaction costs in terms of time, organizational and cognitive effort. In most decision situations, a good result may be expected, and the complete preference ordering is composed directly of the individual preference orderings of the decision committee. *Voting on criteria* certainly involves higher transaction costs, but may still be superior in some decision situations. The following recommendation in respect to the meta-decision problem can be summarized.

According to the simulation results and the lower transaction costs *voting on alternatives* should always be considered the standard procedure. In particular, there is no reason to deviate from this standard, if

- the group faces a *high consensus* in its preferences, or if
- a relatively *large group* votes on *few alternatives.*

Voting on criteria is to be preferred only if

- the group faces *fundamental value conflicts*, and if
- relatively *few people* vote on a *large set of alternatives.*
- Above all, *voting on criteria* is more adequate if special expertise is necessary for the decision and not every group member has this expertise in regard to every criterion. This conclusion cannot be drawn from the simulation, but is easily justified. With *voting on criteria*, an individual may participate in the decisions on some selected criteria and not others, which is obviously not possible with *voting on alternatives.*

In their framework for multicriterion group decisions, Belton and Pictet (1997) reach a similar conclusion. They discuss three different *elementary approaches*, which they call *'sharing'*, *'aggregating'* and *'comparing'*. *'Sharing'* aims to obtain consensus through discussion of the views and reducing differences by explicitly addressing their cause. *'Aggregating'* reaches compromise through a vote or calculation of representative value, and *'comparing'* tries to obtain individual views without necessarily reducing its differences. According to Belton and Pictet, the "costs of *sharing* is high in terms of the time which needs to be committed to the process and the demand for facilitator expertise, but the expectation of a consensual outcome is also higher" (1997: 299). Since in terms of their notation *voting on alternatives* is a method of *aggregating* preferences, and *voting on criteria* is a form of *sharing* preferences, this conclusion endorses the results of our simulation.

It is interesting to consider why *voting on criteria* has this notable opportunity to effect a compromise, which is recognized by Belton and Pictet (1997) as well as by the approach in this paper. If one individual regards one criteria as exceptionally important and another group member regards it as only minor, they may agree on a medium criteria weight, which neither of them initially preferred. This *'creation'* of a compromise is only possible by *voting on criteria*. By *voting on alternatives* this fundamental value conflict will neither be revealed nor reduced, and appropriate compromise alternatives may eventually remain unrecognized. In this aspect, *voting on criteria* resembles a negotiation, because of its high demand of explicitly communicating and discussing values as the basis of making individual choices.

To summarize the results of the simulation and the characteristics of both decision procedures, *voting on alternatives* seems to be appropriate in most standard decision situations. Nevertheless, *voting on criteria* may increase group welfare if a small expert group is facing fundamental conflicts. This seems to be the case for some strategically important, trend-setting decision problems, which frequently emerge in political and industrial decision committees. We may therefore draw the conclusion that it may indeed be reasonable to be concerned about the meta-decision – although it still remains a decision under serious uncertainties. To reduce this uncertainty, the examination of the simulation has to be extended, and more tests have to be made especially regarding the characteristics of the different decision situations, in order to validate the preliminary findings of this paper.

References

Arrow, K.J. (1951), "Social Choice and Individual Values", New York: Wiley.

Arrow, K.J. and Raynaud, H. (1986), "Social Choice and Multicriterion Decision-Making", Cambridge: MIT Press.

Baucells, M. and Sarin, R.K. (2003), "Group decisions with multiple criteria" *Management Science*, Volume 49, Issue 8, 1105–1118.

Belton, V. (1986), "A comparison of the analytic hierarchy process and a simple multi-attribute value function" *European Journal of Operational Research*, 26, 7–21.

Belton, V. and Pictet, J. (1997), "A framework for group decision using a mcda model: Sharing, aggregating or comparing individual information?" *Revue des systèmes de decision*, Volume 6, no. 3, 283–303.

Condorcet, M.J.A.N. de Caritat, Marquis de (1785), "Essai sur l'application de l'analyse à la probabilité des décisions rendues à la pluralité des voix, Imprimerie Royale", Paris [Essay on the application of mathematics to the theory of decision making, in K. M. Baker (ed.), *Condorcet: Selected Writings*, 1976, Indianapolis, IN: Bobbs-Merill].

de Borda, J.C. (1784), "Mémoire sur les élections au scrutin" Histoire de l'Académie Royale des Sciences, Année MDCCLXXXI, Paris, 657–665 [Trans. A de Grazia, *Isis*, 1953, 44, 42–51].

Karni, R.; Sanchez, P. and Rao Tummala, V.M. (1990), "A comparative study of multiattribute decision making methodologies" *Theory and Decision*, 29, 203–222.

Keeney, R. and Raiffa, H. (1976), "Decisions with Multiple Objectives: Preferences and Value Tradeoffs", Cambridge: Cambridge University Press.

Lindstädt, H. (1998), "Qualität von Gruppenentscheidungen (Quality of group decisions)" *OR Spektrum*, Volume 20, 165–177.

Nurmi, H. (1983), "Voting procedures: A summary analysis" *British Journal of Political Science*, Volume 13, Issue 2, 181–208.

Rawls, J. (1971), "Theory of Justice", Cambridge: Harvard University Press.

Sen, A.K. (1970), "Collective Choice and Social Welfare", San Francisco: Holden-Day.

Tarski, A. (1946), "Introduction to Logic and to the Methodology of Deductive Sciences", New York: Oxford University Press, Second Revised Edition.

Thomas, N.P. (2005), "Entscheidungsverfahren und -institutionen zur Lösung gesellschaftlicher Risikoentscheidungen", Marburg: Metropolis.

Young, P. (1995), "Optimal voting rules" *Journal of Economic Perspectives*, 9 (1), 51–64.

Zanakis, S.H.; Solomon, A.; Wishart, N. and Dublish S. (1998), "Multi-attribute decision making: A simulation comparison of select methods" *European Journal of Operational Research*, 107, 507–529.

Part IV

Models of Risk Attitude Modelling and Methodological Issues

Probabilistic Risk Analysis Versus Decision Analysis: Similarities, Differences and Illustrations

Elisabeth Paté-Cornell

Department of Management Science and Engineering, Stanford University, CA 94305, USA
e-mail: mep@stanford.edu

Abstract. The methods of engineering probabilistic risk analysis and expected-utility decision analysis share a common core: a probabilistic model of occurrences of uncertain events. This model is based on systems analysis and on the identification of an exhaustive and mutually exclusive set of scenarios, their probabilities and their consequences. Both methods rely on an assumption of rationality and the use of Bayesian probability, and both assume separation of probability assessments and of preferences among scenarios' outcomes. The major differences are rooted in the nature and the framing of the problems that they address. A risk analysis is often performed before decisions have been fully defined, and one of its objectives is then to identify and characterize risk mitigation options. Furthermore, at the time of the analysis, the decision maker who will eventually use the results is often unknown. Therefore, the definition of Bayesian probability as a degree of belief has to be adapted, for instance, by assuming implicit delegation of the user's judgment to the analyst and the experts, which requires special care in the presentation of the results. Also, a risk analysis is often performed for a single system (e.g., one aircraft) for one unit of time or operation (e.g., one takeoff and landing cycle) when in reality, the analysis may be intended to support risk management decisions that will eventually concern an unknown number of similar systems for an unspecified number of time units. This multiplicity has implications for the treatment of second-level uncertainties (about failure probabilities) and for the need to display these uncertainties in the results. In this paper, the two classical definitions of probability (Bayesian and frequentist) are discussed, focusing on their relevance to both probabilistic risk analysis and decision analysis, when facing aleatory uncertainties (randomness) as well as epistemic uncertainties (limited knowledge about a fundamental phenomenon of interest). The risk and decision analysis methods are then briefly described, along with their similarities and differences. Two illustrations are presented: an analysis, performed in 1990, of the risk of losing a NASA orbiter and its crew due to a failure of the tiles of the thermal protection system, and a method of assessment of the risk of a terrorist attack on the United States in a given time frame, based on available intelligence information (a 2002 study). The latter involves the use of a simple analysis of a game involving alternating decisions and moves by terrorists and the US using a rational model in the descriptive mode. The main conclusion is that whereas the role of the decision analyst is to represent faithfully the beliefs and preferences of a known decision maker in order to identify the preferred alternative, the risk analyst needs to be scrupulous in presenting the model assumptions, as well as the sources and the methods of data processing to allow future decision makers to exercise their own judgments when using the results.

Keywords: risk analysis, decision analysis, utility, probability, Bayesian probability, space shuttle, terrorist attacks

1 Risk Analysis Versus Decision Analysis: Basic Problem And Implications

Probabilistic risk analysis (PRA) is a method widely used in engineering and in other fields to assess the probabilities and consequences of failures of a given system[1]. Developed in large part in nuclear engineering (e.g., Starr, 1969; USNRC, 1975; Kaplan and Garrick, 1981; Henley and Kumamoto, 1992; Bier and Cox, 2006), it has been applied to many other fields e.g., in the medical domain, (e.g., Paté-Cornell, 1999b; Pietzsch and Paté-Cornell, 2004). PRA is based on systems analysis, and it is particularly useful in cases where the evidence available does not include sufficient failure statistics at the system's level, but useful data can be gathered regarding the performance of different components or a hazard in different phases of accident scenarios[2]. Expected-utility decision analysis, by contrast, provides a systematic support to an identified decision maker for consistent, rational decisions under uncertainty[3] (see for example, Raiffa, 1968; Howard and Matheson, 1984). It includes both description of the various scenarios and the preferences among their outcomes.

[1] Risk analysis focuses in general on the downside of a situation because the costs of risk mitigation are often known. The focus of the analysis is thus generally on the probability and the effects of hazardous events, even though if appropriate, the upside can be integrated in the outcome distribution without any fundamental change.

[2] Other areas of risk analysis, e.g., environmental/health risk assessments, are often based on "plausible upper bounds" and on several conservative hypotheses (Paté-Cornell, 1996). Developed mostly for regulatory purposes, they allow only ranking of these upper bounds, but do not provide the costs and benefits of risk mitigation measures

[3] The emphasis, in this paper, is on normative (or prescriptive) methods of risk and decision analysis. There exists a large body of descriptive studies that seek to understand how people actually perceive and accept risks (e.g., Slovic, 1987), or make decisions intuitively without the support of analysis decisions (e.g., Kahneman and Tversky, 1979).

The two methods are normative and share a common core: a probabilistic model of occurrences of uncertain events based on systems analysis, identification of an exhaustive and mutually exclusive set of scenarios, and computation of their probabilities and consequences. In the use of that information, both rely on an assumption of rationality, as defined for decision analysis, by the von Neumann axioms or their equivalents (von Neumann and Morgenstern, 1947) and for risk analysis, by Savage (1954) among others. Both methods generally require the use of Bayesian probability defined as a degree of belief, when a decision needs to be made or a risk estimated in the absence of perfect information (e.g., de Finetti, 1974)[4]. Both methods assume separation of probabilistic assessments on the one hand, and of preferences among scenarios' outcomes on the other hands.

The two methods, however, present major differences rooted in the nature and the framing of the problems that they address. Probabilistic risk analysis is often performed before the decisions to be made and/or the decision makers are known. Indeed, one of the objectives is sometimes to identify and characterize (probabilistically) the effectiveness of possible risk management options. If at that stage, one wants to rank these options, it is sometimes done, *by default*, based on a simple assessment of their expected costs and benefits, which assumes risk-neutrality (Paté-Cornell, 2000, 2002). Obviously, these rankings can change with different risk attitudes, and the results need to be presented so that they can be used in conjunction with other utility functions involving different risk attitudes or even attributes. For example, the failure risks of a chemical facility that can release a toxic substance in the atmosphere may not be valued in the same way by the plant managers, the workers, the neighbors, and the regulators. In that case, the relevant attributes of outcome descriptions and of the utility functions (costs, environmental consequences, effects on human health, etc.) may not even be the same for the different parties involved. The risk analysis should thus be performed so that the results can support decisions that reflect the values and preferences of each of these groups. This implies presenting full joint distributions of outcomes and their possible attributes (including, if relevant, redistribution effects), which can be converted later into probability distributions of multi-attribute utility functions (Keeney and Raiffa, 1976) that can support the decisions of each of the parties involved.

Also, the definition of Bayesian probability as a degree of belief has to be adapted when the user of the results is unknown, for instance, by assuming implicit delegation of probabilistic judgments to the analyst and to the experts involved in the analysis. The decision maker(s), however, may want

[4] Von Neumann and Morgenstern (1947) state clearly that they "The simplest procedure is, therefore, to insist upon the alternative [to probability as a subjective concept], perfectly well founded interpretation of probability as frequency in the long run". In a footnote, however, they concede that "If one objects to the frequency interpretation of probability then the two concepts (probability and preferences) can be axiomatized together." This is what Savage did (Savage, 1954).

to know how the results were obtained in order to use his or her own value judgments in the decision phase. The computation of the risk independently from a specified decision maker thus raises the question of "objective" versus "subjective" probabilistic estimates. Classical statisticians consider that probability cannot be assessed in the absence of an appropriate sample of independent, identically distributed data points. Therefore, in that perspective, the notion of probability is restricted to the description of randomness (aleatory uncertainty) and does not directly apply, for instance, to whether or not a property of interest is true (epistemic uncertainty). Both epistemic and aleatory uncertainties can be measured and combined by Bayesian probability, whereas classical statistics cannot, in principle, address epistemic uncertainties that are at the core of many probabilistic risk analyses.

Furthermore, a risk analysis is often performed for one single system (e.g., one aircraft) for one unit of time or operation (e.g., takeoff-and-landing) when in reality, the analysis is meant to support risk management decisions that will eventually concern an unknown number of similar systems for an unspecified number of time units. This multiplicity has implications for the treatment of second-level uncertainties about failure probabilities and, as shown further, requires the display of these uncertainties in the analytical results.

In this paper, the two classical definitions of probability (Bayesian and frequentist) are discussed, focusing on their relevance – and usefulness – when facing aleatory uncertainties (randomness) as well as epistemic uncertainties (lack of knowledge about a fundamental phenomenon of interest). The risk and decision analysis processes are then briefly described, with their similarities and differences, and illustrated by two past studies performed in the Stanford Engineering Risk Research Group. The first one (performed in 1990) concerns the risk of losing a NASA orbiter and its crew due to a failure of the tiles of the thermal protection system. The second (performed in 2002) is a high-level analytical model of the risks of different types of terrorist attacks on the United States in a given future period, based on available intelligence information at the time of the analysis. That example involves a conflict analysis that relies on an assumption of alternating rational decisions on both sides. It includes probabilities and utilities as assessed by US decision makers and analysts, of their own beliefs and preferences as well as those of various terrorist organizations. The conclusion emphasizes the difference of objectives –as well as the commonality of methods – of decision analysts and risk analysts, who want to provide high-quality results, useful to a known or unknown decision maker.

2 Fundamental Concepts of Probability and Uncertainty

In the course of the analysis of classes of scenarios – whether following a decision option, or leading to the occurrence of an adverse event –, one generally encounters both epistemic and aleatory uncertainties[5]. They characterize respectively the lack of fundamental knowledge e.g., about a still-unknown physical phenomenon, and the simple randomness that is encountered, for example, when throwing a fair dice (e.g., Apostolakis, 1990; Morgan and Henrion, 1990).

Two definitions of probability provide an assessment of aleatory uncertainties: the Bayesian, and the classical frequentist. Bayesian probability, defined as a degree of belief, is fundamentally anchored in the notion of rational choices under uncertainty (e.g., Ramsey, 1926; de Finetti, 1974; Savage, 1954; Press, 1989). It is obtained by combining a prior probability before a new piece of evidence becomes available, the likelihood of these data, and the probability of observing them, which requires considering all possible alternatives to the hypothesis of interest. The priors and the likelihoods all reflect the degrees of belief of the decision maker or the analyst. The likelihood function may represent, for instance, the level of confidence in a particular sensor or a particular source of information.

Classical statisticians, by contrast, define probability as a frequency in a sufficiently large sample of independent and identically distributed trials (e.g., Neyman 1937). It can thus capture aleatory uncertainty, but not epistemic uncertainty, to which they believe the notion of probability simply does not apply. This leaves the decision maker – or the risk analyst who needs to provide information before full knowledge has been obtained – with only one option: the use of Bayesian probability. The question of "subjectivity" versus "objectivity" is often raised in that context. Bayesian probability is by definition subjective. Frequentist probability, however, also includes an element of subjectivity, not only in the choice of the confidence level attached to it, but also in the assumption that past data are relevant to future events. This assumption of a steady state often limits severely the relevance of existing statistics.

The problem – and the tensions – caused by this disagreement about the definition of probability has sometimes been an obstacle to the acceptance of probabilistic methods[6]. Attempts to reconcile the two approaches have been particularly critical in the field of risk analysis where a decision maker, able to provide degrees of belief, has seldom been identified as a unique person. One such attempt to unify the two approaches to probability is based on the consideration of the future frequency of an event as a Bayesian random variable (e.g., Apostolakis, 1990), whose expected value constitutes *the* probability of that event[7]. The absence of a known decision maker in probabilistic risk analysis requires adapting the notion of degree of belief, and carefully

[5] The literature sometimes simply refers to them as "risk" and "uncertainty" but that terminology can be confusing because both involve uncertainties.

[6] In reality, the results of Bayesian and frequentist analyses of the same data sets often converge when the sample size is sufficiently large, as shown for example, in a study of the probability of failure of launch vehicles for space systems (Guikema and Paté-Cornell, 2004).

presenting hypotheses and results to permit the eventual decision maker to adopt – and if needed adapt – the analytical results to represent his or her own degrees of belief.

3 Probabilistic Risk Analysis and Decision Analysis: The Methods and the Processes

Expected-utility decision analysis provides an unambiguous approach to decision making under uncertainty. A known decision maker is generally faced with a set of options among which he or she must choose an optimum. The assumption is that this decision maker follows the axioms of rational choices in one of several equivalent forms (e.g., orderability, continuity, monotonicity, substitutability and decomposability as defined by Howard, 1984). Under each option, the analyst identifies the possible scenarios, structured as a set of collectively exhaustive and mutually exclusive elements. Their probabilities are characterized by a joint probability based on the marginal and conditional probabilities of their components. For each scenario, the analyst then assesses its consequences in terms of magnitudes along various dimensions ("attributes") and the corresponding utility – single- or multi-attribute – of the decision maker. Finally, the expected utility attached to each option is computed; the optimal option is that which maximizes this expected value. The tools of decision analysis involve, for example, decision trees, influence diagrams, and stochastic processes. Problems arise, however, when the decision involves several potential or actual decision makers[8].

This is precisely the issue that the (engineering) risk analyst often faces. His or her task generally starts with a system whose components and functions are often known – although some internal mechanisms may still be uncertain –, that may be subjected to external events (e.g., earthquakes), and that may constitute a hazard for example, because it may release a dangerous substance. The corresponding risk can be defined, for example, by the probability of system failure per time unit if it is a binary situation, or more accurately, by the probability distribution of damage per time unit due to failures of various levels of severity (e.g., losses per year)[9]. When the out-

[7] This definition assumes first that the time unit and the frequency per time unit are sufficiently small to be compatible with the notion of probability, and second, that the concept of repetition applies.

[8] As shown by Arrow (1963) expected-utility decision analysis does not apply in that case. Another problem is that of "collective probabilities", which as discussed further, do not fit directly the Bayesian definition as degree of belief. One can, however, adapt the decision analysis method by assessing what amounts to collective degrees of beliefs and by constructing a valuation function similar to utility for a group of decision makers.

[9] Note that the often-cited definition of risk as "probability times consequence" is generally insufficient to capture the loss distribution and to allow, for instance,

comes involve several attributes (e.g., human casualties and financial losses) the consequences of different scenarios need to be described by a joint distribution of these attributes.

The classic steps of probabilistic risk analysis as used for example, in the nuclear power industry, involve first a functional analysis of the physical system: what functions are needed for operations and what configuration of components – in parallel or in series – fulfill these functions? The second step is to identify the failure modes, i.e., the min-cut sets or minimal sets of events that cause system failure, using a combination of fault trees and Boolean algebra. The third step is the computation of failure probabilities. It involves gathering the data required from all available sources[10] and computing the probabilities of different levels of system failures, accounting for possible dependencies among component failures, in particular those caused by common causes such as external events. The next step is to compute the losses associated to different failure scenarios and the probability distribution of these losses per time unit. The final step is to display and communicate the results, for instance as a risk curve, which represents the probabilities of exceeding each possible level of loss per time unit (i.e., the complementary cumulative distribution function[11]). The description of some of these scenarios and the computations of their probabilities may involve a dynamic analysis and the use of stochastic processes, for example, when the problem involves system deterioration or the evolution of accident sequences as in the case of patient risk in anesthesia (Paté-Cornell, 2000).

Risk analysis, contrary to decision analysis, is not supposed to include preferences for scenario outcomes (in the sense of utility), neither explicitly nor implicitly. As discussed further, this neutrality is not always easy to achieve in practice.

In addition, the risk analyst may want to display in the results the effects of epistemic uncertainties, so that the decision maker can form a better assessment of his or her beliefs regarding the risk. This requires keeping epistemic and aleatory uncertainties separated throughout the analysis, and propagating the uncertainty through simulation (e.g., Monte Carlo or one of its variations, such a the Latin Hypercube method). The results can then be displayed, for example as a family of risk curves, each corresponding to a percentile of the discrete distribution of the future frequency of exceeding any given level of loss in a specified time unit (Paté-Cornell, 1999a, 1996; see Fig. 1). The problem is that the numbers of systems or time units may be unknown at the time of the analysis. As shown further, this propagation and

rational risk-averse decision making. It represents an expected value of the consequences, which fits only the preferences of a "risk-neutral" decision maker.

[10] Different types of information can be used to provide data, including statistics of actual system or component performance, surrogate data (performance in a different environment), test data, engineering models, and expert opinions.

[11] Again, if several attributes are involved in the description of the outcomes, the results may include their joint distribution.

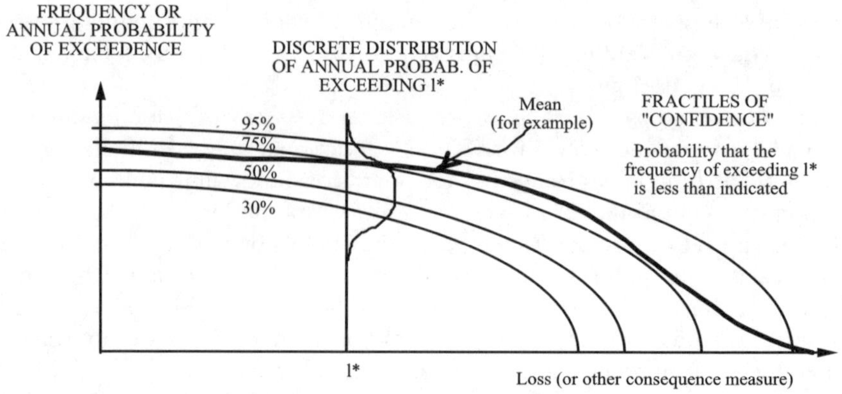

Fig. 1. Risk analysis results as a family of risk curves (second-order uncertainty analysis) (Source: Paté-Cornell M.E. 1999a)

representation of uncertainties allows computing the probabilities of failures of multiple systems and multiple independent failures in future time periods.

At the end of this exercise, the risk analyst is often in a position to identify the system's weakest points, different reinforcement options, and their costs and benefits. As mentioned earlier, recommending a course of action presumes knowing the risk attitude of the decision maker, but by default, can be based at first, on simple expected values of costs and benefits. Yet, as shown further in the NASA example, these recommendations may not be those that the actual decision makers will want to implement, for various reasons that can involve different risk attitudes.

4 Risk Analysis and Decision Analysis: Difference and Similarities

The two methods are complementary and address problems of uncertainty in decision, one focusing on information, the other also involving preferences. Therefore, they show both similarities and differences. By definition, decision analysis, in addition to a preference model, generally includes a risk analysis describing both upsides and downsides, embedded in its structure. This implies that the structure of the "factual" part of both analyses is the same; it is based on the same construction of scenarios representing conjunctions of events and random variables. Whether these variables are discrete or continuous, the computation of scenario probabilities is based on two laws of logic: Bayes theorem and the total probability theorem[12].

[12] Note that these two theorems of logics hold regardless of the definition of probability adopted.

Both decision analysis and engineering risk analysis are based on an axiomatic foundation of logics and rationality[13]. They involve full distributions of the consequences of a set of possible scenarios that must be structured as collectively exhaustive and mutually exclusive to be amenable to probability computations. In both methods, one encounters similar questions such as: to what depth should the analysis be performed in different parts of the problem? In other words, when does one have enough information about a phenomenon or the performance of a subsystem, so that further decomposition is not necessary? As it will become clear further through examples, the optimal decomposition level varies across the different parts of the problem. It has a critical effect on the results, and it depends on the choice of the data that constitute the best evidence base. It is that choice of variables, among other factors, that makes decision and risk analyses both an art and a science and does not make them amenable to cookie-cutter procedures that could be automatically implemented by untrained operators or entirely performed by a computer.

The fundamental premise of rationality also implies that both methods of analysis rely on Bayesian probability, at least when facing epistemic uncertainties, which cannot be captured through classical statistics[14]. Classical statistics, of course, are useful in risk analysis in the context of a steady state, and they provide estimates that are close to those that are obtained through Bayesian statistics, which are often more complex.

The definition of Bayesian probability as a degree of belief makes its use relatively simple in decision analysis where one can "encode" the assessments of the decision makers or of the experts that he or she has explicitly chosen. By contrast and as mentioned above, even though risk analyses are explicitly performed to support risk management decisions, the actual decision maker (user, customer) is generally unknown at the time of the exercise. Furthermore, one performs such an analysis *because* sufficient statistical data are generally not available at the system level at the time when a decision has to be made, thus precluding the use of classical statistics and dissipating the illusion of objectivity that they are assumed to provide. The use of Bayesian probability, however, prompts the question: whose degree of belief? There may be several users of the analysis, at different times, under different circumstances, or in different parts of an organization. Therefore, one has to rely on an assumption of implicit delegation of the degree of belief of decision makers to the analyst and to the experts with whom he or she has chosen to work. The degrees of belief of the analyst and of experts in different fields then become the basis of the probabilities used in

[13] The rational foundation of Bayesian probability can be found in the works of Ramsey (1926) and Savage (1954) among others.

[14] This is true unless one stretches the notion of frequencies to include the proportion of times when experts have been right in a particular field. This approach, however, is seldom specific enough to be useful, especially when one can instead, analyze in greater details the potential causes of different types of failures.

different parts of the problem. For decision makers to adopt the same views, the analyst must thus be scrupulous in his or her description of assumptions, of the sources of data, and of the methods by which they were processed.

In both risk and decision analyses, the structure of the "factual" model – as opposed to the representation of preferences – and the choice of its components (random events and variables) reflect the skills of the analyst, but also the constraints of the resources allocated to that task. Another issue is the choice of models. In decision analysis, it is sometimes stated that there is no such thing as uncertainty about a model, presumably because it constitutes the best representation of the decision maker's view of a phenomenon of interest[15]. By contrast, in risk analysis, different experts may rely upon different representations of the world. Different models may thus have to be considered and included so that decision makers who have not participated in the analysis get the full spectrum of information. In addition, uncertainties about the parameters of each model have to be included in the computations. To the extent that the "experiment" will be repeated (multiple similar systems and time periods), these uncertainties have to be propagated throughout the computations and represented in the results.

Indeed, uncertainties about probabilities are at the heart of Bayesian computations. A classic computation involves a random variable and assumes a particular form for its probability distribution. The uncertainties about the value of one or several parameters can in turn, be represented by a probability distribution (e.g., Normal), and this distribution updated with every new piece of evidence. Similarly, Howard (1970) showed, on a specific example, how one can start from an uninformative distribution (i.e., uniform on the $[0, 1]$ interval) about the probability of an event and proceed to sequential updatings of that distribution with every new piece of information. That computation can be simplified greatly by using "conjugate priors", involving for example, binomial and Beta distributions.

At any given time, if one wishes to use expected-utility decision analysis to choose among options involving an event whose probability has been updated in that way, it is the *mean* of that distribution that is to be used in the computations as *the* probability of that event. In risk analysis, if eventual decisions involve several similar systems and several time periods, the full distribution of that probability (per time unit) cannot be reduced to its expected value and used as if there was no uncertainty about it. This is true because if $p(F)$ is the probability of one failure in a given time period, and if there is no uncertainty about it, the probability of n independent failures in the same time period is simply $[p(F)]^n$. But if there are uncertainties about

[15] Note, however, that for a rational decision maker, uncertainty about the nature of a fundamental phenomenon may well result in model uncertainty, for example, whether or not there is "memory" in the occurrence of earthquake, therefore if a Poisson model of occurrence is appropriate or not.

$p(F)$ (e.g., because of uncertainties about different parameter values), one can represent this probability as a random variable as described above. Although the Bayesian probability of one event is the expected value of that random variable ($EV[p(F)]$), the probability of several such independent events is generally not $[EV(p(F))]^n$ – i.e., the value that one would obtain by simply raising to the nth power the probability of a single event – but $EV[p(F)^n]$[16]. Therefore, in that case, one cannot simply compute the probability of an event (or of a known number of events) as a single number (a mean) and raise it to the power n, even if one knew that number at the time of the analysis. In a well-defined decision analysis, by contrast, one can compute directly the probability of the scenarios of interest (for a known number of "experiments"), given each of the identified options. The multiplicity of identical systems and time periods, thus has implications for the display of uncertainties about probabilities in risk analysis, but not in decision analysis with known time frames, options, systems, etc.

Another key difference concerns preferences. Decision analysis includes preferences in a utility function. Risk analysis does not, even though some recommendations that come out of it may be based by default on implicit risk neutrality with respect to costs and benefits, or sometimes on worst-case scenarios[17]. These are only assumptions regarding preferences, which have to be presented as such. In reality, as mentioned earlier, the analysis must be performed so that the probability distribution of outcomes can be further transformed into a probability distribution of utilities, which may not be a linear function. It is on the basis of that utility distribution that the decision will be made. This requires full display of the risk analysis results as opposed to a simple expected value of the outcomes[18].

The separation of preferences and probability assessment can be trickier than it appears at first. First, the choice of the attributes that describe the scenario outcomes in a risk analysis may limit the inclusion of some preferences in the decision phase. Second, it is often tempting for experts with strong preferences about the decision that the analysis will support to inject these preferences in their probabilistic assessments. As an illustration,

---

16 For example, if one represents the absence of information about the probability of an event in a given time period by an uninformative uniform distribution on the $[0, 1]$ interval, the probability of that event is the expected value of that distribution, i.e, 1/2. The probability of two such independent events, however, is not $(1/2)^2 = 1/4$, but 1/3 because $EV(p^2(F)) = [1/3x^3]$ between 0 and 1.

17 The only common ground for firm recommendations is the case of dominance in which the analyst has identified an option under which one is better off under all possible scenarios that may unfold.

18 This is particularly critical when computing the risk of severe events with extreme consequences (e.g., large natural catastrophes or large-scale terrorist attacks). Whereas the expected value of the outcomes is indeed influenced by these extreme values, a risk-averse utility function will allocate even greater weight to the tail of the outcome distribution.

consider the case of a seismologist whose first question, when asked for the probabilities of earthquakes of different characteristics in the Eastern US, was: what are they going to be used for? The local seismicity, of course, is the same whether the site is that of a nuclear reactor or a chicken coop. What varies with the criticality of the facility is the value of information, thus the amount of resources that should be allocated to the seismic risk analysis. Keeping preferences and probability separated and persuading experts that conservatism should be put in the decision criteria and not in the probabilistic estimates is one of the features of a credible analysis of either type.

In decision analysis, the preference criteria are squarely represented by utility functions, which can be encoded by presenting the decision maker with hypothetical lotteries and asking him or her to identify (for instance) their certain equivalents. If he or she is risk-neutral, the utility of an outcome is simply proportional to the measure of its consequences. In addition, in the classical definition of rationality, the decision maker is by definition ambiguity-neutral, i.e., he/she is assumed to be indifferent between two identical lotteries (same probabilities, same consequences) regardless of differences in the information bases that support the probabilities in the two lotteries. In reality, people often are and *want to be* (e.g.,) ambiguity-averse (the Ellsberg "paradox"; Ellsberg, 1988), and the definition of rationality itself can be revisited to allow for other attitudes towards ambiguity (e.g., Paté-Cornell and Davis, 1994).

The results of a risk analysis, when used in a decision analysis context, can be coupled with an explicit utility function applied to the distribution of the outcomes, reflecting the relevant risk attitude (and in general, its variations along the outcome axis). More often, however, the risk analysis can be used to check that the probability of a system's failure is below a given threshold of risk tolerance, or to rank threats and risk mitigation options, for example, on the basis of expected costs and benefits, or of worst-case assumptions. Note that the use of a threshold of risk tolerance does not satisfy the von Neuman's axioms of rationality, but is another expression of objectives and preferences. Since in addition, the decision maker(s) may not be indifferent to uncertainties about failure probabilities, it is often *required* that risk analysis results include a display of these uncertainties, not only in a single risk curve but in a family of risk curves as shown in Fig. 1.

Similarities between risk and decision analyses thus come from a common assumption of rationality, and the same framework for the factual and probabilistic part of the analysis. The differences stem mainly from the fact that in a decision analysis, the options and the decision maker are known, whereas risk analysis must provide more detailed information to decision makers who are often unknown at the time of the computations. What follows is a limited illustration of these concepts.

5 Illustrations: The Tiles of the NASA Shuttle and the Risks of a Terrorist Attack on the US

5.1 The Risk of Losing an Orbiter Due to a Failure of the Tiles of the Space Shuttle

In that study, which was performed in the late 1980's based on the first 33 shuttle flights, one objective was to assess the probability of losing an orbiter and its crew due to a loss of tiles of the thermal protection system (Paté-Cornell and Fischbeck, 1993a and 1993b). It was also to allocate this risk among the tiles in different locations on the orbiter's surface to set maintenance priorities based on risk-criticality. Another objective was to make some recommendations to NASA, including organizational adjustments, to improve the maintenance of the tiles. The study was performed on a very low budget, which did not permit a second-level analysis of uncertainties, only a discussion of their potential effects on the recommendations. However, the first-level analysis (based on single values of failure probabilities) was sufficient to yield specific conclusions. The risk analysis model was based on a partition of the orbiter's surface into min-zones, each defined by a range of values of four parameters: the heat load, the aerodynamic forces, the density of debris hits, and the criticality of the subsystems under the aluminum skin. Two main failure modes of the tiles were identified: loss of tiles under regular loads of vibration and aerodynamic forces, and debonding under a hit, especially by a piece of insulation detached at takeoff from the surface of the external tank. This second failure mode was the cause of the loss of the orbiter Columbia in February 2003, when such a piece of debris hit the panels that protected the edge of the left wing. The possibility of that failure mode became clear when the authors superposed maps of debris hits, showing a higher density of tile damage under the right wing of the orbiter. That "cloud" of hits corresponded to the attachment of a fuel line on the external tank, which in turn appeared to weaken the bond of the insulation on the tank's surface. The risk computation was based on the influence diagram shown in Fig. 2.

The total probability of a shuttle accident due to tile failure was shown to be in the order of 1/1000 per flight, or about 10% of the overall probability of an accident. It was also shown that about 15% of the tiles contributed about 85% of the probability of failure, therefore, that all tiles were not equally critical. In the process of gathering data, it had become clear that a small number of tiles had not been properly bonded, either because of poor maintenance, or because they had been poorly installed in the first place. About twelve such tiles had been found after processing of about half of all tiles. Therefore, one could reasonably suspect that approximately twelve others remained with a weak bond. The study thus emphasized the possibility of a combination of high loads (from a poor bonding of the insulation of the external tank) and low capacity (from weak bonds on some tiles).

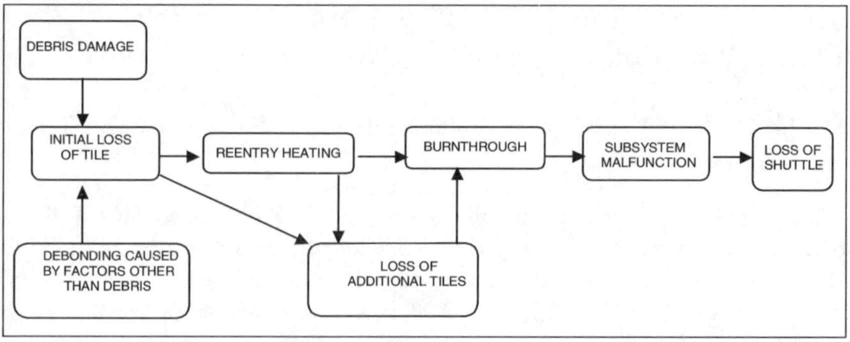

Fig. 2. Influence diagram for the PRA of the tiles of the space shuttle (Source: Paté-Cornell and Fischbeck, 1993a);
Legend: Oval nodes: Uncertainties about events and random variables. White nodes: Uncertainties about terrorist groups and their activities, including (striped) the elements of an attack scenario. Grey nodes: U.S. side. Square node: decision node. Hexagonal node: Consequences to the U.S. of an attack scenario given countermeasures. Arrows: Probabilistic dependencies

Several recommendations were communicated to NASA orally or in writing: that a more elaborate risk analysis with an adequate budget be performed, that special attention be given to the bonds of the most risk-critical tiles, that the time constraints on the tile maintenance be relaxed, that an in-flight repair kit be provided to the astronauts, and that the bonding of the insulation of the external tank be improved. Interestingly, only the recommendations that concerned Kennedy Space Center – which had funded the study – were seriously considered. They involved mainly increasing the attention given to the bonding of the most risk-critical tiles. The recommendations that had to be implemented by Johnson Space Center, where tile maintenance procedures were set, were apparently ignored. For example, the in-flight repair kit was called too expensive, unfeasible, or a potential cause of additional problems. Yet, after the death of seven astronauts, such a kit became available. Also, the bonding of the insulation of the external tank was modified, but apparently, only to make its application more environmentally friendly. In any case, the debonding of a large chunk of it during the launch of the doomed Columbia flight showed that it has not received sufficient attention, which may be attributable to the dispersion of space centers that seemed to care mostly to the subsystem for which they were responsible.

One of the features of the analysis was the choice of its depth to permit the best use of available information. At a first level of statistical observation, one could simply have relied on the number of shuttles lost thus far due to the tiles (none). At a second level, one could have relied on the number of tiles that had debonded in flight: two at the time of the 1990 analysis, which was too small a sample (given that there are in the order of 25,000 tiles per orbiter) to permit a relevant estimation of the probability of an accident.

This is why we chose a deeper level of analysis based on loads and capacities, which lead us to the analysis of the number of poorly bonded tiles and of the density of debris hits (we assumed that the density of poor bonds was uniform across the surface). This information base provided us with a larger and more stable database than the very small sample of lost tiles. Interestingly, one of the contractors, charged with the task of reducing the failure probability that we had found, did choose to rely on the number of tiles lost in flight. After a total of about 60 flights without additional loss of tiles since our study, the probability had been reduced to 1/3000 per flight (instead of 1/1000 as we had found). When a few flights later, several tiles were lost, their probability had to be raised and the contractor found a figure that was again in the order of 1/1000 and consistent with our findings a decade or so earlier. Finally, Columbia was destroyed by one of the failure modes that we had identified, and our original recommendations may now have a greater chance of being implemented in the remaining life of the Shuttle program.

This story illustrates one of the fundamental caveats to risk and decision analysts, which is to *not* perform an analysis for an organization or an individual who is not willing to use the results for decision support (or wants to influence them), but only to justify decisions already made. In the case of Columbia, the failure could have been anticipated and prevented; but the "proof" of its vulnerability (the accident itself) came too late. The same can be said of the 9/11 attack on the United States in 2001 which had several precursors such as the 1993 attack on the World Trade Center. More attack attempts have occurred, and more are to be expected – and they might well succeed. The US Department of Homeland Security needs to allocate its resources according to risk-reduction priorities (as opposed to political pressures). The following illustration shows how a risk analysis model can be constructed to allow assessing the risk of different attack scenarios.

5.2 A Risk Analysis Framework for Terrorist Threats on the United States

The risk of a terrorist attack on the US involves two main factors: the threats of different types of attack scenarios, and the vulnerabilities of the potential targets of such attacks[19]. Reinforcing all potential targets or addressing all potential threats is unfeasible and it is imperative to allocate resources efficiently. To structure large quantities of data of very different nature, and to set priorities among risk reduction measures, a risk analysis model was developed at a very high level of aggregation (Paté-Cornell and Guikema, 2002b). It is designed to be dynamically implemented and continuously updated on the basis of new intelligence information.

[19] Threats and vulnerabilities, in this case, are the equivalents of the classic loads and capacities used to compute the risk of failure of engineered systems ($p(\text{Failure}) = p(\text{Loads} > \text{Capacity})$).

The overarching model is based on two interrelated decision analyses that can be viewed as a sequential game of alternate moves. The first one concerns the decisions of the different terrorist groups based on their preferences, as revealed through intelligence sources or through their own statements, and on the resources available to them[20]. The second is that of the United States authorities, based on information that has been gathered regarding terrorist intents. The latter is described in the influence diagram shown in Fig. 3. A key element of that model is an assessment of the terrorists' "supply chain", which involves people and skills, communications, weapons and materiel, transportation, and cash. Each of these components can provide clues to terrorist activities, especially air transportation. An attack scenario is characterized by a terrorists' choice of weapon, target and means of delivery. The two models are run based on probabilities and utilities assessed by the US for both sides. A hypothetical application of the model was published

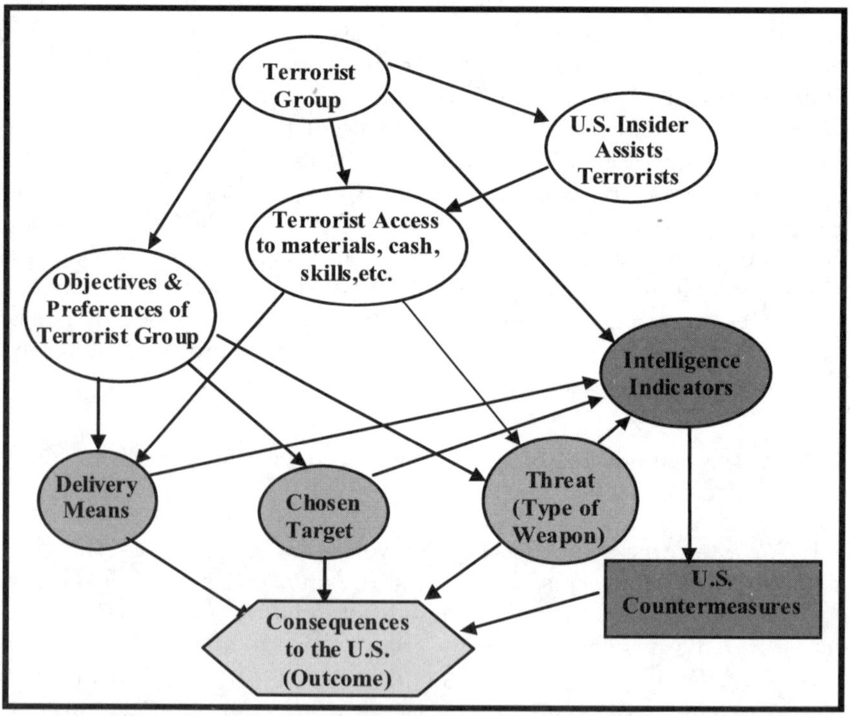

Fig. 3. Influence diagram representation of an overarching model for the prioritization of the risks of a terrorist attack on the United States (Source: Paté-Cornell and Guikema, 2002)

[20] Note that in this case, we used the rational decision analysis model in the descriptive mode because, in our judgment, it was an acceptable approximation of what we could anticipate given what we had observed.

in the unclassified literature (ibid.). For illustrative purposes, two terrorist groups were considered (extremist Islamists and disgruntled Americans), as well as four broad classes of attack scenarios: the detonation of a nuclear warhead, a "dirty bomb" made of nuclear spent fuel and conventional explosives, a biological attack such as small pox, and repeated conventional attacks on urban centers e.g., with improvised explosive devices (IED's). The probability of a successful attack involves a sequence of events: planning of the attack, acquisition of the weapon, introduction of that weapon on the US soil (if needed), successful implementation steps, and non-detection by the US, or US failure to act in time to stop the attack. Unsurprisingly given the potential consequences, the detonation of a nuclear warhead tops the priority list of this illustrative example; dirty bombs that can cause a real scare but much less damage, are considerably lower on the threat scale. The point however, is that such an analysis needs to be run in real time to guide the search for additional information and timely counter-terrorism measures, and that Bayesian updating is the best way to capture the evolution of the state of information[21].

This is a case where a default expected value of the losses (e.g., average number of human casualties in an attack on a specific target) is not a sufficient representation of the risk. Repeated conventional urban attacks, for example, are more likely and easier to execute than some other types of attacks, even though each of them may not cause as much damage. Therefore, the expected value of the associated losses may exceed, for instance, that of an improbable but much more destructive detonation of a nuclear warhead. A full probability distribution of the damage must be part of the risk results if the decision makers who wish to use them are risk-averse and want this risk aversion to be reflected in a non-linear utility function.

6 Conclusions

The two fields of probabilistic risk analysis and expected-utility decision analysis are tightly linked. Both rely on an assumption of rationality and in general, on Bayesian probability. A model of the risk analysis type is often at the core of a decision analysis. But whereas risk analysis focuses exclusively on event scenarios, their probabilities and their consequences, decision analysis involves, in addition, the alternatives that are considered and the preferences of a decision maker. The main differences thus lie in the problems' structures. In a decision analysis, the decision to be made, the decision maker,

[21] The probabilities used in the assessment of the risks of different types of terrorist attacks are clearly unstable. Intelligence information is collected every day. Surprises (small and large) occur all the time. Some of the factors of the model, however, are more stable than others for example, the existing vulnerabilities in our systems, the stated preferences of the various groups, or the existence of some known terrorist supply networks.

and the options considered are known *a priori*. By contrast, risk analysis is often performed without knowledge of who are the ultimate users of the results, or which options will be considered in the risk management phase. Indeed, the objectives of such an analysis often include finding the weak points of a system, identifying risk reduction options, and presenting information that allows setting priorities among them if the risk is found to be intolerable. A default option is to do so on the basis of expected values of cost and benefits. In all cases, one of the analytical requirements is to keep the probabilities and preferences separated. In decision analysis, risk attitudes must be captured by utility functions; in risk analysis, the probabilities represent the degrees of belief of analysts and experts, and presumably, will be adopted by the decision maker(s). One of the challenges is to keep these risk estimates free of attempts to influence the results and the decisions that they will support. It is essential in that context to represent scrupulously the underlying assumptions, the sources of data and the processing methods, and to present the results at the level of detail that the problem requires. This may involve representation of second-degree uncertainties (about event probabilities themselves), especially if the analysis is applied to several systems over several time units.

The role of the decision analyst is clear: to represent faithfully the beliefs and preferences of the decision maker in order to identify the preferred alternative. That of the risk analyst is more complex. It is to present as exactly as possible the state of knowledge, i.e., the assumptions of the model, the sources of information and the processing of the data, in order for future decision makers to be able to exercise their own judgments when using the results.

In all cases, a probabilistic analysis, of a risk or a decision, should not be performed for people who do not want to know the results, who try to influence them to serve their own purposes, or who have already made up their mind and simply seek justification. Probabilistic analyses are generally performed before complete knowledge is available and as such involve a degree of subjectivity. They are only tools and not something that one "believes in" as one would in religious dogmas. If potential users – and the people who are subjected to their decisions – prefer to rely on their instincts, so be it. The complexity of many systems, however, does not permit that luxury, when the stakes are high and human intuition insufficient to capture the intricacies of the problem at hand.

References

Apostolakis G. (1990), "The concept of probability in safety assessments of technological systems", *Science*, Vol. 250, pp. 1359–1364.

Arrow K.J. (1963), *Social Choices and Individual Values*, Wiley, New York (2nd Edition).

Bier V.M., Cox, L.A. (2006), "Probabilistic Risk Analysis for Engineered Systems", Chapter 16 in Advances in Decision Analysis, Edward, Miles and von Winterfeldt Eds., Cambridge University Press.

de Finetti B. (1974), *Theory of Probability*, Wiley, New York.

Ellsberg, D. (1988), "Risk, ambiguity and the Savage axioms" Chapter 13 in *Decision, Probability, and Utility*, P. Gärdenfors and N.E. Sahlin, Eds, Cambridge University Press.

Garrick, J.B. (1984), "Recent case studies and advancements in probabilistic risk assessment", *Risk Analysis*, Vol. 4, No.4, pp. 267–279.

Guikema, S.D., Paté-Cornell, M.E. (2004), "Bayesian analysis of launch vehicle success rate", *Journal of Spacecraft and Rockets*, Vol. 41, No. 1.

Henley E.J., Kumamoto, H. (1981), *Reliability Engineering and System Safety*, Prentice Hall, Englewood Cliffs, NJ.

Howard R.A. (1970), "Decision analysis: Perspectives on inference, decision and experimentation", *Proceedings of the IEEE*, Vol. 58, No.5, pp. 632–643.

Howard, R.A. (1984), "Risk preference", in *The Principles and Applications of Decision Analysis* Howard and Matheson, Eds. SDG, Pub. Menlo Park, CA.

Howard R.A., Matheson, J.E., Eds. (1984), *The Principles and Applications of Decision Analysis*, SDG Pub., Menlo Park, CA.

Kaplan S., Garrick, B.J. (1981), "On the quantitative definition of risk", *Risk Analysis*, Vol. 1, No. 1, pp. 11–27.

Keeney R., Raiffa, H. (1976), *Decisions with Multiple Objectives*, Wiley, New York.

Kahneman D., Tversky, A. (1979), "Prospect theory: An analysis of decision under risk", Econometrica, Vol. 47, 263–291.

Morgan M.G., Henrion, M. (1990), *Uncertainty*, Cambridge University Press, New York, NY.

Neyman, J. (1937), "Outline of a theory of statistical estimation based on the classical theory of probability", *Philosophical Transactions of the Royal Society*, Ser. A, Vol. 236, pp. 333–380.

Paté-Cornell, M.E., Davis, D.B. (1994), "A challenge to the compound lottery axiom: A two-stage normative structure and comparison to other theories," *Theory and Decision*, Vol. 37, No. 3, pp. 267–309.

Paté-Cornell, M.E., Fischbeck, P.S. (1993a), "Probabilistic risk analysis and risk-based priority scale for the tiles of the space Shuttle," *Reliability Engineering and System Safety*, Vol. 40, No. 3, pp. 221–238.

Paté-Cornell, M.E., Fischbeck, P.S. (1993b), "PRA as a management tool: Organizational factors and risk-based priorities for the maintenance of the tiles of the space shuttle Orbiter," *Reliability Engineering and System Safety*, Vol. 40, No. 3, pp. 239–257.

Paté-Cornell, M.E. (1996), "Uncertainties in risk analysis: Six levels of treatment", *Reliability Engineering and System Safety*, Vol. 54, pp. 95–111.

Paté-Cornell, M.E. (1999a), "Conditional uncertainty analysis and implications for decision making: The case of the waste isolation pilot plant," *Risk Analysis*, Vol. 19, No. 5, pp. 995–100.

Paté-Cornell M.E. (1999b), "Medical application of engineering risk analysis and anesthesia patient risk illustration," *American Journal of Therapeutics*, Vol. 6, No. 5, pp. 245–255.

Paté-Cornell, M.E. (2000), "Greed and ignorance: Motivations and illustrations of the quantification of major risks", *Proceedings of the Study Week on "Science for*

Survival and Sustainable Development" (March 12–16, 1999), pp. 231–270, Pontificiae Academiae Scientiarum Scripta Varia (Report of the Pontifical Academy of Sciences), The Vatican.

Paté-Cornell, M.E. (2002), "Finding and fixing systems weaknesses: Probabilistic methods and applications of engineering risk analysis," *Risk Analysis*, Vol. 22, No. 2, pp. 319–334.

Paté-Cornell, M.E., Guikema, S.D. (2002), "Probabilistic modeling of terrorist threats: a systems analysis approach to setting priorities among countermeasures," *Military Operations Research*, Vol. 7, No 4.

Pietzsch, J.B., Paté-Cornell, M.E., Krummel, T.M. (2004). "A Framework for Probabilistic Assessment of New Medical Technologies", Proceedings of PSAM7/ESREL04, Berlin, Germany, Springer-Verlag Pub. London UK: 2224–2229.

Press, S.J. (1989), *Bayesian Statistics: Principles, Models, and Applications*, John Wiley and Sons, New York, NY.

Raiffa H. (1968), *Decision Analysis*, Addison Wesley, Cambridge, MA.

Ramsey F.P. (1926), "Truth and probability" in *The Foundations of Mathematics and Other Logical Essays*, Kegan, London.

Savage, L.J. (1954), The Foundations of Statistics, Wiley, New York.

Slovic, P. (1987), "Perception of risk", *Science*, Vol. 236, pp. 280–285.

Starr, C. (1969), "Social benefit vs. technological risk", *Science*, Vol. 165, pp. 1232–1239.

Tversky A., Kahneman, D. (1974), "Judgment under uncertainty: Heuristics and biases", *Science*, Vol. 185, pp. 1124–1131.

U.S. Nuclear Regulatory Commission (USNRC) (1975), *Reactor Safety Study: Assessment of Accident Risk in U.S. Commercial Nuclear Plants*, WASH-1400 (NUREG-75/014). Washington, D.C.

von Neumann J., Morgenstern, O. (1947), *Theory of Games and Economic Behavior*, Princeton University Press.

A Conjoint Measurement Based Rationale for Inducing Preferences

Heiko Großmann[1,2], Michael Brocke[2], and Heinz Holling[2]

[1] School of Mathematical Sciences, Queen Mary, University of London, UK
[2] Psychologisches Institut IV, Westfälische Wilhelms-Universität Münster, Germany
e-mail: h.grossmann@qmul.ac.uk, brockem@psy.uni-muenster.de, holling@psy.uni-muenster.de

Abstract. For additive models of preferences or choices among multi-attribute options an approach to inducing preferences is presented which offers new opportunities for empirical investigations of choice behavior as well as the validation of preference elicitation techniques. The approach is founded in measurement theory and draws on finite conjoint measurement and random utility theory. Preferences are induced by teaching respondents to choose among multi-attribute options in accordance with a weak order that can be represented by a unique set of utility values. It is shown that the utility values can be recovered with estimation procedures for probabilistic choice models. As the utility values are known a priori, they can serve as a standard of comparison for estimates in empirical investigations. The measurement theoretic background and a procedure for teaching preferences are described in detail. Data from two experiments provide evidence that accurate numerical utility values can be induced with tasks that require only qualitative judgments but do not reveal any numerical information.

Keywords: preference inducement, finite conjoint measurement, random utility theory, discrete choice analysis

1 Introduction

The concept of utility is a cornerstone of most theories of decisions among multi-attribute options. Standard conceptions of multi-attribute decision making, such as multi-attribute utility theory (Keeney and Raiffa, 1976), conjoint measurement (Luce and Tukey, 1964), or multi-attribute extensions of random utility theory (Luce and Suppes, 1965), presume that the utility of options determines preferences and choices. Widely used techniques for eliciting preferences in multi-attribute settings, like conjoint analysis (Green & Srinivasan, 1978, 1990), policy capturing (Karren and Barringer, 2002), and discrete choice analysis (Louviere et at., 2000), also rest on the concept of utility and yield utility values as a crucial part of their results. However, there seems to be no adequate criterion available for making direct evaluations at the level of utility values when assessing the validity of such methods is of interest or when the question of how preferences are affected by the experimental manipulation of variables of the decision context is an issue. Researchers have therefore usually employed indirect measures like holdout choices or preference reversals. In the present paper we offer an alternative criterion that represents a theoretically-founded standard of comparison at the level of utility values.

The endeavor to establish a direct standard of comparison at the level of utility values confronts a number of impediments. First, utilities reside at the construct level and are not directly observable. Hence, the utility values of a person are unknown. How then should one be able to compare different preference elicitation techniques or appraise the effect of experimental manipulations at this level of analysis? Second, preferences are usually heterogeneous. That is, different individuals have different sets of utilities and this exacerbates the problem of establishing some common ground on which utilities can be compared.

A potential solution to the first problem would be to elicit preferences with some technique and use the resulting utilities as the standard of comparison. However, any conclusions reached by using this approach would critically depend on the specific method used for eliciting the reference values. For instance, if one was interested in assessing the validity of different elicitation techniques the problem would become circular since the results would depend on the validity of the method that has been singled out to obtain the benchmark values.

The issue of preference heterogeneity might be dealt with by actively reducing heterogeneity to at least alleviate the problem. For example, a sample of respondents could be investigated which can be assumed a priori to have comparatively homogeneous preferences. Of course, this method requires at least partial knowledge about the utilities in the sample. A more systematic approach to reduce heterogeneity in experimental settings would be to ask participants to adopt the preferences of a fictitious person and to perform some task vicariously on behalf of that person. This approach, known as the principal-agent paradigm, has been applied in several recent studies (Ariely, 2000; Huber et al., 2002).

In order to overcome the difficulties implied by the nature of preferences and utilities we propose to generate samples of respondents with homogeneous preferences which can be represented by a known set of utility values. These utility values then constitute the proposed standard of comparison. A sample with homogeneous preferences is established by means of an inducement procedure. In short, this procedure takes the form of a computer administered training during which respondents are gradually led to evaluate multi-attribute options in accordance with a certain preference structure. In this training, the options are described by several attributes with a finite number of levels each. The preference structure is defined by a weak order that can be represented additively by a unique set of utility values. Weak orders with this property are characterized by a uniqueness theorem in finite conjoint measurement (Fishburn and Roberts, 1988) which will be explained shortly. We demonstrate that even though the inducement of preferences is based on the deterministic theory of conjoint measurement, the choices are consistent with a certain class of probabilistic choice models. Consequently, statistical estimation procedures for these models can be used to recover the induced utility values.

The following features of the approach are worth noting. First, participants have to internalize only the qualitative information represented by the weak order. That is, for every pair of options they have to learn which option is preferred or whether the options are equivalent. Second, the qualitative information is sufficient to determine a unique set of numerical utilities and, hence, to establish an unambiguous standard of comparison. Third, no numerical information regarding the utility values of the attribute levels is revealed at any stage of the training. Moreover, participants receive no information as to the additive rule relating overall and marginal utilities. Fourth, the approach offers the opportunity to systematically vary utility structures because (within certain constraints) the experimenter is free to choose a set of utilities for constructing the weak order which has to be internalized. Finally, since all participants are taught the same weak order, and hence the same set of utilities, the heterogeneity problem is eliminated provided that participants successfully master the learning task.

The application of the method requires that participants are first trained to evaluate a set of multi-attribute options according to a specified weak order. After a test of learning success an experimental manipulation can be introduced that is hypothesized to alter the utility structure. The effectiveness of the manipulation can then be evaluated, for instance, by comparing utility values elicited after the manipulation with the utilities underlying the weak order.

In what follows we describe the measurement theoretic underpinnings of the proposed approach in more detail. Subsequently, a training procedure is presented which implements the approach. Finally, we report the test results for the training procedure's effectiveness in two studies.

2 Measurement Theoretic Background

The approach briefly described in the previous section posits an additive relationship between the overall utility of an option and the utility values of its levels. Since we want to employ the utility values associated with the attribute levels as the standard of comparison in empirical investigations, our goal is to teach participants to apply an additive rule using a given set of level utilities when evaluating options. However, as we do not want our respondents to actually compute their responses, numerical values are not presented and it is not explained how these values add up to yield the overall utility of an option. Instead, participants have to interiorize or implicitly learn a preference relation defined by a weak order which uniquely determines the set of utility values. We now address the question how weak orders with this property can be constructed.

2.1 Conjoint Measurement

The question when a weak order (complete and transitive relation) $\succsim$ on a product set $A = A_1 \times \ldots \times A_K$ can be additively represented is the subject of conjoint measurement. In other words, conjoint measurement is concerned with the problem under which conditions there do exist real-valued utility functions u_k defined on the set of attribute levels A_k,$k = 1, \ldots, K$, such that for any two options $a = (a_1, \ldots, a_K)$ and $b = (b_1, \ldots, b_K)$ in A we have

$$a \succsim b \Leftrightarrow \sum_{k=1}^{K} u_k(a_k) \geq \sum_{k=1}^{K} u_k(b_k) \,. \tag{1}$$

For preferences among options that are characterized by two attributes, Luce and Tukey (1964) were the first to state a set of purely algebraic conditions, so-called axioms, for a weak order which are sufficient to prove the existence of an additive representation as in (1). Their results were generalized by Luce (1966) who presented a corresponding representation theorem for three or more attributes. Alternative formulations of this theorem can be found in Krantz et al. (1971, p. 302) and Fishburn (1970). As a particular feature, the representation theorem not only states conditions under which a weak order can be additively represented but at the same time characterizes the degree to which such a representation is unique. In particular, under the conditions of the theorem, if there is another set of functions $\tilde{u}_k$ defined on A_k, $k = 1, \ldots, K$, for which (1) holds, then there exist numbers $\alpha > 0$ and β_k such that

$$\tilde{u}_k = \alpha u_k + \beta_k \tag{2}$$

for all $k = 1, \ldots, K$. Equation (2) means that the function u_k that assigns utility values to the levels of attribute k is unique up to a positive similar transformation, that is, a linear transformation with an attribute specific intercept term β_k and a slope parameter α which is the same for all attributes. The important consequence of this uniqueness property is that any difference (or contrast) of utility values within attribute i can be compared with any difference of utility values within attribute j. In other words, within-attribute differences are measured on a common ratio scale.

The representation theorem of conjoint measurement provides the motivation for our idea to teach respondents a preference structure in order to induce utility values that are unique up to the degree specified in (2). However, it is well-known that the representation theorem referred to here applies only when every attribute possesses an infinite number of levels. This fact is a consequence of the solvability axiom (Krantz et al., 1971, p. 301) used in its proof. Consequently, when only attributes with a finite number of levels are considered and an additive representation as in (1) exists, it cannot be claimed on grounds of the representation theorem that the utility functions are unique. In fact, in the case of finite conjoint measurement – that is, when all attributes have a finite number of levels – the utility functions possess only rather weak uniqueness properties if additional restrictions on the preference relation are not imposed (Krantz et al., 1971, p. 431).

Additional conditions that are necessary and sufficient to proof the uniqueness of utility functions for finite conjoint measurement were presented by Fishburn and Roberts (1988). It is worth noting that any attempt to use conjoint measurement in an empirical study deals with a finite number of levels for each attribute. In a strict sense, all empirical studies therefore represent instances of finite conjoint measurement. It is surprising then, that Fishburn and Roberts's paper did not receive much attention.

As weak orders with a unique additive representation as defined by (1) and (2) are central to our approach, we briefly describe how they can be constructed following Fishburn and Roberts (1988, Corollary 1).

To this end, consider K attributes with $n_k + 1$ levels each. Denote the set of levels for attribute k by $A_k = \{a_{k,1}, \ldots, a_{k,n_k+1}\}$. Let $d = (d_{1,1}, \ldots, d_{1,n_1}, \ldots, d_{K,1}, \ldots, d_{1,n_K})$ be a vector of size $\sum_{k=1}^{K} n_k$. Furthermore, consider homogeneous linear equations of the form

$$\sum_{k=1}^{K} \sum_{j=1}^{n_k} \varepsilon_{i,k,j} d_{k,j} = 0\,, \tag{3}$$

with coefficients $\varepsilon_{i,k,j}$ in $\{-1, 0, 1\}$. The existence of a weak order $\succsim$ on $A = A_1 \times \ldots \times A_K$ with a unique additive representation according to (1) and (2) then depends on finding a strictly positive solution d (i.e., a vector d with exclusively positive components) to a system of $\sum_{k=1}^{K} n_k - 1$ linearly independent equations as in (3) for which the coefficients $\varepsilon_{i,k,j}$ in the i-th equation satisfy the following three additional conditions:

P1: For fixed i and every k all $\varepsilon_{i,k,1}, \ldots, \varepsilon_{i,k,n_k}$ are either in $\{0, 1\}$ or $\{0, -1\}$.
P2: For fixed i and every k all nonzero $\varepsilon_{i,k,j}$, $j = 1, \ldots, n_k$, are contiguous or form an interval of consecutive j values.
P3: For fixed i both $+1$ and -1 appear among the coefficients $\varepsilon_{i,k,j}$.

Now starting from a strictly positive solution $d = (d_{1,1}, \ldots, d_{1,n_1}, \ldots, d_{K,1}, \ldots, d_{1,n_K})$ a utility function u_k for attribute k, $k = 1, \ldots, K$, can be defined by fixing $u_k(a_{k,1})$ at some arbitrary value and letting $u_k(a_{k,j}) = u_k(a_{k,1}) + \sum_{i=1}^{j-1} d_{k,i}$ for $j = 2, \ldots, n_k + 1$. Finally, the weak order $\succsim$ on A is defined by (1).

In practice, one begins with a tentative set $d_{k,j}$, $j = 1, \ldots, n_k$, $k = 1, \ldots, K$, of positive integers that correspond to differences between adjacent attribute levels in the utility structure one intends to induce in the participants. That is, for every k and $j = 1, \ldots, n_k$ the number $d_{k,j}$ can be interpreted as the difference $u(a_{k,j+1}) - u(a_{k,j})$. If a system of $\sum_{k=1}^{K} n_k - 1$ linearly independent equations as in (3) that satisfy P1, P2, and P3 can be established and if this system can be solved by the $d_{k,j}$ values, then the desired weak order can be constructed as outlined above. Otherwise, the $d_{k,j}$ have to be modified until a respective system of equations exists.

It should be noted that the above construction implies indifference (i.e., $a \succsim b$ and $b \succsim a$) for at least $\sum_{k=1}^{K} n_k - 1$ pairs of options a and b in A. Conversely, it can be shown, that any weak order with a unique additive representation as in (1) and (2) necessarily yields at least $\sum_{k=1}^{K} n_k - 1$ equivalent pairs (Fishburn and Roberts, 1988). Thus, indifference is crucial in finite conjoint measurement for establishing the uniqueness of an additive representation. Consequently, special attention has to be paid to indifference when respondents are trained to evaluate options according to a weak order satisfying (1) and (2).

In principle, a test of learning success can be performed by asking the respondents to produce a ranking of all options in which ties are explicitly permitted. Yet, within the theoretical framework of conjoint measurement a single error in such a ranking would mean that the training procedure has not perfectly induced the desired utilities. The reason is that conjoint measurement is a deterministic theory without a concept of response errors. From our point of view it is conceivable, however, that by and large a respondent has internalized the preference structure in spite of ordering some options incorrectly on occasion. Hence, we suggest that the evaluation of learning success should be based on a model that is compatible with conjoint measurement, but can tolerate response errors. In the following, we establish a connection between conjoint measurement and random utility theory. In this context, specific predictions will be derived as to what results are to be expected when the inducement of preferences has been successful and respondents are tested with choice tasks.

2.2 Connection with Random Utility Models

As a deterministic theory, conjoint measurement would require an individual to always select the same option when repeatedly presented with a fixed set of choice options. Clearly, observed choices violate this strong requirement. To account for random fluctuations, which at different times may yield different choices from the same set, choices are usually treated within a probabilistic framework. In general, deterministic and probabilistic theories cannot be reconciled easily. In the following we show, however, that within a certain class of probabilistic choice models observed choices of a person, who has been successfully trained to evaluate multi-attribute options according to a weak order with a unique additive representation, can be fully explained by the utility values underlying that order.

In particular, we are interested in a special type of independent random utility models known as Thurstone models (Yellott, 1977). Given a set $C_n = \{o_1, \ldots, o_n\}$ of n (not necessarily multi-attribute) options, a Thurstone model posits the existence of scale values $v(o_1), \ldots, v(o_n)$ and n independent, identically distributed continuous random variables $X_1, \ldots, X_n$ with a known cumulative distribution function F such that the probability $p_S(o_i)$ of choosing option o_i from a subset $S \subseteq C_n$ is given by

$$p_S(o_i) = P\left[v(o_i) + X_i = \max\{v(o_j) + X_j | o_j \in S\}\right] \tag{4}$$

for every $o_i \in S$ and all $S \subseteq C_n$.

It is well-known (Yellott, 1977) that if F is the cumulative distribution function of the (standard) double exponential distribution defined by $F(x) = \exp(-\exp(-x))$ for all real x, Equation (4) is equivalent to

$$p_S(o_i) = \frac{\exp(v(o_i))}{\sum_{o_j \in S} \exp(v(o_j))} \,. \tag{5}$$

In other words, the Thurstone model corresponding to the double exponential distribution is the multinomial logit model, which represents the most prominent model of discrete choice analysis (Louviere et al., 2000). Moreover, if F is the cumulative distribution function of some other member from the location-scale family of double exponential distributions, that is, $F(x) = \exp(-\exp(-(ax+b)))$ for some constants $a > 0$, b and all x, it follows that the corresponding Thurstone model (4) is equivalent to a multinomial logit model (5) with scale values $v'(o_1), \ldots, v'(o_n)$ in place of $v(o_1), \ldots, v(o_n)$ which are related by $v'(o_i) = av(o_i)$ for all $i = 1, \ldots, n$.

As an aside, we note that in using the name double exponential distribution we follow Yellott (1977). This distribution is usually referred to as the Gumbel distribution or extreme value distribution of Type 1 in the statistical literature, and the members of the corresponding location-scale family are known as Fisher–Tippett distributions.

We now turn to the case of multi-attribute options and establish the connection between conjoint measurement and the multinomial logit model. To this end, we assume that every option $o \in C_n$ is characterized by K attributes with a finite number of levels each. More precisely, we consider the set $C_n = A_1 \times \ldots \times A_K$, where as before $A_k = \{a_{k,1}, \ldots, a_{k,n_k+1}\}$ denotes the set of levels of attribute k. Suppose that based on utility values $u_k(a_{k,j})$, $j = 1, \ldots, n_k + 1$, $k = 1, \ldots, K$, a weak order on C_n that possesses a unique additive representation has been constructed as described in the previous section. If a respondent has internalized this order, then the overall utility $\tilde{u}(o_i)$ of option $o_i = (o_{i,1}, \ldots, o_{i,K})$ with $o_{i,k} \in A_k$ can be expressed as

$$\tilde{u}(o_i) = \tilde{u}_1(o_{i,1}) + \ldots + \tilde{u}_K(o_{i,K}) \,, \tag{6}$$

where according to (2) $\tilde{u}_k(o_{i,k}) = \alpha u_k(o_{i,k}) + \beta_k$ for every i and k. Consequently, for every i we have $\tilde{u}(o_i) = \alpha u(o_i) + \beta$ where $u(o_i) = u_1(o_{i,1}) + \ldots + u_K(o_{i,K})$ and $\beta = \beta_1 + \ldots + \beta_K$.

If the respondent is now asked to choose among options from subsets $S \subseteq C_n$, and if in addition it is assumed that the corresponding choice probabilities $p_S(o_i)$ can be described by a Thurstone model, where the scale value for every option $o_i \in C_n$ is given by $\tilde{u}(o_i)$ according to (6) and if the random variables $X_1, \ldots, X_n$ are distributed according to *some* double exponential distribution with cumulative distribution function $F(x) = \exp(-\exp(-(ax + b)))$, then the choice probabilities can be represented by

$$p_S(o_i) = P\left[\alpha u(o_i) + \beta + X_i = \max\{\alpha u(o_j) + \beta + X_j | o_j \in S\}\right] . \quad (7)$$

Since for any random variable X in (7) the shifted variable $\beta + X$ is again distributed according to some member of the family of double exponential distributions, it follows that for every S the choice probabilities $p_S(o_i)$ in (7) can be equivalently expressed by (5) with the scale values $v(o_1), \ldots, v(o_n)$ replaced by $cu(o_1), \ldots, cu(o_n)$, where $c = \alpha a$. More explicitly, the choice probabilities are given by

$$p_S(o_i) = \frac{\exp(cu_1(o_{i,1}) + \ldots + cu_K(o_{i,K}))}{\sum_{o_j \in S} \exp(cu_1(o_{j,1}) + \ldots + cu_K(o_{j,K}))} . \quad (8)$$

Notice that generally the components α and a of c cannot be disentangled.

Thus far we have only considered a single respondent. When we turn to the entire sample, Equation (8) holds for every individual who has acquired the preference structure, but possibly with a different value of c for each person. In the following, we nevertheless assume to the contrary that c is the same for all respondents. That is to say that the ratio of α to the standard deviation of the random variables in (7) is constant throughout the sample, whereas the value of α does not have to be.

Under the assumption that c does not depend on the individual respondents, model (8) is appropriate for the whole sample. The model parameters are the values $cu_k(a_{k,j})$, $j = 1, \ldots, n_k + 1$, $k = 1, \ldots, K$, which can be estimated by maximum-likelihood procedures for the multinomial logit model using a suitable coding of the attribute levels or, equivalently, suitable identifiability constraints. The constraints imply that for every attribute k only n_k parameters can be estimated. Moreover, each of those estimates corresponds to a difference in the parameters $cu_k(a_{k,j})$.

Consequently, when respondents have acquired a unique weak order based on the conjoint measurement approach, we expect every parameter estimate obtained from the multinomial logit model (8) to be a multiple of a corresponding difference in the original utility values $u_k(a_{k,j})$ that have been used for constructing the weak order. This prediction will be tested in the empirical studies reported later. If it can be substantiated, we conclude that the attribute-level utilities underlying the weak order fully account for the choice probabilities even though they have been induced according to the deterministic conjoint measurement model.

In concluding this section we want to mention that instead of the Thurstone model based on the double exponential distribution we could also have used the respective model for the normal distribution to establish a similar connection with conjoint measurement. We concentrated on the double exponential distribution since the corresponding multinomial logit model is most widely used in applications. The discussion in this section was predicated on the assumption that respondents have been trained successfully to apply a weak order with a unique additive representation. We now address the question how this goal can be achieved.

3 Training Procedure

The goal of the training procedure is to teach respondents to evaluate multiattribute options according to a weak order with a unique additive representation. To construct such an order, we apply the method described in the section on conjoint measurement. As was already noted, the weak order then necessarily contains a certain number of equivalent objects for which no order of preference exists. In the following such objects will be referred to as tied objects. Since tied objects are crucial for the uniqueness of the representation, special attention has to be paid to teaching indifference relations.

We adopt the so-called principal-agent paradigm (e.g., Huber et al., 2002). In this paradigm, respondents represent agents who have to perform tasks vicariously on behalf of a fictitious person, called the principal. At the beginning of the training respondents receive a written cover story that explains the attributes and levels used for characterizing the options. The story also presents a rough sketch of the principal's preferences in narrative form. The purpose of this description is to provide some ordinal information concerning the attributes and levels that can serve as a starting point when working on the subsequent tasks. However, since we only want to teach relations among options, the description does not contain any numerical values.

The tasks are grouped into several modules to be described below with one module per type of task. Preceding each task, respondents receive a written instruction in which it is explained and an example is provided. The modules are administered computerized. Every module comprises a certain task in several trials. After each trial respondents automatically receive textual feedback as to whether their response was right or wrong. Additionally, the correct response is indicated in case a wrong answer is given. Furthermore, the feedback is enhanced by highlighting correct responses in green and errors in red color. We assume that by using the feedback respondents can gradually revise their initial beliefs about the principal's preferences and finally arrive at evaluating the options in accordance with the weak order.

The construction of the tasks was guided by several principles. First, no numerical information was presented. This was motivated by the goal to teach respondents preferences in an ecologically valid manner instead of just having them carry out simple arithmetic with the numbers presented. We assume that the use of numbers in the training would have resulted in much faster and easier learning, but at the same time it would have seriously undermined generalizability of the procedure and results to real-world settings. Second, the training had to employ multiple tasks. One reason for this was that multiple tasks allowed us to focus the respondent's attention on certain aspects of the weak order as tied objects or relations among only two or several options. Another reason was to keep the participants' task engagement at a high level. Third, task complexity was intended to increase during the training. That is, whereas early modules only asked comparatively simple questions, later modules required the integration of a larger amount of information and also a more complex kind of response. These principles gave rise to a total of five modules shown in Table 1.

The first two modules represent paired comparison tasks. List descriptions of two options in terms of attribute levels are simultaneously presented on the computer screen. In Module I respondents have to indicate which of the two options would be chosen by the principal or whether she would be indifferent. Since this kind of task might cause respondents to focus on positive features of the alternatives (Meloy and Russo, 2004), Module II was designed to counteract this potential bias. Here, respondents are asked to indicate which option would be rejected by the principal or, again, whether she would be indifferent.

While Modules I and II offer indifference as a response category, in consideration of the importance of tied objects, Module III is targeted at systematically training indifference relations. To this end, descriptions of four options are simultaneously presented on the screen. These options are chosen in such a way that every set contains two pairs of equivalent options. The task is to mark two options between which the principal would be indifferent, thereby identifying both pairs.

Table 1. Training modules

Module	Options per trial	Task
I	2	Choose preferred option
II	2	Reject less preferred option
III	4	Identify equivalent options
IV	5	Order first two options, then pick intermediate option from remaining three options
V	3	Order first two options, then decide position of third option

The most demanding tasks are presented in Modules IV and V. In Module IV every trial displays a set of five options. Two of these options are shown at the top of the screen whereas the remaining options are displayed at the bottom. The response requires two judgments. First, respondents have to indicate which of the two options at the top would be preferred by the principal. Subsequently, the option that lies (strictly) between those at the top has to be identified among the options at the bottom. The five options in a trial are chosen in such a way that each subtask has a unique solution. Fig. 1 presents an example of this kind of task in which the first subtask is already completed.

Finally, Module V presents a set of three options and also employs two subtasks. The first subtask is the same as in Module IV. That is, among two options displayed at the top of the screen the most preferred option has to be identified. The second subtask then asks how a third object shown at the bottom of the screen relates to the options at the top. Here respondents have to choose among five response categories represented by arrows: a) less preferred than both options, b) equivalent to top left option, c) situated between both options d) equivalent to top right option, and e) more preferred than both options. The three options for every trial are again chosen in such a way that both subtasks have a unique solution. An example is shown in Fig. 2.

We now turn to empirical studies in which the practicability of our approach to inducing preferences was investigated.

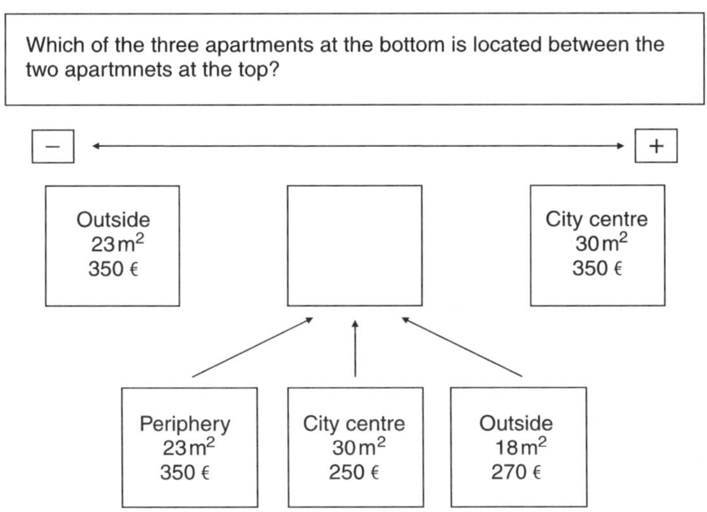

Fig. 1. Sample task for Module IV with first subtask already completed

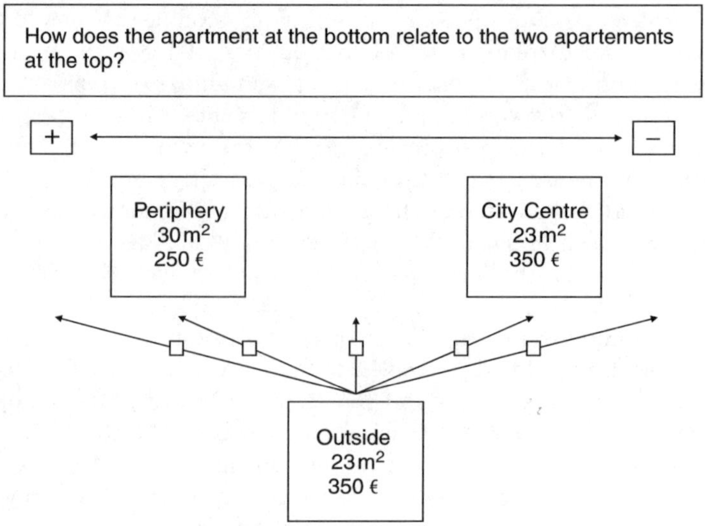

Fig. 2. Sample task for Module V with first subtask already completed

4 Empirical Test

In the following we describe a study in which the proposed approach was used to teach preferences for apartments. Subsequently, we present the results of a replication study with a different sample of respondents. In both studies, apartments were described by three attributes with three levels each yielding a total of 27 options. The attributes and levels are shown in Table 2. Utility values were worked out and used to construct a weak order with a unique additive representation as described in the section on conjoint measurement. The resulting values are also shown in Table 2. Since within each attribute the utility of one level can be chosen arbitrarily, we fixed the value for the lowest level of each attribute at zero.

In the first study, seventeen undergraduate students volunteered to participate in the study. Every respondent was paid 10 Euros. The schedule of the study was as follows: First, respondents were asked to rank order the 27 apartments according to their own preferences. To this end, each respondent received a pile of 27 cards representing the apartments and a large sheet of cardboard. The ranking task including the option to produce ties was explained in detail. Respondents were then given 25 minutes to generate the ranking that was fixated on the cardboard. Following the ranking task, the computerized training procedure was administered as described in the previous section. Between Modules III and IV a one hour interval was inserted. Subsequently, Modules IV and V were administered. Altogether the training procedure took 2.5 hours. Following a short break of 15 minutes the respondents were asked next to rank order the apartments again, but this time according to the preferences of the principal. As this task was deemed to be rather demanding, 40 minutes were allocated for this purpose. Finally, each respondent had to perform 27 computerized choice tasks in 20 minutes, where they had to choose from among three apartments in accordance with the preferences of the principal. In the following, we only present the analysis of the choice data.

The choice data were analyzed within the multinomial logit model. Parameter estimates were computed by maximum-likelihood using the SAS software. The attribute levels were dummy coded using for each attribute the level with the highest utility value in Table 2 as the reference category. Thus, the parameter estimates represent contrasts between levels and are expected to be negative. Table 3 presents the results. The table also shows the true values for the contrasts derived from Table 2. Based on the theoretical derivations presented in the previous sections we expect the estimates to coincide with the true contrasts (i.e., differences) up to a positive factor. Fig. 3 shows a scatterplot of the estimated against the true contrasts. The points fall almost perfectly on a straight line.

Table 2. Attributes, levels, and utilities of apartments

Attribute	Levels	Utility values
Rent	350 €	0
	270 €	4
	250 €	5
Location	Outside	0
	Periphery	2
	City centre	4
Size	18 m^2	0
	23 m^2	1
	30 m^2	2

Table 3. True contrasts and parameter estimates

			Parameter estimates	
Attribute	Contrast	True value	Original study	Replication
Rent	350 €–250 €	−5	−16.36	−9.64
	270 €–250 €	−1	−2.58	−1.20
Location	Outside – city centre	−4	−12.43	−7.98
	Periphery – city centre	−2	−5.77	−3.57
Size	18 m^2–30 m^2	−2	−5.74	−2.96
	23 m^2–30 m^2	−1	−3.15	−1.21

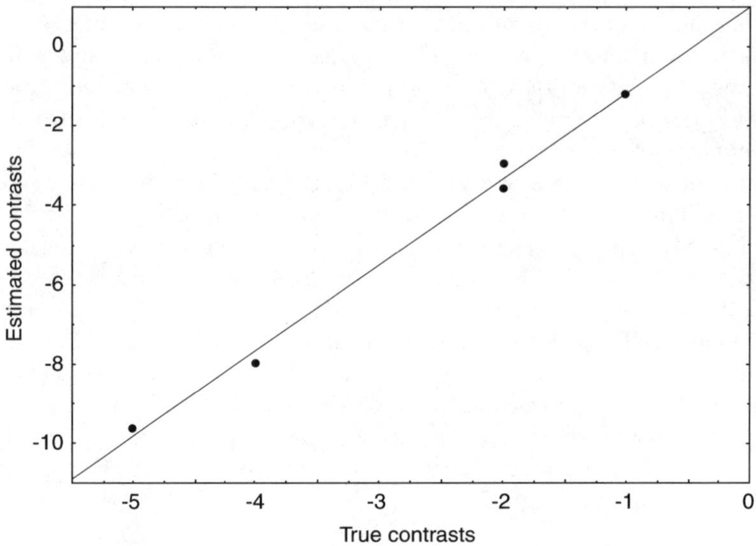

Fig. 3. Estimated and true contrasts in original study

A linear regression of the estimated on the true parameters yields a highly significant slope parameter of $\beta_1 = 3.35$ ($p < .001$) whereas the intercept of $\beta_0 = 0.70$ does not reach statistical significance ($p = .08$). An analysis of the residuals reveals some deviations from normality, however, so that the latter test should not be overemphasized. The same analysis without the intercept term results in a slope parameter of $\beta_1 = 3.14$. Moreover, the coefficient of determination for that model is equal to $R^2 = .991$ indicating that over 99% of the variation in the parameter estimates can be explained by a simple multiplication of the true contrasts. Based on these findings we conclude that respondents have successfully learned the weak order and hence the underlying utility values.

The replication study took place one year after the original study. The sample consisted of forty-six undergraduate students none of which had participated in the original study. Every participant was paid 10 Euros. The design of the replication study was identical to the one of the original study except that the order of the second ranking task and the task requiring choices among three apartments was reversed. Moreover, each respondent had to make choices from 54 sets whereas the original study used only 27 sets. Maximum-likelihood estimates of the parameters in the multinomial logit model were estimated using the SAS software. The results are also shown in Table 3. A scatterplot of the estimated and true contrasts is exhibited in Fig. 4. A linear regression of the estimated on the true parameters yields a highly significant slope parameter of $\beta_1 = 2.16$ ($p < .001$) and a significant intercept of $\beta_0 = 0.98$ ($p = .01$). As in the original study, not too much emphasis should be placed on the test for the intercept because an analysis of

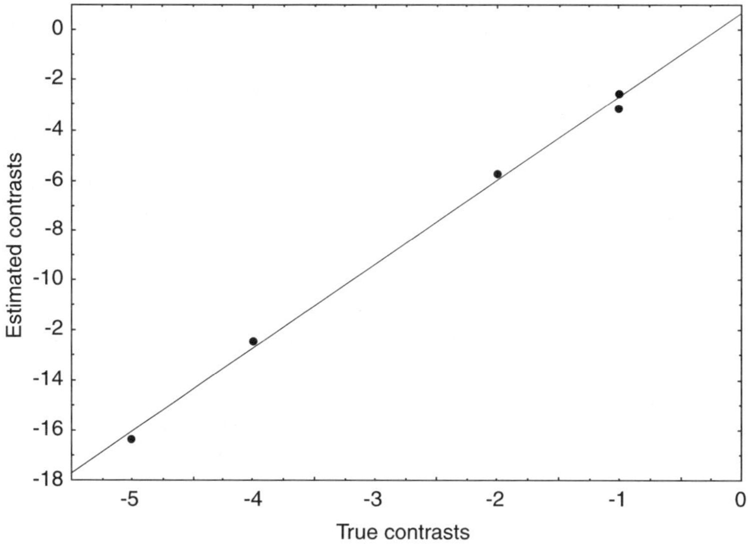

Fig. 4. Estimated and true contrasts in replication study

the residuals again indicates some deviations from normality. The regression without an intercept term yields a slope parameter of $\beta_1 = 1.87$ and a coefficient of determination of $R^2 = .971$. In terms of goodness of fit the results of the replication are thus similar to those of the original study.

5 Discussion

The main goal of the present work has been to establish a coherent framework for inducing preferences among multi-attribute options. By tying together results from finite conjoint measurement and random utility theory we have elaborated the basis of a method for teaching accurate numerical utility values without disclosing any numerical information. A training procedure that implements the method was presented and evaluated in two empirical studies. The results of these studies provide unequivocal evidence that preferences can be successfully and very accurately induced with our approach. The hypothesis that preferences are indeed acquired is further corroborated by the fact that the task for testing learning success differed from the ones used during the training phase. Consequently, respondents had to generalize their learning experience, a goal they achieved remarkably well with our training.

The inducement of preferences is an important step for empirical research on preference assessment methods. The significance of a corresponding procedure derives from its manifold prospects for empirical research. As was mentioned previously, accurately induced utility values can serve as a standard of comparison for appraising the effects of experimental manipulations on preferences as well as for evaluating preference elicitation techniques. The usefulness of being able to perform comparisons at the level of utilities has been recognized recently (Ariely, 2000; Huber et al., 2002). For example, in their study of preference assessment methods Huber et al. (2002) acknowledged that using an agent task "can tell us not only how the methods differ from each other but also how they differ from the true preference structure which the agent seeks to emulate" (p. 67). However, the cited works differ from our usage of the principal-agent paradigm in that they do not strive to induce preferences. Instead, information regarding the utility values is presented overtly in the form of bar graphs that the respondents have at their disposal during the entire experiment. We consider it a virtue of our approach that we do not use the more or less overt numerical information contained in such displays but rely on qualitative information only. Nevertheless, we are able to make accurate numerical predictions as to what results have to be expected when learning success is tested with choice tasks.

We emphasized earlier that an inducement procedure eliminates the problem of preference heterogeneity. It should be noted, however, that our procedure can also be used to introduce heterogeneity in a perfectly controlled manner. In the simplest case, this can be achieved by applying our training procedure in two groups but using a different weak order for each group. It is thus possible, for example, to generate experimental conditions with a linear or nonlinear preference structure (Huber et al., 2002). Another application would be the manipulation of choice difficulty by varying the range of utility levels (Stone and Kadous, 1997).

As exemplified by the preceding discussion, our procedure has a number of benefits to offer and a broad range of potential applications. Nonetheless, there are also some challenges and limitations. First, additive models have been criticized for not isomorphically representing judgment and decision processes. Previous research has shown that the assumption of additivity is not tenable under all circumstances (e.g., Payne et al., 1993) and it has also been claimed that simple heuristics can account for evaluations of multi-attribute options almost as well or even better than additive models (Gigerenzer and Goldstein, 1996). Studies of experienced decision makers in the field of naturalistic decision making (Zsambok and Klein, 1997) also indicate that additive models are not always appropriate. Notwithstanding these findings, there seems to be a consensus that additive models are often good paramorphic representations of judgment and decision making processes (Doherty and Brehmer, 1997). To the best of our knowledge, almost all applied approaches to preference measurement rely on the principle of additivity in one form or another.

Second, another stream of research has demonstrated that utility functions and preferences are not stable, but are rather evoked or constructed in a given context (Fischhoff, 1991; Slovic, 1995). At first glance, the goal to induce preferences may seem to be at odds with this position. Yet, we consider our approach to be perfectly compatible with a constructive preferences perspective. Even though the inducement procedure aims at conveying a stable preference structure, insomuch that respondents can reliably apply the learned preferences after completion of the training, this structure is intended to serve primarily as a baseline against which effects of experimental manipulations can be judged. For example, in a test of the compatibility hypothesis (e.g., Meloy and Russo, 2004) – which states that under instructions to select positive attributes receive higher weight, whereas under reject instructions more weight is placed on negative attributes – we would expect that respondents modify the acquired utility values accordingly.

Third, we believe that our approach is most suitable for group-level investigations. In our empirical studies the inducement procedure worked well at the level of groups. By contrast, additional analyses of individual respondents show that those data contain more noise. This is not surprising, however, since it cannot be expected that all people master the learning task equally well.

In conclusion, although more applications of the method to substantial research questions are certainly needed before firmer conclusions can be drawn, the approach presented here offers many advantages. It is an innovative approach for the inducement of preferences that has an underlying theoretical rationale and empirical support from the evaluation studies.

Acknowledgement. This research was supported in part by the Deutsche Forschungsgemeinschaft under grant Ho 1286/2-2.

References

Ariely, D. (2000), Controlling the information flow: Effects on consumers' decision making and preferences. *Journal of Consumer Research*, 27, 233–248.

Doherty, M.E., Brehmer, B. (1997), The paramorphic representation of clinical judgment: A thirty-year retrospective. In Goldstein, W.M., Hogarth , R.M.(Eds.), *Research on judgment and decision making: Currents, connections, and controversies* (pp. 537–551). Cambridge, UK: Cambridge University Press.

Fischhoff, B. (1991), Value elicitation: Is there anything in there? *American Psychologist*, 46, 835–847.

Fishburn, P.C. (1970), *Utility theory for decision making*. New York: Wiley.

Fishburn, P.C., Roberts F.S. (1988), Unique finite conjoint measurement. *Mathematical Social Sciences*, 16, 107–143.

Gigerenzer, G., Goldstein, D.G. (1996), Reasoning the fast and frugal way: Models of bounded rationality. *Psychological Review*, 103, 650–669.

Green, P.E., Srinivasan, V. (1978), Conjoint analysis in consumer research: Issues and outlook. *Journal of Consumer Research*, 5, 103–123.

Green, P.E., Srinivasan, V. (1990), Conjoint analysis in marketing: New developments with implications for research and practice. *Journal of Marketing*, 54, 3–19.

Huber, J., Ariely, D., Fischer, G. (2002), Expressing preferences in a principal-agent task: A comparison of choice, rating, and matching. *Organizational Behavior and Human Decision Processes*, 87, 66–90.

Karren, R.J., Barringer, M.W. (2002), A review and analysis of the policy-capturing methodology in organizational research: Guidelines for research and practice. *Organizational Research Methods*, 5, 337–361.

Keeney, R.L., Raiffa, H. (1976), *Decisions with multiple objectives: Preferences and value tradeoffs*. New York: Wiley.

Krantz, D.H., Luce, R.D., Suppes, P., Tversky, A. (1971), *Foundations of measurement. Vol. I: Additive and polynomial representations.* New York: Academic Press.

Louviere, J.J., Hensher, D.A., Swait, J.D. (2000), *Stated choice methods: Analysis and application.* Cambridge, UK: Cambridge University Press.

Luce, R.D. (1966), Two extensions of conjoint measurement. *Journal of Mathematical Psychology*, 3, 348–370.

Luce, R.D., Suppes, P. (1965), Preference, utility, and subjective probability. In Luce, R.D., Bush, R.R., Galanter, E.(Eds.), *Handbook of mathematical psychology, Vol. III.* New York: Wiley.

Luce, R.D., Tukey, J.W. (1964), Simultaneous conjoint measurement: A new type of fundamental measurement. *Journal of Mathematical Psychology*, 1, 1–27.

Meloy, M.G., Russo, J.E. (2004), Binary choice under instructions to select versus reject. *Organizational Behavior and Human Decision Processes*, 93, 114–128.

Payne, J.W., Bettman, J.R., Johnson, E.J. (1993), *The adaptive decision maker.* New York: Cambridge University Press.

Slovic, P. (1995), The construction of preference. *American Psychologist*, 50, 364–371.

Stone, D.N., Kadous, K. (1997), The joint effects of task-related negative affect and task difficulty in multiattribute choice. *Organizational Behavior and Human Decision Processes*, 70, 159–174.

Yellott, J.I. (1977), The relationship between Luce's choice axiom, Thurstone's theory of comparative judgment, and the double exponential distribution. *Journal of Mathematical Psychology*, 15, 109–144.

Zsambok, C.E., Klein, G. (Eds.). (1997), *Naturalistic decision making.* Mahwah, NJ: Erlbaum.

Active Search for Probability Information and Recall Performance: Is Probability an Outstanding Element in the Mental Representation of Risky Decisions?

Odilo W. Huber

Department of Psychology, University of Fribourg, Rue de Faucigny 2,
CH-1700 Fribourg, Switzerland
e-mail: odilo.huber@unifr.ch

Abstract. Recall performance of (1) actively searched information vs. information prestructured by the experimenter and (2) actively searched probability information vs. non-probability information in quasi-realistic risky decisions was investigated. 42 subjects decided in 2 scenarios each. Information presentation was varied within subjects by means of the Active Information Search method (O.Huber, Wider & O.W.Huber, 1997). After a frugal basic description of the scenario (prestructured information) subjects ask questions and receive answers until they subjectively have enough information to decide (active search). In only 55% of the tasks probability was searched. Recall was measured 48 hours afterwards as surprise task, showing (1) actively searched probability being on average better recalled than other actively searched information and (2) actively searched information being better recalled than prestructured information. It is argued that (1) actively searched information in general is of greater importance, because it matches the specific information needs of the decision maker during the construction of the mental representation and (2) the role of probability information in the mental representation differs between subjects and tasks. Actively searched risk control information is proposed as one reason for this difference. If risk control is expected to be successful, probability information is of limited value.

Keywords: recall-performance, probability information, quasi-realistic risky decision, active information search, mental representation

1 Introduction

Research on risky decisions is dominated by models specificating concepts originating from SEU-Theory (Edwards, 1961). Research in this tradition has been extremely fruitful. Only to mention one of the most influential models, Kahneman and Tversy (1979) proposed the prospect theory which solves a number of problems for the original SEU model, e.g. preference reversals. These models share a set of basic assumptions, of which for the present paper one is most relevant: The knowledge of the probability of possible consequences is necessary in all cases to compute the expected utility of the alternatives.

However, experimental research on risky decisions has shown that decision makers not in all or even in only a minority of risky decisions are actively interested in probabilities, but instead are trying to control possible future risks: O. Huber, Wider and O.W. Huber, 1997; O. Huber, Beutter, Montoya and O.W. Huber, 2001; O. Huber and O.W. Huber, 2004; O. Huber and Macho, 2001a; Ranyard, Williamson and Cuthbert, 1999; Ranyard et al., 2001; Schulte-Mecklenbeck and O. Huber, 2003; Williamson et al., 2000a, b. These papers investigate quasi-realistic risky decision scenarios and use the Active Information Search (AIS) method as basic experimental paradigm (O. Huber et al., 1997) which allows the decision maker actively to inspect only the information he subjectively needs for his decision (see method section for details). The fact that not all decision makers are interested actively in probabilities questions the role of probability information as a general component of the subjective representation of every risky decision problem. O. Huber (2004, in press) suggests the concept of Risk Defusing Operators (RDOs) to explain these results. RDOs are actions intended additionally to a specific alternative with the aim to control negative outcomes. He proposes a model of the decision process specificating a classification of RDOs and the conditions for their application. For decision makers deciding without probability information, the model proposes, that if decision makers are able to control a risk, probability information is of very limited value. Research on real decisions yield similar results regarding the role of probabilities and risk control (see, e.g. Lipshitz and Strauss, 1997; Shapira, 1994, for managerial decisions, or Shiloh et al., 2004, who investigate decisions in genetic counseling).

However, regarding that decision makers are not interested in probabilities in many risky decisions, decision makers who actively ask for probabilities in a specific risky decision situation should consequently attach special importance to this information. That does not mean that actively searched probabilities must influence decisions, because risk defusing may play a role simultaneously. Probability may e.g. affect the intensity of search for RDOs or the maximal cost for an RDO that will be accepted.

These results point in two directions, on the one hand questioning the importance of probability information for all risky decisions, and on the other hand letting the role of actively searched probability information unclear. In the vast majority of experiments that present probability information to the decision maker without active search it is processed (see, e.g., O. Huber and O.W. Huber, 2003). However, the role and importance of actively searched probability information in the mental representation of the decision problem has not been investigated directly. The main objective of this paper therefore is to investigate the importance of actively searched probability information for the subjective representation. This is not possible with research paradigms using complete information presentation. I suppose this to be one reason that research on that problem to my knowledge has not been done. However, the AIS method gives at hand a possibility to address this question.

It is expected (1) that probability information is not searched in many decisions, which would replicate the findings cited above, and (2) that because the search of probability information is a nonstandard and willful process decision of the decision maker, actively searched probability information is of especially high importance for the construction of the subjective representation. This should result in higher importance of actively searched probability information in comparison to the importance of other actively searched non-probability information.

A second objective of the paper is the evaluation of the subjective importance of actively searched information vs. information that is prestructured by the experimenter. Paradigms that use complete information presentation expose decision makers to information that is chosen and structured by the experimenter. It can be presumed that the presented information is chosen according to a decision theory of the experimenter and does not correspond to information needs of the decision maker spontaneously emerging during the construction of the subjective representation of the decision problem (for a discussion, see O. Huber, 2004, in press). If information can actively be chosen, it can be expected that information subjectively needed will be inspected. It therefore can be expected that actively searched information is of higher importance for the construction of the mental representation than prestructured information.

Weber, Weber, Goldstein & Busemeyer (1991; see also Weber, Goldstein and Barlas, 1995) propose a non-reactive way to investigate decision processes. They emphasize the possibilities that emerge if different cognitive processes are analyzed that contribute to the decision process. They propose the investigation of memory processes to analyze the decision process and discuss relevant results.

This paper will follow this suggestion and assess the subjective importance of (1) actively searched information vs. information prestructured by the experimenter and (2) actively searched probability information vs. actively searched non-probability information by measuring memory performance. The AIS method will be used to operationalize both dichotomies. Because there is no knowledge concerning the investigation of memory processes in combination with the AIS method, recall performance will be used as a simple operationalization of the proposition of Weber et al. (1991).

2 Method

42 undergraduate students of psychology participated in the experiment (mean age 25.15 years). None of them had participated in a similar experiment before. Two independent variables were varied, scenario and information mode.

2.1 Scenario

Two quasi-realistic risky decision scenarios were used, railway fanclub and turtles. Both are slightly modified versions of scenarios used in O. Huber and O.W. Huber (2004). These were designed as to ensure that decision makers were lays in the respective domains and thus to minimize background knowledge. In both tasks one safe and one risky alternative was presented in order to investigate the role of probabilities for their choice or rejection.

2.2 Railway Fanclub

The subject takes the role of the president of a railway fanclub. The club owns some passenger railway cars and of late a matching steamer. The club wants to operate the train on public rails. The public railway company offers two possible packages:

Operations Three Days Per Year (Safe Alternative)

The club may operate the train three days per year on a light railway. The days can be chosen by the club and the club bears all operation expenses.

Operations at Regular Intervals (Risky Alternative)

The club may operate the train on the light railway each weekend as a substitute for modern trains of the regular schedule . The club would get considerable profit and the regular operation would fulfill a dream of many members. However, in case of breakdown of the steamer the club had to pay high penalties to the railway company.

2.3 Turtles

The subject takes the role of a executive of an environmental program with the aim to safe a species of sea turtles from extinction. There are only a few individuals left, which are placed in a lab. However, they do not breed in the lab. An investigation for natural habitats where the turtles could breed resulted in two possible options:

Beach (Safe Alternative)

A suitable beach is located close to the lab. There are no predators but the water quality is not optimal. This will result in a low breeding frequency of the turtles.

Island (Risky Alternative)

An island provides optimal conditions for the breeding of the turtles. There are no predators for adult turtles. However, in the sea beneath the island sometimes a very small plankton species (saltwater mites) may occur. If the mites occur, they invade the brood of the turtles and kill it.

Mode of Information Acquisition and Actively Searched Information Content

The main objectives of this study are (1) to investigate differences between actively searched information and information prestructured by the experimenter (factor mode of information acquisition) and (2) to investigate the difference between actively searched probability information and actively searched non-probability information (factor actively searched information content).

Information presentation was operationalized by means of the AIS method (O. Huber et al., 1997). Here, the subject initially gets only a scarce description of the decision scenario and the alternatives like the ones presented above. Then, the subject may ask questions. After each question, the experimenter presents the answer on a printed card. Printed presentation was chosen in order to minimize non-verbal effects. When the subject feels to have enough information to decide, he decides by marking the chosen alternative. In order to find possible questions and to construct standardized answers, extensive pre-experiments are necessary. The AIS method allows the decision maker to inspect only the information which is subjectively needed. Thus, information search allows inferences on the subjective representation of the decision problem by measuring the elements the representation is constructed of. Additionally, the AIS method increases the validity of the results for most real complex decisions, in which information search is a standard element of the decision process. It has to be noted that the task descriptions given above are not identical to the subjective representations of the decision problems, because most decision makers search for additional information and therefore the representation is in general based on more complex and detailed information.

In summary, there are two basic modes of information presentation, the basic description that is prestructured by the experimenter and actively searched information, which depends of the specific information needs of a decision maker. Within the latter, there is probability information and non-probability information.

Of special relevance for this investigation is the probability of the negative event. Subjects who asked in the railway fanclub scenario were presented a probability of 32% for steamer failure or in the turtles task of 36% for the occurrence of the mites.

2.4 Design and Procedure

Each subject decided in both scenarios. The two possible orders of scenarios were counterbalanced between subjects. The independent variable mode of information acquisition was varied within each task by means of the AIS method. Actively searched information content was a random variable depending on the search behavior in each specific task.

The procedure consisted of two sessions, a decision session and a recall session two days later.

Decision Session

After an instruction explaining the AIS procedure, a training task (travel scenario, O. Huber and O.W. Huber, 2003) was performed. Then the basic description of one of the tasks described above was given, followed by questions of the subject and the corresponding answers of the experimenter. E.g., the subject would ask: 'How many turtles are in the lab?' The experimenter would place the corresponding answer card in front of the subject: 'There are 20 turtles in the lab.' After the decision, the basic description of the second task was given to the subject, followed by questions and answers. In this task, the subject additionally made a preliminary decision after each answer. This did not result in any differences in the dependent variables and will therefore not be discussed any further. As in the preceding tasks, the subject stopped information search and decided definitively, when she felt to have subjectively enough information for a decision. Finally, the subject was instructed that in the second session, she would be given new decision scenarios.

Recall Session

As subjects expected new scenarios in the recall session, the instruction to recall all information they had got concerning their first task (the task was named) two days before was surprising for the subjects. After recalling the first task, the procedure was repeated for the second task. The recall session was audiotaped.

In the decision session two provisions ensured that intentional memory processes play no role:

(1) During the whole decision process, the subject kept the description of the scenario and all given answers in front of her. So, memorization was not necessary.
(2) The final instruction prevented memorization, because subjects did not expect any memory task in the recall session.

Therefore it can be expected that the memory traces of the decision process measured in the recall session are due to the decision task itself and not due to an interfering memory task.

All subjects were run individually. The decision session lasted about 40 minutes, the recall session about 15 minutes.

3 Results

The material of one subject was incomplete and a second subject missed the appointment for the recall session, so 40 subjects with two tasks each were available for analysis. Former experience with the AIS method showed that decision tasks of one decision maker were independent of each other (e.g. O. Huber et al., 2001; O. Huber and O.W. Huber, 2003). This paper again will treat the decision tasks of each decision maker as independent.

3.1 Decision Session

The reliability of the assignment of the answers to the questions regarding the independent variable of interest actively searched information content was very high (two trained coders, Cohens $\underline{\kappa} > .98$). Descriptive results for this section can be found in Table 1.

General Information Search and Decisions

The mean number of questions was 4.31. Table 1 shows the mean questions regarding the safe alternative, the risky alternative and the decision situation. Two subjects did not ask any questions in the railway fanclub scenario. There were no differences in the mean number of questions between the two scenarios ($F(1, 78) = .10$).

In 66.3% of the tasks the risky alternative was chosen. This was independent of the scenarios ($\chi^2(1) = 2.74$, $p > .05$).

Active Search for Probability Information

In 55% of the tasks subjects asked for the probability of the negative outcome of the risky alternative. This was independent of scenarios ($\chi^2(1) = .81$, $p > .05$). It has to be noted, that a question did not have to contain the term probability to be coded as probability question. Questions were coded as probability questions regardless if they were in frequentistic or probabilitistic format (see, e.g. Cosmides and Tooby, 1996; Gigerenzer and Hoffrage, 1995). E.g. if a subject asked 'How often do the mites occur?', she go the answer: 'The mites occur in 36% of the breeding periods.' This coding is conservative as it favors the alternative hypothesis that probability information is always searched.

Table 1. Number of questions for the safe alternative, for the risky alternative and for the decision situation by scenario. Furthermore, choices of the risky alternative and search for probability information are reported

	Turtles	Railway fanclub	Total
n	40	40	80
Number of questions concerning the			
safe alternative	1.20	.38	.79
Risky alternative	2.42	2.28	2.35
Decision situation	.78	1.88	1.32
Total	4.40	4.22	4.31
Rate of tasks with decision for the risky alternative	.58	.75	.66
Rate of tasks with a question for the probability of the negative outcome	.50	.60	.55

A logit analysis (see, e.g. Agresti, 1990; DeMaris, 1992) showed that decisions were independent of the questions for the probability of the negative outcome of the risky alternative and the scenario. Logit analyses fit a hierarchy of logit models to the data. They use the G^2 statistic, which is approximately χ^2 distributed. The independence of decisions of the factors mentioned above was evidenced by the fitting of the null model which assumes that no predictor has an effect ($G^2(3) = 5.88$; $p > .1$). The inclusion of the predictors scenario ($G^2(2) = 3.11$; $p > .1$; $\Delta G(1) = 2.77$; $p > .05$) or probability questions ($G^2(2) = 2.53$; $p > .1$; $\Delta G(1) = 3.35$; $p > .05$) did not result in a significantly better fit.

3.2 Recall Session

The audiotapes of the recall session were coded according to the following procedure:

1. The basic descriptions of the scenario were separated into information items (for details, see Mayring, 2003). One information item contained one single fact given in the basic description. The reliability of this coding was very high (two trained coders, Cohens $\underline{\kappa} > .97$). For the turtles scenario, a total of 21 information items resulted vs. 19 information items for the railway fanclub scenario.
2. The item pool for a specific task of a specific subject consisted of the information items of the respective scenario and the answers to the questions the subject was given in the decision session.
3. All information in the audiotapes was separated into distinctive information items. The individual information items were assigned to the corresponding items of the respective item pool. For the factor contents of active search (probability / non-probability) coding reliability was very high (two trained coders, Cohens $\underline{\kappa} > .97$).
4. Repetitions of the same information item in recall were coded only once.
5. Information concerning the procedure or the communication between subject and experimenter was ignored (e.g. '... and then I asked...').
6. Information concerning the decision problem that was not listed in the specific item pool was coded as false memories (M=.02). In most cases false memories consisted in general background knowledge that was introduced into the representation.

To analyze recall performance, the rate of remembered information was computed by dividing the number of remembered items by the number of information items available in the specific item pool for each category of interest. Due to the division two tasks without information search in the decision session had to be eliminated from the analysis.

Mode of Information Acquisition

Mode of information acquisition had a strong effect. 32.2% of the information from the basic description prestructured by the experimenter was remembered compared to 48.8% of the actively searched information. Analyzed with a MANOVA with these variables as dependent variables, scenario as fixed factor and decision as random factor confirmed this result: The repeated measures factor was highly significant ($F(1, 74) = 14.38$, $p < .001$, partial $\varepsilon^2 = .163$), whereas all other factors or interactions had no influence (all other F values were smaller than 1). Descriptive values for the scenarios can be found in Table 2.

Actively Searched Probabilities

Of the actively searched probabilities 65.9% were remembered. A logit analysis investigating the 44 tasks with probability searched in the decision session revealed an effect of the predictor scenario. The null model fitted the data ($G^2(3) = 4.44$; $p > .2$). The model containing the predictor scenario fitted significantly better ($G^2(2) = .27$; $p > .8$; $\Delta G(1) = 4.17$; $p < .05$). The effect size of scenario was $w = .31$. A significant increase of the fit of this model by inclusion of the factor choice is not possible (safe alternative chosen 63.6% vs. 66.7% risky alternative chosen). In the turtles scenario more probability information was remembered than in the railway fanclub scenario. Descriptive values can be found in Table 2.

Actively Searched Probabilities vs. Non-probabilities

Recall performance was tested by means of an ANOVA including all 78 tasks with questions in the decision session. The dependent variable was computed in the following way: because the factor scenario had a significant effect in the analysis reported above, recall rate for actively searched probabilities and non-probabilities was computed for each scenario separately. Subsequently,

Table 2. Percentage of recalled actively searched information, prestructured information, actively searched probability information and actively searched non-probability information by scenario

Rate of recall of	Turtles	Railway fanclub	Total
Prestructured information of the basic description	.34	.31	.33
Actively searched information	.52	.47	.50
Actively searched non-probability information	.47	.50	.4.48
– of n tasks with questions in the decision session	40	36	78
Probability information (of tasks with probability questions in the decision session)	.79	.50	.66
– of n tasks with probability questions in the decision session	24	20	44

the average recall rate for probability information was subtracted from the recall rate of non-probability information. The resulting value is the difference in actual recall of non-probability information from the expected value of probability recall. The expected value had to be used instead of the actual, because probability recall was nominal scaled an therefore could not be used directly in an ANOVA. Between subjects factor was scenario.

The results show a strong difference of the recall rates (the constant is different from zero: $F(1,76) = 12.73$, $p < .001$, partial $\varepsilon^2 = .1$) and an effect of scenario ($F(1,76) = 17.85$, $p < .001$, partial $\varepsilon^2 = .190$). This result was confirmed by a second ANOVA analyzing only the 44 tasks with probability search in the decision session. Both factors remained significant. These results indicate (1) the scenarios differ, in the turtles scenario recall being better than in the railway club scenario, where recall rate for probabilities is not superior to non-probabilities and (2) that the overall recall rate for probability information is higher than for non probability information. Table 2 shows descriptive values.

4 Discussion

The results confirmed the hypotheses:

1. Actively searched information is recalled better than information given to the subject that is prestructured by the experimenter.
2. Search for probability information occurs only in about half of the tasks.
3. Actively searched probability information is on average recalled better than actively searched non-probability information. However, only in the turtle scenario probability recall is much better, in the railway fanlub scenario it is only slightly higher.

The first result shows the importance of actively searched information for the construction of the mental representation of the decision problem. It justifies the concerns of O. Huber et al. (1997) that decision research based on information completely prestructured by the experimenter is of only limited value for the explanation of decision processes in complex real decisions where information search in most cases is part of the decision process.

The result that not in all tasks probability information is searched replicates the findings of other investigations with the AIS method referred to in the introduction and confirms the assumption that probability information is not relevant for all risky decisions. However, because in this experiment a safe and a risky alternative is presented, it could be argued that probability information is not necessary if a subject rejects any risky alternative and chooses the safe one. This can be ruled out because search for probability information is independent of choice. Second, in the railway fanclub scenario in 87.5% and in the turtles scenario in 90% of the tasks information regarding the risky alternative was searched, indicating that the risky alternative had been taken into consideration in most decisions. Therefore it can be concluded that information search for probabilities is not a standard element of the decision process for all risky decisions. This could be due to the representation of probabilities itself or the importance of probabilities in the mental representation of the specific decision problem.

Probability representation could be the cause for the non-search if representation is not assumed to be interval scaled but ordinal scaled of the type safe–risky–impossible. Kunreuther et al., (2001) presented subjects low probabilities and found that subjects did not evaluate them differently even if they differed extremely. However, this explanation is not very probable for the data of this experiment, if the pattern of probability search over subjects is analyzed: 30% of the subjects never searched, 30% searched in one task and 40% in both. This pattern does not differ from a pattern that would result from a random process ($\chi^2(2) = 3.37$, $p > .05$)[1]. This result strengthens the alternative explanation, that probability search depends on the specific mental representation of the decision problem and not of general strategies of the decision maker, because in the latter case it would be expected that subjects do or do not search in all tasks but do not show both types of behavior.

The finding that actively searched probability information overall is better recalled than actively searched non-probability information indicates the special role of probability information in the mental representation of some scenarios if it is actively searched. This can be explained in two ways: As result of the structure of information or because probability information is in some cases of special importance for the construction of the mental representation. Most approaches to risky decision making suppose that utilities are computed on the basis of probabilities and values of the consequences (e.g. Kahneman and Tversky, 1979). Probabilities and consequences are of equal importance. Then, the better memory performance for actively searched probability information in quasi-realistic risky tasks could be explained by its structure. Probability information here consists of one item of numerical information but consequence information consists of different dimensions which must be integrated in one value. Due to this consideration, the memory performance could be better because probabilities are directly accessible and do not need further processing. However, the recall performance differs in the two tasks. In the turtles task performance is clearly better than in the railway fanclub task. This contradicts the consideration above, because the type of probability information and non-probability information given to the subjects did not differ between the scenarios. If the structure of the information would have caused recall differences, the performance in both tasks should be the

[1] This result was obtained by the comparison of the distribution of the number of tasks in which subjects searched for probability (0 to 2) with the distribution that would be produced by a Bayesisan random process with two levels, where at each level the probability for a search is .55 (overall probability of search for probability). The expected distribution is 20.25% search in no task, 49.5% search in one task, and 30.25% search in two tasks.

same. Probability information search therefore strengthens the considerations based upon probability information search: The search and recall of probability information seems to be dependent of the specific representation of the decision problem.

For descriptive approaches to risky decision making the mental representation of risky decisions not containing probability information poses a challenge. That must not mean that these decisions are to be classified as nonrational, e.g. because probabilities were just ignored due to unpleasant feelings connected with the perception of risk (for the discussion of the transformation of probabilities into feelings, see Loewenstein, Weber, Hsee and Welch, 2001). Alternative concepts as O. Hubers RDO concept (2004a, b) can provide a solution to this problem. Many decision makers do not bear risk just passively but try to control the risk actively by integrating an additional action into the mental representation. If a decision maker finds or constructs a possible action to control the risk contained in the alternative, probability information is of minor importance, because a negative event does not lead to an negative outcome. E.g., the railway fanclub could find an insurance against the financial consequences of the breakdown of the steamer or ask other railway clubs if they could provide a substitute steamer. In this case, the breakdown of the steamer leads to no (insurance) or only minor (transport of the substitute) negative consequences. O. Huber and O.W. Huber (2003) propose a model to explain the search for probabilities and RDOs. They suppose, that the search for these categories is dependent on expectations of the decision maker. These are supposed to be dependent on background knowledge and local cues. E.g., most people expect based on background knowledge, that in general if a newly constructed car is invented, there are no good probability information available for malfunctions of the engine and therefore will not search. If, however, in a specific situation a local cue is available, that an information source is accessible, search would be initiated. E.g., if people would know that research has been done on the occurrence of plankton close to the island of the turtles task, they would ask the biologist in charge if he has collected probability information about the occurrence of the mites.

In this paper the investigation of decision processes by investigating information needs has proven to be able to provide fruitful contributions to research on risky decision making. This approach allows to address research questions that cannot be targeted at with paradigms that expose the decision maker to what is defined by the experimenter as relevant information. Incomplete information presentation opens a way to investigate the construction of mental representation. The capabilities of that method are further enhanced by the measurement of memory performance as proposed by Weber et al. (1991), yielding results that could not be accomplished with the investigation of information search alone. The results of this paper indicate that there is more research to be done on the problem of probability processing in complex decisions and the role of probability information in the construction of the mental model of the decision situation, especially concerning the factors that initiate probability search and control the integration of probabilities into the mental representation. This paper does not address the topic of decision domains (e.g. Blais and Weber, 2001) which could play a role here, especially in interaction with the expectations the model of O. Huber and O.W. Huber (2004) suggests. This model could be used as a starting point to investigate the questions emerging from the results of this paper, especially why and when subjects do or do not search for probability information in a specific decision and why probability is integrated more or less deep into the mental representation.

References

Agresti, A. (1990), *Categorical data analysis.* New York: Wiley.

Blais, A.R. and Weber, E.U. (2001), Domain-specifity and gender differences in decision making. *Risk Decision and Policy,* 6(1), 47–69.

Cosmides, L., Tooby, J. (1996), Are humans good intuitive statisticians after all? Rethinking some conclusions from the literature on judgment under uncertainty. *Cognition,* 58, 1–73.

DeMaris, A. (1992), *Logit Modeling: Practical Applications.* Newbury Park, CA: Sage.

Edwards, W. (1961), Behavioral decision theory. *Annual Review of Psychology,* 12, 473–498.

Gigerenzer, G., Hoffrage, U. (1995), How to improve Bayesian reasoning without instruction: Frequency formats. *Psychological-Review,* 102, 684–704.

Goldstein, W.M., Weber, E.U. Content and discontent: Indications and implications of domain specificity in preferential decision making. In J.Busemeyer, R. Hastie, D.L. Medin (Eds.), *Decision Making from a Cognitive Perspective* (pp. 83–136). San Diego: Academic Press, 1995.

Huber, O., Macho, S. (2001), Probabilistic set-up and the search for probability information in quasi-naturalistic decision tasks. *Risk Decision and Policy,* 6, 1–16.

Huber, O., Huber, O.W. (2003), Detectability of the negative event: Effect on the acceptance of pre- or post-event risk defusing actions. *Acta Psychologica,* 113, 1–21.

Huber, O., Huber, O.W. (2004), What makes gambles special? – Active search for probability and risk defusing information in gambles and quasi-realistic risky tasks. *Manuscript submitted for publication.*

Huber, O. (2004), Entscheiden unter Risiko: Aktive Risiko-Entschärfung. *Psychologische Rundschau,* 55, 127–134.

Huber, O. (in press). Risky decision making: Focus on risk defusing behavior. In: F. Columbus (Ed). *Advances in Psychology Research.* New York: NovaSciences Publishers.

Huber, O., Beutter, C., Montoya, J., Huber, O. W. (2001), Risk defusing behavior: Towards an understanding of risky decision making. *The European Journal Of Cognitive Psychology,* 13, 409–426.

Huber, O., Wider, R., Huber, O.W. (1997), Active information search and complete information presentation in naturalistic risky decision tasks. *Acta Psychologica,* 95, 15–29.

Kahneman, D., Tversky, A. (1979), Prospect theory: An analysis of decision under risk. *Econometrica*, 47, 263–291.

Kunreuther, H., Novemsky, N., Kahnemann, D. (2001), Making low probabilities useful. *Journal of Risk and Uncertainty*, 23, 103–120.

Lipshitz, R., Strauss, O. (1997), Coping with uncertainty: A naturalistic decision-making analysis. *Organizational Behavior and Human Decision Processes*, 69, 149–163.

Loewenstein, G.F., Weber, E.U., Hsee, C-K., Welch, N. (2001), Risk as feelings. *Psychological Bulletin*, 127, 267–286.

Mayring, P. (2003), Qualitative Inhaltanalyse: Grundlagen und Techniken. Weinheim; Basel: Beltz.

Ranyard, R., Hinkley, L., Williamson, J. (2001), Risk management in consumers' credit decision making. *Zeitschrift für Sozialpsychologie*, 32,152–161.

Ranyard, R., Williamson, J., Cuthbert, L. (1999), The development of conversation-based process tracing methods for the study of naturalistic decision making. In Green, B.(ed), *Risk Behavior and Risk Management in Business Life* (pp. 127–136). Dordrecht: Kluwer.

Schulte-Mecklenbeck, M., Huber, O. (2003), Information search in the laboratory and on the Web: With or without the experimenter. *Behavior Research Methods, Instruments & Computers*, 35, 227–235.

Shapira, Z. (1994), *Managerial risk taking.* New York: Russell Sage.

Shiloh, S., Gerad, L., Goldman, B. (2004), Patients' information needs and decision-making processes: What can be learned from genetic counseling? *Paper submitted for publication.*

Weber, E.U., Goldstein, W.M., Barlas, S. (1995), And let us not forget memory: The role of memory processes and techniques in the study of judgement and choice. In Busemeyer, J.R., Hastie, R., Medin, D.L. (Eds.). *The Psychology of Learning and Motivation: Vol. 32. Decision Making from a Cognitive Perspective* (pp. 31–81). San Diego: Academic Press.

Weber, E.U., Goldstein, W.M., Busemeyer, J.R. (1991), Beyond strategies: Implications of memory representation and memory processes for models of judgment and decision making. In: Hockley, W.E., Lewandowsky, Stephan (Ed). (1991), *Relating theory and data: Essays on human memory in honor of Bennet B. Murdock.* (pp. 75–100). Hillsdale, NJ: Lawrence Erlbaum Associates.

Williamson, J., Ranyard, R., Cuthbert, L. (2000 b). A conversation-based process tracing method for use with naturalistic decisions: an evaluation study. *British Journal of Psychology*, 91, 203–221.

Williamson, J., Ranyard, R., Cuthbert, L. (2000a), Risk management in everyday insurance decisions: Evidence from a simulation study. *Risk Decision and Policy*, 5, 19–38.

Evolutionary Analysis of Risk Attitudes in Competitive Bidding Environments Using Simulation

Jayavel Sounderpandian

University of Wisconsin – Parkside Kenosha, WI, 53141-2000, USA
e-mail: sounderp@uwp.edu

Abstract. Exactly how risk-averse a person should be is an elusive question that may be answerable only through evolutionary analysis. The literature on the evolution of risk attitudes is quite diffuse. Researchers have taken widely varying approaches. A survey of these approaches is presented. One approach to study the evolution is to simulate a small society of individuals whose risk taking behaviors are interrelated according to simple rules. The aim of this paper is to introduce a few different ways to conduct such simulations and visualize the results. These approaches can be extended in diverse ways. While this type of simulation is unlikely to produce normative or prescriptive results for an individual, it may reveal some facts about the collective fate of a society. Simulation codes written in *Mathematica* are included.

Keywords: risk aversion, competitive bidding, evolutionary game theory, memetic analysis, simulating a society

1 Introduction

Normative and prescriptive theories of decision making assume that the risk attitude of a decision maker (DM) is exogenous, and derive or compute the DM's optimal choice among risky alternatives based on that assumption. Questions regarding how the DM came to possess a particular risk attitude were largely ignored until the 1990s. But the questions are quite tantalizing and have therefore attracted a lot of attention in recent years. Some questions to ponder are: Is the risk attitude of a person controlled by genes, and if so, is a person born with and destined to die with the same risk attitude? Contrarily, if a person makes a conscious choice of one particular risk attitude, how should that person make that choice? If a person wants to change his/her risk attitude based on life experiences to date, is there an optimal way to do it? Does the risk taking behavior of an individual depend on environmental parameters such as how the neighbors behave, and if so, how?

Such questions point to a biological and evolutionary process of individual behavior rather than a formal optimization process. Unfortunately, biological or evolutionary approaches are unlikely to give rise to normative or prescriptive theories of decision making, despite the fact that Cooper (1987) has shown how the Savage axioms can be derived from biological principles. Such derivations still cannot prescribe an optimal level of risk aversion for a given situation, because Savage axioms only prove the existence of a utility function or risk attitude and not the optimality of a risk attitude. A more recent and more impressive research along this line is that of Robson (1996). In this line of research, assumptions are made about the number of offsprings produced by an individual being a function of the commodity bundle the individual consumes. But this function need not be constant over time, and will depend heavily on what the other competing individuals consume. Allowing for such dependencies make the problem intractable. Similar problems apply to the results presented in Sinn (2003). It is no wonder, therefore, that economists took risk attitudes as exogenous and proceeded from there. But differing views of rationality have prompted a second look at evolutionary processes.

Herbert Spencer (Spencer 1890, 1892) was probably the first to propose biological origins for economic behavior, and he did influence several economists of his time and in later years. Alfred Marshall wrote a "biological" chapter in his *Principles* (Marshall 1961). An analysis of this chapter can be found in Hodgson (1993a,b). But Marshall did not go any further than mentioning the biological nature of certain branches of economics. Only recently have many researchers paid some attention to evolutionary analysis of decision making behavior in competitive environments. A milestone is the work by Maynard Smith (1982) who wrote about evolutionary aspects of repeated games. His idea of evolutionarily stable strategies in non-cooperative game theory led to better characterizations of Nash equilibria than other approaches.

Since evolution involves the interaction of numerous individuals who might alter their rules of behavior based on their experience, formal analysis of evolution becomes quite complex and intractable. Savage (1972, p. 16), for instance, envisaged that a person has to make just one decision, namely, how to live the rest of one's life. Citing the intractability of that decision, he calls the approach "ridiculous." The complexity of decision analysis is determined by

- the numerosity of the individuals involved
- the degree of interaction among the individuals
- diversity of individual behaviors
- complexities of how individuals change their behavior

Simplifications that make the analysis tractable inevitably entail corresponding degrees of limitations in the applicability of the results. But one has to make a start somewhere, and hope that advances in large scale simulation technology will lead to more impressive results. This paper lists a few interesting environments that are worthy of simulation. These environments involve a combination of players taking risks in a competitive environment. One can extend the approach in many diverse ways, some of which might even yield prescriptive results for particular scenarios.

An interesting application of simulation is to conduct contests, such as student teams managing virtual businesses competing with one another. Similar contests can be conducted using the simulation approaches mentioned here. A virtual, risky environment is specified and the contestants submit behavior rules for their virtual players in that environment. The evolution of the virtual players can then be simulated for several periods. According to some criteria, such as final wealth, winners may be determined. It is indeed the intention of the author to conduct such contests in the near future.

In the next section, we review the literature on evolutionary aspects of risk taking based on game theory, utility functions and game playing automata. After that, a few simulation approaches are described in a competitive bidding environment. *Mathematica* codes for these approaches are given in the appendix.

2 Evolution and Game Theory

The evolutionary game theory approach was given a boost by Maynard Smith (1982). A compilations of results along this line can be found in Weibull (1997). Given an underlying game and continuous mixed strategies for its players, among many of its Nash equilibria, only a subset might be evolutionarily stable. It is instructive to see how, starting from one combination of strategies, the players move along an *orbit* toward a stable equilibrium. While an orbit describes one particular path starting at a given point, a *vector field* plot describes the direction of movement from any given point. Some plots are illustrated in Weibull (1997). This line of research gave rise to concepts such as *stability* and *basins of attraction.* Stability concerns the effect of a small perturbation in the current position. If the evolutionary forces bring the system back to the starting position, then that position is stable. A basin of attraction of a stable position is the set of positions from where evolutionary forces lead to that stable position. Besides Weibull's book, Ellison (2000) and Binmore and Samuelson (1999) contain many results along this line. Such results are possible when the underlying game is crisply defined in normal form, because the orbits then follow well-defined first-order differential equations that satisfy Lipschitz continuity condition. Plots of orbits and vector fields for these cases are possible when the dimension of the strategy space is two or three.

An offshoot of this approach is the analysis of the distribution of different types of individuals in the population being analyzed, and how the distribution varies with time. Taylor and Jonker (1978) put forth the idea of *replicator dynamics* which determines, in a sense, the orbit of the distribution. While this idea was put forth in biological context, it adapts well to game theoretic situations. Weibull (1997) contains an extensive analysis of replicator dynamics in game theoretic contexts. Oechssler and Riedel (2002) show how evolutionary robustness, which is a stronger condition than evolutionary stability, imply the stability of many additional types of replicator dynamics.

The idea of repeated games gave rise to a wealth of literature. Among the papers written, many focused on evolutionary concepts. Kandori et al. (1993) analyze, among other things, how risk taking behavior affects evolution. Dekel and Scotchmer (1999) analyze the case where a particular risk taking behavior can be modeled as adopting a particular lottery. They then derive the implications of first and second order stochastic dominance of lotteries on the evolution. In addition, they define and analyze a "tail dominance" criterion.

During the course of a multiperson repeated game, some players may form a coalition. Coalition is very well studied in the mainstream literature in game theory. In addition, the evolution of coalitions has also been studied. Keenan and O'Brien (1993) report some results on coalition formation using simulation.

3 Evolution and Utility Functions

Bernoulli (1738) proposed that a person should have a logarithmic utility function for wealth because the pleasure derived from an additional dollar should be inversely proportional to current wealth. This helped explain such phenomena as the St. Petersburg paradox. Kelly (1956) showed that in a repeated betting environment based on a faulty information source will attain maximum expected final wealth only if a logarithmic utility function is followed. Sinn's (2003) argument in favor of logarithmic utility is no surprise because it uses the same underlying mathematics. As already pointed out, neither Kelly nor Sinn consider competition from others and assume a stable process. Even if we are granted a stable process the troubles of logarithmic utility do not go away. Using the appealing *single-switch property*, Bell (1988) argued that logarithmic utility functions are not reasonable. A single-switch property can be explained as follows: In a choice between two gambles A and B, suppose a DM prefers A at the current wealth level. As the wealth of the DM increases, the DM switches from A to B at some point due to decreasing risk aversion. Single switch property requires that the DM never switch back to A as the wealth continues to increase. With a logarithmic utility function, the DM can switch back to A. Bell (1995) introduces a *contextual uncertainty condition* which can be explained as follows: Suppose a DM is facing currently two uncertainties A and B in his wealth, where A is a larger (riskier) uncertainty than B. Against this contextual uncertainty, the DM has to choose between two gambles X and Y. Bell proposes that the resolution of the riskier uncertainty A should have a greater bearing on the choice between

X and Y than the resolution of the less risky uncertainty B. Bell then shows that the only type of utility function that satisfies both the single-switch and contextual uncertainty conditions is

$$u(w) = aw - be^{-cw}$$

where w is the wealth, and a, b and c are constants satisfying $a \geq 0$, $b, c > 0$. Indeed, Bell and Fishburn (2000) compile a list of utility functions for wealth that satisfy four appealing rationality conditions.

Certain types of utility functions can make an unattractive gamble attractive when taken along with other gambles. This has some troubling consequences because a DM does not know what gambles the future might bring in. The DM might reject a gamble which looks like the correct decision today. But it could turn out to be a wrong decision tomorrow when further gambles become available. Pratt and Zeckhauser (1987) called the type of risk aversion that avoid this problem, *proper risk aversion.* They proved that the utility functions that are either linear or are sums of exponential functions are the only ones that have proper risk aversion.

But utility functions for wealth can lose their rationality in evolutionary contexts because wealth may not necessarily be the relevant criterion. Often the rank of the wealth of a person in a society, for example, could be the relevant criterion. Pollock and Lewis (1993) show how wealth utility functions can fail in some simple game theoretic evolution. They claim that their results can explain phenomena such as the Allais paradox. McCardle and Winkler (1992) show how the implied utility functions of optimal strategies in repeated gambles are discontinuous, implying the failure of utility function approach in such circumstances. Levy (2003) argues with supporting empirical evidence that in the long term, it is chance rather than utility functions that make people rich. Degeorge et al. (2004) also use empirical evidence to show that in evolutionary contexts an individual will at times take risks, which are disallowed by utility functions.

The idea of defining the utility function itself as a probability or survival has been proposed by some researchers. Borch (1968) suggests that the evaluation of risky cash flows over time seems to follow a "utility function"

$$u(w) = \text{the probability that the DM loses not more than } w \text{ in the near future}$$

so that $u(w)$ is the probability of survival through the near future. The same idea is implicit in Bordley and Li Calzi (2000). Defining the utility function as probability of survival, Karni and Schmeidler (1986) derive that in sequential decision making situations, the independence axiom must be satisfied in order to maximize expected utility. This meant that the Allais paradox cannot be justified in evolutionary contexts. Hagen (1992) pointed out some situations in which one has to violate the independence axiom, giving legitimacy to the Allais paradox. In any case, the idea that utility can be conceived as the probability of survival hints at an evolutionary basis for risk attitudes.

4 Evolution and Memory

Harsanyi (1967) explains how a DM may base his decisions on Bayesian revision of probabilities in repeated incomplete information games. Needless to say, memory is needed to store and recall the prior probabilities and thus memory can improve a player's chances of survival. To analyze the role of memory, some researchers have used the theory of game-playing automata. Abreu and Rubinstein (1988) take this line of inquiry and trace the behavior of the automata. Ben-Porath (1993) proves that when two automata A and B play a zero sum game with a saddle point, A can do better than the saddle point payoff if and only if its memory is exponentially larger than that of B.

Memory can also help to signal one's intention to other players. The role of signaling in repeated games has been extensively studied both in theoretical and empirical contexts. Kreps (1990) contains a good summary of the findings.

In sum, memory plays an important role in decision making behavior. Unfortunately, the inclusion of memory increases the complexity of evolutionary analysis. In the simulation approaches mentioned in the remaining half of this paper, memory has been ignored.

5 Evolution and Simulation

When a system's behavior depends on many interacting random variables, simulation has been the common approach to study the behavior of that system. In what follows, three simple approaches to simulation are presented.

To bring about interaction among the players, a competitive bidding environment is used in the simulation. Competitive bidding environments do not have simple solutions and thus it is unlikely that anything better than simulation can be done to analyze them. Rothkopf and Harstad (1994) confirm that "elegance and powerful theorems are insufficient to obtain practical advice" in competitive bidding environments.

6 The Scenario

A society of n players competitively bid for m ($m < n$) identical lottery tickets. Each lottery has probability p of winning a prize of $\$ X$ and $(1-p)$ probability of winning nothing. The top m bidders are awarded one lottery each and their wealth is reduced by their respective bid amounts. These awardees play out the lottery and winnings are added to their respective wealths. The only source of income for any individual is the income from winning the bid and winning the lottery. Each player is endowed with an initial wealth of W_0 and incurs a living expense of $\$ 1$ per period. It is therefore necessary to win the bid and the lottery to survive beyond W_0 number of periods. The winning bid amounts are not common knowledge.

Any of the above parameters can be changed in the simulations. For instance, different individuals may be endowed with different amounts initially. The advantage of equal endowment is that it validates the comparison of different behaviors.

6.1 Behavioral Analysis

Each player is assigned a random positive value d which is the amount by which the player will increase or decrease the bid amount next period. The player will increase the bid amount if he does not win the bid or after winning the bid wins the lottery as well; he will decrease the bid amount if he wins the bid and the lottery. In the results presented here, $n = 12$ players and $m = 10$ lotteries. The code will admit other values for these parameters. We track and plot the wealth, bid amount and the quantity d of each player at each period. The *Mathematica* code used for this simulation is in Appendix A. The graphics output from this analysis are grayscale drawings of wealths, bid amounts and ds. A sample output for 100 simulated periods is shown in Figs. 1, 2 and 3 below. A nonlinear grayscale has been used to enhance the visibility of the changes.

6.2 Genetic Analysis

In many environments, the consequences of a given level risk aversion may become apparent only over a long period of time, over many generations. Accordingly, in this approach we simulate the society over an extended number of periods. Over time, many individuals will be ruined as their wealths become zero, and they "die" and exit the system. When an individual exits the system he is replaced by an "offspring." There are many schemes for creating

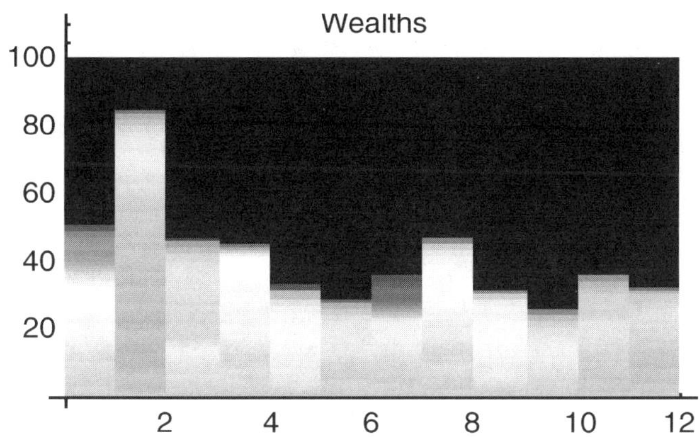

Fig. 1. The wealths of 12 interacting individuals. Lighter shade indicates greater wealth

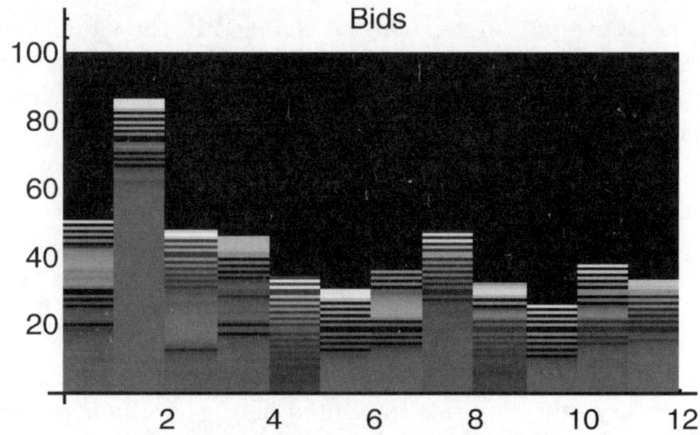

Fig. 2. The bid amounts of the 12 interacting individuals. Lighter shade indicates larger bids. When an individual is about to be ruined, he becomes desperate and bids very high

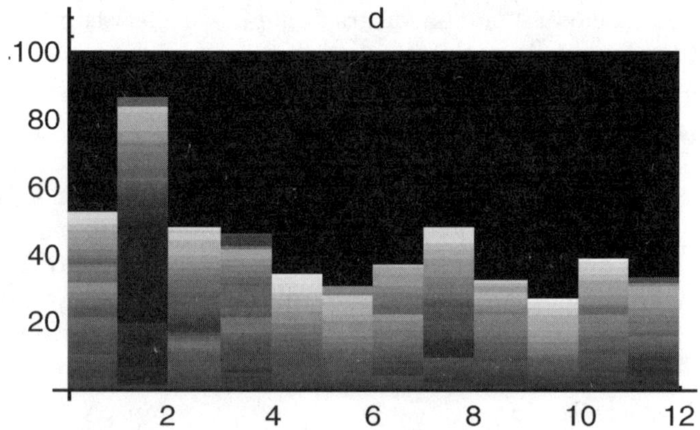

Fig. 3. The amounts by which the 12 individuals change their bids. Lighter shades indicate larger values

and introducing offsprings into the system. For example, an offspring may be endowed with the average characteristics of the society, or with those of the most successful individual in the society. In the results presented here, the offspring is created with the characteristics of the wealthiest individual. A justification for this scheme is that it most likely that the wealthiest person has the best survival strategy and is worth replicating. Other schemes, if desired, can be easily coded into the simulation.

In the simulation presented here, the behavior of an individual is characterized by three values $d1$, $d2$ and $d3$, defined as follows:

$d1 =$ increase in bid amount if the individual loses the bid

$d2 =$ increase in the bid amount if the individual wins the bid and wins the lottery

$d3 =$ increase in the bid amount if the individual wins the bid but loses the lottery.

All three values have been restricted to $\{0, +1, -1\}$ for simplicity. Once again, it is not difficult to remove this restriction in the code.

A *Mathematica* program that simulates 20 individuals over 200 periods is in Appendix B. Its output is reproduced below. It first shows the initial status of the 20 individuals. These are, for the most part, randomly generated. This table is iteratively updated at the end of each period, and the iteration continues for 200 periods. After 200 periods, the status of each individual is once again tabulated as output.

At the end of the simulation, the program outputs the following tabulation.

We note from the ID column that only the offsprings of original individuals with IDs 7, 15, 17 and 20 have survived. We also see that among the survivors, $d1$ is predominantly negative, $d2$ is predominantly positive and $d3$ is predominantly negative indicating that this pattern is more likely to survive.

6.3 Memetic or Spatial Analysis

How a person should change his degree of risk aversion after an experience is almost the same as asking what is the optimal degree of risk aversion in a given situation, and therefore just as difficult to answer. In the previous approaches the adjustment behavior is random (or an inherited random behavior that was successful). Another approach is to assume that every individual adjusts his risk attitude depending on the actions of all of his neighbors in the last period. This approach has some appeal because most people have only their neighbors' actions as a guide in this type of situations. We then need to define who the neighbors are and assume some rules for adjusting the risk attitude based on the actions of those neighbors. To be able to define neighbors, we assume that each individual is a lattice point in a 2-dimensional grid. We assume each person has eight neighbors who are at the lattice points immediately to the north, northeast, east, ... northwest from his own place. In the cellular automata literature, such a neighborhood is called a *Moore neighborhood.*

In sociological simulations a behavior pattern is called a *meme* and a meme may spread form one player to another through proximity. Morris (2000) calls the meme that spreads from one player to another a contagion. Gaylord and D'Andria (1998) present detailed guidelines for simulating a society for memetic analyses.

As another simplification that makes the simulation quick, the degree of risk aversion of any individual is assumed to be a number from 0 to 4, 0

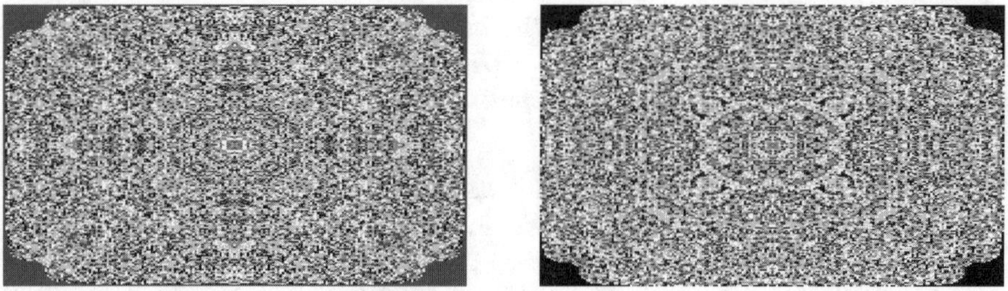

Fig. 4. Spatial Analysis of Risk Taking Behavior, shown after 99 periods (left) and 100 periods (right). Each dot is an individual, and there are 40,000 of them. A lighter shade indicates smaller risk aversion

denoting low risk aversion and 4 high. The behavior modification rule is assumed to be *totalistic,* meaning the new risk aversion level is a function of the total of the eight neighbors and one's own. The total can be a maximum of $9 \times 4 = 36$ and a minimum of $9 \times 0 = 0$. Thus, we use a function $f : \{0, 1, 2, \ldots, 36\} \rightarrow \{0, 1, 2, 3, 4\}$ which specifies the risk aversion level of the individual under consideration for the next period.

Mathematica has a built-in command called `CellularAutomaton[]` which is well suited for this type of simulation. The code used for this simulation is in Appendix C. The output of this code is shown in Figure 4. The instance of the code can be explained as follows: Initially, all individuals are assumed to have an average level of risk aversion, namely, 2. The individual at the center of the figure is assumed to increase his level to 3. According to a behavior modification function f,the details of which can be seen in the code [the code shows $f(36) = 3, f(35) = 1, \ldots, f(0) = 2$], the program simulates the risk taking behavior of his neighbors, and then their neighbors, and so on... After 100 periods, a total of 40,000 individuals would have changed their behavior as seen in the figure. Each dot in the figure is an individual. A lighter shade of gray for the dot indicates a smaller risk aversion (0 = white and 4 = black). The figure shows the risk aversion levels of the individuals after 99 periods on the left and after100 periods on the right. The four corners of the left figure display individuals with a risk aversion level (gray level) of 3. In the right, they have a level of $f(27) = 4$.

7 Conclusion

The first half of this paper presents a literature survey that brings to light the usefulness of evolutionary analysis of risk attitudes. Unfortunately, evolutionary analyses can easily become too complex and intractable. An approach to analyze complex cases is simulation. The second half of this paper presents three simulation approaches. The presented simulations are too simple to yield normative or prescriptive results. But by extending these simulations in different ways, and perhaps with further advances in simulation technology and computer power, larger projects that yield useful results can be undertaken. Even some prescriptive results might then be obtained.

Simulation can also be used to conduct evolution contests. It is the intention of the author to design and conduct such contests in the near future.

The appendices contain simulation codes written in *Mathematica* version 5.0. The codes may not run properly in earlier versions of the software. Readers may write to the author for details regarding the codes.

A Appendix

(All codes are in Mathematica 5.0. They may not run properly in earlier versions)

```
<<Graphics‘Graphics‘
nPlayers = 12;
Endowment = 100;
Prize = 10;
nPrizes = 3;
Pwin = 0.8;
nPeriods = 100;
EV = Prize*PWin;
Players =  Table[{Endowment,(EV-1)*Sqrt[Random[]],
   (EV-1)/nPeriods},{nPlayers}];
FindWinningBid[Players_]:=
   Max[1, Min[Take[Sort[Players /. {_, B, _}ƒB], -nPrizes]]];
Rules  ={{0,_,_}ƒ{0,0,0},{W_,B_,d_}/;
   (B<WinningBid)ƒ{Max[0,W-1], B+d,d}, {W_, B_, _} /;
   (B>=WinningBid) ƒ Won = (Random[] < PWin);
   {Max[0, W - 1 - B + If[Won, Prize, 0]],
   Max[0, B + If[Won, -d, d]], If[Won, d*(1 + Random[]),
   d*Random[]]})};
History = Players;
Do[(WinningBid = FindWinningBid[Players];
Players = Players /. Rules;
History = Join[History, Players]), {nPeriods}];
Whistory = History /. {W_, _, _}ƒg;
Bhistory = History /. {_, B_, _}ƒg;
dHistory = History /. {_, _, d_}ƒg;
Btable = Table[BHistory[[i + (Range[nPeriods]-1)*nPlayers]],
   {i, 1, nPlayers}];
Wtable = Table[WHistory[[i + (Range[nPeriods] - 1)*nPlayers]],
    {i, 1, nPlayers}];
dTable = Table[dHistory[[i + (Range[nPeriods] - 1)*nPlayers]],
```

```
    {i, 1, nPlayers}];
R[i_,t_]:= Rectangle[{i -1, t -1}, {i, t}];
MW:= Max[WHistory];
MB:= Max[BHistory];
Md:= Max[dHistory];
Show[Table[Graphics[{Text["Wealths", {nPlayers/2,nPeriods*1.1}],
   GrayLevel[Sqrt[WTable[[i,t]]/MW]], R[i,t]}], {i,1,nPlayers},
   {t,1,nPeriods}], Axes→True];
Show[Table[Graphics[{Text["Bids",{nPlayers/2, nPeriods*1.1}],
   GrayLevel[(BTable[[i, t]]/MB)^0.3], R[i, t]}],{i, 1,
   nPlayers},{t, 1, nPeriods}], Axes→True];
Show[Table[Graphics[{Text["ds",{nPlayers/2,nPeriods*1.1}],
   GrayLevel[(dTable[[i,t]]/Md)^0.2], R[i,t]}], {i,1,nPlayers},
   {t, 1, nPeriods}], Axes→True]
```

B Appendix

```
(* Memetic Analysis *)
<<Statistics`DescriptiveStatistics`
(* Initialize *)
nPlayers=20;
Endowment=1300.;
Prize=20;
nPrizes=4;
pWin=0.7;
nPeriods=200;
(* MIntense = Mutation Intensity *)
MIntense = 0.2;
RandomSeed[3];
(* Create Players *)
Players = Table[{Endowment, Prize/2, Round[3 Random[]-1.5],
   Round[3Random[]-1.5], Round[3 Random[] - 1.5], i},
   {i, 1, nPlayers}];
(* Print initial values *)
Print["Initial Values"]
Print[TableForm[N[Players], TableHeadings→{Automatic,
   {"Wealth","Bid","d1","d2","d3","ID"}}]]
Rules ={
(* Dead Players *)
{W_, _, d1_, d2_, d3_, i_}/; (W <= 0):>{0.,
   0., d1, d2, d3, i},
(* Bid losers *)
{W_,B_,d1_,d2_,d3_,i_} /;
```

```
   (B<WinningBid):>{Max[0., W-1.],
Max[1.,B+d1],
d1, d2, d3, i},
(* Bid Winners *)
{W_, B_, d1_, d2_, d3_, i_} /; (B >= WinningBid)
   :> (Won = (Random[] < pWin);
{Max[0., W -1 - B + If[Won, Prize, 0]],
Max[1, B + If[Won, d2, d3]],
d1, d2, d3, i}
)};
(* Iterate *)
nRep = 0;
Clock = 0;
Do[(Clock++; WinningBid = Max[1, Min[Take[Sort[Players /.
   {_, B_, _, _, _, _}:> B], -nPrizes]]];
Players = Players /. Rules;
Players = Sort[Players];
(* Reproduce *)
While[Players[[1,1]]==0,
(nRep++;
Players[[1]] = Players[[nPlayers]];
(* Endow the newborn *)
Players[[1,1]] = Endowment;
Do[Players[[i,1]] -= Max[0,Players[[i,1]]-Endowment/nPlayers],
{i, 2, nPlayers}];
(* Mutate *)
Players = Sort[Players])
]),
{nPeriods}];
(* Print output *)
Print["Clock = ", Clock]
Print[TableForm[N[Players], TableHeadings →{Automatic,
   {"Wealth","Bid","d1","d2","d3", "ID"}}]]
Print["Expected Value = ", Prize*PWin]
Print["Sustainable bid = ", (Prize*nPrizes*Pwin - nPlayers) /
   nPrizes]
Print["Average Bid = ", Mean[Players/.
   {_, B_, _, _, _, _}:> B]]
Print["Standard Deviation of Bids = ",
   StandardDeviation[Players /.{_, B_, _, _, _, _}:> B]]
Print["Last WinningBid = ", WinningBid]
Print["Lottery Won? ", Won]
Print["nRep = ", nRep]
```

C Appendix

```
RasterGraphics5[x__]:=Graphics[Raster[1-Reverse[x/4]]];
Show[GraphicsArray[Map[RasterGraphics5,CellularAutomaton[
{FromDigits[{
(* Totalistic Transition Rule ordered f(36) to f(0) *)
3,1,4,0,3,4,
3,4,1,4,0,2,0,2,3,2,
4,2,1,3,1,0,1,1,0,4,
1,4,3,2,2,1,4,3,4,1,
2
},5],
{5,1},{1,1}},
(* Disturbance and Background *)
{{{3}},2},
(* Number of periods *)
100,
-2]]],AspectRatio→Automatic]
```

References

Abreu, D., Rubinstein, A. (1988), "The structure of nash equilibrium in repeated games with finite automata." *Econometrica*, Vol. 56, No. 6, pp. 1259–1281.

Ben-Porath, E. (1993), "Repeated games with finite automata." *Journal of Economic Theory.* Vol. 59, pp. 17–32.

Bell, D.E. (1988), "One-switch utility functions and a measure of risk." *Management Science*, Vol. 34, pp. 1416–1424.

Bell, D.E. (1995), "A contextual uncertainty condition for behavior under risk." *Management Science.* Vol. 41, pp. 1145–1150.

Bell, D.E., Fishburn, P.C. (2000), "Utility functions for wealth." *Journal of Risk and Uncertainty.* Vol. 20, pp. 5–44.

Bernoulli, D. (1738), "Specimen theoriae novae de mensura sortis." *Commentarii Academiae Scientiarum Imperialis Petropolitanae.* Vol. 5, pp. 175–192. Translated in L. Sommer (1954), "Exposition of a new theory on the measurement of risk." *Econometrica*, Vol. 22, pp. 23–36.

Binmore, K., Samuelson, L. (1999), "Evolutionary drift and equilibrium selection." *Review of Economic Studies*, Vol. 66, pp. 363–393.

Borch, K. (1968), "Decision rules depending on the probabilty of ruin." *Oxford Economic Papers*, Vol. 20, No. 1, pp. 1–10.

Bordley, R., Calzi, L.M. (2000), "Decision analysis using targets instead of utility functions." *Decisions in Economics and Finance* Vol. 23, pp. 53–74.

Cooper, W.S. (1987), "Decision theory as a branch of evolutionary theory: A biological derivation of the Savage axioms". *Psychological Review,* Vol. 94, No. 4, pp. 395–411.

Degeorge, F., Moselle, B., Zeckhauser, R. (2004), "The ecology of risk taking." *The Journal of Risk and Uncertainty.* Vol. 28, No. 3, pp. 195–215.

Dekel, E., Scotchmer, S. (1999), "On the evolution of attitudes towards risk in winner-take-all games." *Journal of Economic Theory*, Vol. 87, pp. 125–143.

Ellison, G. (2000), "Basins of attraction, long-run stochastic stability, and the speed of step-by-step evolution." *Review of Economic Studies*, Vol. 67, pp. 17–45.

Friedman, D. (1996), "Equilibrium in evolutionary games: Some experimental results." *The Economic Journal*, Vol. 106, pp. 1–25.

Gaylord, R.J., D'Andria, L.J. (1998), *Simulating Society: A* Mathematica *Toolkit for Modeling Socioeconomic Behavior.* Springer, Berlin Heidelberg New York.

Hagen, O. (1992), "Survival through the Allais paradox." *Theory and Decision*, Vol. 32, pp. 209–217.

Harsanyi, J. (1967), "Games with incomplete information played by Bayesian players, Part I: The basic model," *Management Science*, Vol. 14, pp. 159–82.

Hodgson, G.M. (1993a), "The Mecca of Alfred Marshall," *The Economic Journal*, Vol. 103, No. 417, pp. 406–415.

Hodgson, G.M. (1993b), *Economics and Evolution: Bringing Life Back into Economics.* Polity Press, Cambridge, MA.

Kandori, M., Mailath,G.J., Rob, R. (1993), "Learning, mutation, and long run equilibria in games." *Econometrica*, Vol. 61, No. 1, pp. 29–56.

Karni, E., Schmeidler, D. (1986), "Self-preservation as a foundation of rational behavior under risk." *Journal of Economic Behavior and Organization*, Vol. 7, pp. 71–81.

Keenan, D.C., O'Brien, M.J. (1993), "Competition, collusion and chaos." *Journal of Economic Dynamics and Control*, Vol. 17, pp. 327–353.

Kelly Jr., J.L. (1956), "A new interpretation of information rate." *The Bell System Technical Journal*, Vol. 35, pp. 917–926.

Kreps, D.M. (1990), A Course in Microeconomic Theory. Princeton University Press, Princeton, NJ.

Levy, M. (2003), "Are rich people smarter?" *Journal of Economic Theory.* Vol. 11, pp. 42–64.

Lippman, S.A., Mamer, J.W. (1988), "When many wrongs make a right: An asymptotic analysis of risk aversion and additive risks." *Probability in the Engineering and Information Sciences*, Vol. 2, pp. 115–127.

Marshall, A. (1961), *The Principles of Economics.* 9th edition, Macmillan, London.

Maynard Smith, J. (1982), *Evolution and the Theory of Games.* Cambridge University Press, New York.

McCardle, K.F., Winkler, R.L. (1992), "Repeated gambles, learning and risk aversion." *Management Science*, Vol. 38, No. 6, pp. 807–818.

Morris, S. (2000), "Contagion." *Review of Economic Studies*, Vol. 67, pp. 57–78.

Oechssler, J., Riedel, F. (2002), "On the dynamic foundation of evolutionary stability in continuous models." *Journal of Economic Theory.* Vol. 107, pp. 223–252.

Pollock, G.B., Lewis, K. (1993), "Gambling in a Malthusian universe." *Rationality and Society*, Vol. 5, No. 1, pp. 85–106.

Pratt, J.W., Zeckhauser, R.J. (1987), "Proper risk aversion." *Econometrica*, Vol. 55, No. 1, pp. 143–154.

Robson, A.J. (1996), "A biological basis for expected and non-expected utility." *Journal of Economic Theory*, Vol. 68, pp. 397–424.

Rothkopf, M.H., Harstad, R.M. (1994), "Modeling competitive bidding: A critical essay." *Management Science*, Vol. 40, No. 3, pp. 364–384.

Savage, L.J. (1972), *The Foundations of Statistics.* Dover, New York.

Sinn, H.-W. (2003), "Weber's law and the biological evolution of risk preferences: The selection dominance of the logarithmic utility function", 2002 Geneva Risk Lecture. *The Geneva Papers on Risk and Insurance Theory*, Vol. 28, No. 2, pp. 87–100.

Spencer, H. (1890), *First Principles*, 5th edition, Williams and Norgate, London.

Spencer, H. (1892), *Essays Scientific, Political and Speculative.* Appleton, New York.

Taylor, P., Jonker, L. (1978), "Evolutionarily stable strategies and game dynamics." *Mathematical Biosciences*, Vol. 40, pp. 145–156.

Weibull, J.W. (1997), *Evolutionary Game Theory.* MIT Press, Cambridge, MA.

Behavior in Risky Decisions: Focus on Risk Defusing

Oswald Huber

University of Fribourg, Department of Psychology, Route de Faucigny 2, CH-1701 Fribourg, Switzerland
e-mail: oswald.huber@unifr.ch

Abstract. In experiments on risky decisions with gambles as alternatives the central factors determining decision behaviour are: The subjective values of the outcomes, and their subjective probability. The present paper first reports results of a number of experiments indicating that this central result cannot be generalized. In quasi-realistic risky scenarios, many decision makers are not interested in probability information and many search actively for risk-defusing operators (RDOs). An RDO is an action intended by the decision maker to be performed additionally to a specific alternative in order to decrease the risk. The paper also gives an overview about experimental research with RDOs. Topics include the factors that determine the search for RDOs and the factors affecting the acceptance of an RDO. Finding an acceptable RDO has a distinct effect on choice: If for a specific risky alternative an RDO is available, this alternative is chosen most often. The consequences of the concept of RDOs on theories about decision behaviour and on aiding decision making are discussed. Expectations to find usable probability information and to find information about the existence of an RDO are also discussed as factors explaining differences between different types of decision situations and gambles.

Keywords: risky decision behaviour, risk defusing, risk, decision process, information search, probability

1 Introduction

Decision theory has experimentally investigated risky decisions since its early days, about 50 years ago. An impressive number of empirical results have been gathered, and important theories have been developed, see, for example, Lopes (1995). The most prominent among these theories are subjectively expected utility (SEU) theory (Edwards, 1961) and prospect theory (Kahneman and Tversky, 1979; Tversky and Kahneman, 1992).

According to the experimental results of decision theory and the prevailing theories, the central factors determining decision behaviour are:

(1) the subjective values of the outcomes, and
(2) their subjective probability.

One has to take into account, however, that most of these experiments use (simple) gambles as alternatives or alternatives that are prestructured by the experimenter like gambles. Therefore, it is not clear whether this result can be generalized to all decision situations. The central role of gambles in psychological decision research is emphasized by Lopes' (1983) statement that gambles are as indispensable to research on risk as the fruitfly is to genetics.

Despite the advantages of using gambles there is a nagging suspicion that the essence of what naive decision makers often do when making a risky decision often is missed by models based on gambles. Everyday decision tasks differ in a variety of aspects from gambles, for example, in the topics of control, background knowledge, and the construction of a subjective representation of the task. Discussions of such differences can be found, for example, in Crozier and Ranyard (1997), Dörner and Wearing (1995), Goldstein and Weber (1995), Hogarth and Kuenreuther (1995), Huber (1997), Huber and Kühberger (1996), Lipshitz and Strauss (1997), Ranyard and Craig (1995) or Wagenaar (1988). Because of these differences, the hypothesis that the results from experiments with gambles can be generalized to other decision situations cannot be taken for granted, but has to be tested. Furthermore, there are methodological characteristics of experiments with gambles that may hamper generalizability.

2 Peculiarities of Experiments with Gambles or Similarly Structured Tasks

Two main problems may limit the generalizability of results obtained from experiments with gambles: task characteristics of gambles and specific properties of the experimental procedure.

2.1 Task Characteristics of Gambles

In contrast to many everyday risky decisions, in a decision task involving gambles (or tasks that are prestructured like a gamble by an experimenter) it is clear what consequences an alternative has, on which events the occurrence of the consequences depends and what the probabilities are.

Another central difference is control, which is usually excluded deliberately in experiments with gambles. In choices among gambles, besides choosing one alternative, the decision maker cannot exert any control at all. Consider in contrast the situation of a person who has to decide whether or not to travel into a country where an epidemic infectious disease rages. This person may not only contemplate on the probability to become infected, but may look actively for ways to prevent an infection (e.g., by cooking water before drinking it, wearing protective gear, etc.) or inquire whether a vaccination exists.

This neglect of the topic of control is probably the most serious deficiency of traditional risky decision theory[1].

2.2 Characteristics of the Experimental Procedure

With very rare exceptions in experiments on risky decision behaviour the following procedure is adopted:

An experimenter explains to the subject (orally or in writing) which alternatives exist, what events are relevant (for example, drawing a *green* ball from an urn), which consequences may occur with which probabilities. Thus, all the information considered to be relevant by the experimenter is presented whereas all information considered as irrelevant is not presented. In experiments with alternatives other than gambles (for example, in a framing task) the situation is prestructured like a gamble by the experimenter.

This experimental procedure has two potential reactive effects:

(1) The situation is prestructured as a gamble by the experimenter. Therefore, there is the danger of *forcing* the structure on the subject. Furthermore, the experimenter has no opportunity to learn how a decision maker would have spontaneously structured a situation, without the experimenter's prestructuring.
(2) Because all informations are presented, the experimenter cannot find out in which information the subject is genuinely interested. If a presented item of information is used by the decision maker, the experimenter does not know whether the subject would have searched for this item and used it if it had not been presented.

3 Are Subjective Probabilities and Subjective Values Central in all Risky Decision Situations?

Our first series of experiments aimed to answer the question whether the main result from risky decision theory (subjective probability and subjective value as the central components) can be generalized to situations that are not prestructured as a gamble. These experiments differ in the type of tasks and in the method of information presentation from standard experiments.

---

[1] Cohen and Hansel (1959) introduced the element of control already at the beginning of risky decision research by using gambles with a skill component. This approach, however, was not followed up systematically in decision research.

3.1 Quasi-Realistic Risky Tasks

In a quasi-realistic task, a scenario with at least two alternatives is described to the decision maker. Such a description leaves to the decision maker the task to construct a subjective representation of the scenario. This construction is considered as an essential part of the decision process.

An example of a quasi-realistic task is the *post office* task:

The subject is in the role of the head of the post office in the village H. The post office has very cramped conditions, and has faced the following problem for several years: In November and December, it has to deal with many parcels. If the number of parcels to be handled is too large, conditions may become unbearable.

The village administration offers the manager now (in spring/summer) the opportunity to rent the local meeting hall for CHF 12,000 (about $10,000, €8000). This hall is located besides the post office and would solve the problem.

Whether it pays off to rent the hall depends on the amount of parcels to be handled before Christmas. If the amount of parcels to be handled is high, then renting the hall is necessary, because otherwise the delivery of the parcels would be delayed, and complaints of customers would have to be expected. Furthermore, working conditions for the employees would become very bad. If, however, the amount of parcels to be handled were low, then it would not be necessary to rent the hall. The CHF 12,000 would have been wasted.

The subject has to decide whether to rent the hall or not.

3.2 A Non-reactive Method of Information Search: Method of Active Information Search

If we want to find out which information the decision maker is genuinely interested in, a methodology has to be employed that forces the decision-maker to actively search for information. The *method of active information search* (AIS) has been developed in order to attain this objective (Huber et al., 1997).

In the basic version of this method, the procedure of information acquisition is as follows:

(1) The subject is given a short description of the decision situation. This description is the same for all subjects and explicitly mentions the possibility of a negative consequence.
(2) Then, he or she can ask questions in order to obtain more information from the experimenter. Note that the *subject* asks the questions, not the experimenter.
(3) For each question, the experimenter presents a prepared answer, printed on a card or shown on a computer monitor. The reason for presenting printed answers is to avoid non-verbal influences.

(4) The participant may ask any and as many questions as wanted. If he or she thinks to have sufficient information he or she can make the choice.

Elaborate pre-experiments are necessary to optimize the short description and to find out the questions people ask in order to prepare the answers.

Different variations of the AIS-method have been developed. In the *list-version* (Huber et al., 2001), the participant can select one at a time from a list of questions. This version is especially useful to present the task on the computer or in the Internet (Schulte-Mecklenbeck and Huber, 2003). In the conversation-based version (Raynard, Williamson and Cuthbert, 1999) AIS is combined with think-aloud instructions and spoken answers are given. The different versions evoke a different number of questions (more questions in the list-version than in the basic version), but the distribution of types of questions (e.g., questions requesting information about the probability of the negative outcome, questions about the availability of risk defusing actions, see next section) is not affected.

3.3 Summary of Experimental Results

Several experiments have been performed which enable a test of the hypothesis that in quasi-realistic risky tasks the same factors are central as in gambles, namely, subjective values and subjective probabilities: Huber et al. (1997), Huber et al. (2001), Huber and Huber (2003), Huber and Huber (2004), Huber and Macho (2001), Ranyard et al. (1999), Ranyard et al (2001), Schulte-Mecklenbeck and Huber (2003), Williamson et al. (2000a,b).

In all these experiments, the subjective values of the outcomes are a central factor in the decision process. This finding is in agreement with results from experimental research on gambles.

However, there are important behavioural differences between choices in quasi-realistic tasks and choices among gambles:

(1) Most decision-makers in most quasi-realistic decision tasks are not interested actively in probability information. It seems sufficient for them to know that a negative consequence *may* happen.
(2) Often, risk-defusing behaviour plays a central role in the decision process. If decision-makers realize that an otherwise positive alternative may lead to a negative outcome, they search for an additional action (risk-defusing operator, see below) that reduces the involved risk.

These two main results are discussed in the following sections.

Low Interest in Probability Information

This general result contradicts one of the fundamental assumptions of traditional decision theory. Therefore, it is necessary to rule out alternative explanations.

(1) Could it be that people do not search for probability information because they choose a non-risky alternative anyway? This alternative explanation can be ruled out because the non-interest for probability information is independent from whether a decision maker chooses a risky alternative or a non-risky one. Furthermore, in several of the tasks *all* alternatives were risky.
(2) Could it be that people do not search for probability information because they *generally* introduce or infer probabilities from their background knowledge? Huber and Macho (2001) showed that in four quasi-realistic decision tasks subjects did not introduce or infer usable probability information. Therefore, this alternative explanation can be discarded. It is nevertheless possible that in *specific* decision situations the decision maker can introduce or infer probability information from background knowledge. For example, if the negative event is the loss of one's job, the decision maker may have a feeling how certain or uncertain her or his job is.
(3) In experiments using the *basic* AIS version, questions have to be coded. Could it be that in the coding process, questions searching for probability information were not coded as such because, for example, the participant did not use the word "probability"? In the experiments investigating information search (e.g., Huber et al., 1997; Huber et al., 2001; Huber and Huber, 2003) a question needed not necessarily contain the word "probability" or "uncertainty" in order to be classified as a probability question. All questions indicating a search for information about probability or uncertainty were coded as probability questions. In cases of doubt the item was coded as question for probability information, so that an error of assignment worked against our hypothesis. Thus one can rule out the alternative explanation that probability questions were overlooked. In the list-version of the AIS (e.g., Huber et al., 2001; Schulte-Mecklenbeck and Huber, 2003), the term "probability" was used explicitly and no coding was necessary.

In concluding this short discussion it should be noted that in the search for probability information there is a large difference between tasks. The difference ranges from 6% to nearly 60%. Furthermore, even if decision makers search for probability information, most are satisfied with imprecise answers (e.g., "the probability is small") and only few search for precise probability values (Huber et al., 1997).

Risk Defusing Operators

A *risk-defusing operator* (RDO) is an action intended by the decision-maker to be performed additionally to a specific alternative and is expected to decrease the risk. In the post office task, for example, many decision makers inquired whether it was possible at short notice to lease bureau containers or similar devices. For the risky alternative *not* to rent the hall leasing bureau containers is an RDO. It is intended for the case that the number of parcels turns out to be large.

RDOs are common in everyday risky decision situations. Typical examples are: taking out insurance, getting a vaccination, making a backup copy of computer files, or wearing protective gear in order to avoid contact with a toxic substance. In Sect. 5, some basic types of RDOs are defined.

In many quasi-realistic scenarios the majority of decision-makers actively searches for RDOs. Finding an RDO has a distinct effect on choices: if a decision-maker finds an RDO or an RDO is available, he or she chooses the risky alternative in question much more often than without finding an RDO. Also in a multistage investment task, people were willing to buy control with the help of an RDO, when they had the opportunity to do so (Huber, 1996). It should be noted that also in the search for information concerning RDOs there are large differences between tasks.

Conclusion

To sum up the answer to the question posed at the beginning of this chapter: The available studies clearly reveal that the results from experiments with gambles *cannot* be generalized to all risky choice situations. It should be noted that this conclusion concerns decision making behavior only. On a formal level, RDOs can well be incorporated into a gambling framework (see the discussion section).

4 Risk Defusing in the Decision Process

In experiments with quasi-realistic scenarios – as usually in real decision situations – the decision task is not prestructured as a gamble. It is rather an important part of the task for the decision maker to find a structure by constructing a subjective representation. Some theories explicitly take into account structuring (e.g., prospect theory). Most often, however, this aspect is ignored. The search for RDOs shows that the mental representation is not static but dynamic and may be altered in the course of the decision process.

For those decision makers who search for an RDO, the decision process can be sketched as follows:

(1) The decision maker detects that an otherwise attractive alternative x *may* also lead to a negative consequence.
(2) Search decision: The decision maker has to decide whether to search for an RDO or not (local process decision).
(3) Acceptance decision: If the decision maker decides to search for an RDO and the search is successful, a decision has to be made whether or not to *accept* the detected RDO (local process decision). The introduction of an

RDO changes the mental representation. If a promising RDO is found, the causal path leading to the negative outcome can be eliminated.
(4) If the RDO is acceptable, the decision maker chooses alternative x.

In the next sections, the available experimental results concerning search, acceptance and choice are summarized.

4.1 Search for RDOs

For the search decision, the attractiveness of the alternative and the expectation to find an RDO have been identified as relevant factors. Moreover, control beliefs as a personality variable has an effect. Persons with internal control belief search more often for RDOs than those with external control belief (Huber and Bartels, 2005).

The search for an RDO often gives rise to cost (money, time, effort...) and at the beginning of the search it is not clear whether the search will be successful or not. A decision maker should be willing to bear these cost more likely when the alternative is attractive. A first indication for this effect of attractiveness is the finding that RDOs are searched more often for the alternative that later is chosen (Huber et al., 2001). However, further experiments are necessary where the attractiveness of an alternative is varied deliberately.

Decision makers should search for an RDO more likely if their expectation to find an RDO is higher. This hypothesis was confirmed, for example, in Huber and Huber (2004). M. Bär (2002) measured participants' expectations to find an RDO and found an attractiveness bias. If an alternative was more attractive (and everything else was equal) the expectation to find an RDO was higher than when the alternative was less attractive.

4.2 Acceptance of RDOs

Once an RDO is found it is not necessarily acceptable. There are desirable effects of the RDO (prevention or compensation of the negative outcome), but there may be costs of the application of an RDO (e.g., the premium of an insurance, time). The decision maker has to weigh up the positive effect against the costs. The effect of both factors, costs and effect, has been investigated experimentally.

The higher the cost of an RDO the less likely it is accepted (Williamson et al., 2000a). Huber and Huber (2003) varied whether the cost of the RDO had to borne with certainty or probabilistic. The majority of subjects chose the alternative with an RDO with probabilistic cost.

The goal of applying an RDO is to prevent the negative outcome or to compensate for the negative outcome. The effect of an RDO indicates to what degree the goal is attained. For example, a vaccination against an infectious disease may prevent the outbreak of the disease completely or it may alleviate the disease, but not prevent it. As might be expected, RDOs having more effect are preferred to those having less (provided everything else is equal). An alternative with an RDO with partial effect is chosen more often than one without RDO. The latter was also observed in Huber (1995).

It should be noted that the effect of an RDO must be distinguished from the probability of its success (e.g., the probability that the vaccination is successful). The success (i.e., the intended effect) of an RDO may be only probabilistic, not certain. For example, a vaccination may not be successful for a minority of people who have been vaccinated. If success probability is higher, the RDO is accepted more often. Huber and Macho (2001b) found that an alternative with an RDO with probabilistic success was chosen more often than an alternative with no RDO, and that an alternative with an RDO with certain success was chosen more often than an alternative with an RDO with probabilistic success. Huber and Huber (2003) varied success probability via the probability to detect the occurrence of a negative event in good time to initiate an RDO. The higher the success probability was the more often the alternative with the RDO in question was chosen.

4.3 RDOs and Choice

Finding an acceptable RDO has a distinct effect on choice: If for a specific risky alternative an RDO is available, this alternative is chosen more often than if no RDO is available: Huber (1995), Huber et al. (2001), Huber and Huber (2003), Huber and Macho (2001 b), Ranyard et al. (2001) and Williamson et al. (2000a).

Huber and A. Bär (2005) varied the success of the search for an RDO. In most cases, decision makers searched an RDO only for one alternative. If the search was not successful, the choices were on chance level. If the RDO search was successful, in 98% of the cases the alternative with the RDO was chosen.

5 Types of RDOs

In the following sections three criteria are introduced that enable a useful classification of RDOs: (i) the target of the RDO, (ii) whether the RDO has to be applied before or after a negative event has taken place, and (iii) whether the RDO prevents the negative event or interrupts the causal chain between negative event and negative outcome. Thus, criteria (ii) and (iii) are applicable in situations where one can identify a specific negative event (e.g., an infection with a germ) that leads to a negative outcome (e.g., a dangerous disease).

5.1 Target of RDOs

The criterion *target* classifies an RDO according to the global component it attempts to change. There are two possibilities that are both spontaneously used by subjects: (1) the RDO prevents the occurrence of the negative outcome, or at least makes it less likely (outcome prevention), or (2) the RDO does not prevent the occurrence of the negative outcome, but introduces a compensation instead (outcome compensation). Thus, outcome prevention influences the probability, Outcome Compensation the utility of the outcome. A typical example of an outcome prevention RDO is a vaccination against an infectious disease. A typical outcome compensation RDO is taking out insurance or making a backup copy of one's computer files.

In the theory of individual choice under risk, Ehrlich and Becker (1972) introduced a similar distinction: self-protection and self-insurance. Self-protection[2] concerns the decrease of probability and self-insurance the decrease of the severity of risk. Even if relatively low attention is given to the empirical investigation of these concepts, there are some studies. For example, Shogren (1990) tests the reaction of subjects to risks that are reduced either through private or collective self-protection or self-insurance.

5.2 RDO Application Before or After the Negative Event

Dependence classifies RDOs according to the occurrence of a negative event, which leads to the negative outcome. For example, an infection with a specific germ may be the negative event, which leads to the disease. It should be noted that often there is no *objective* way to determine which event is the negative event.

Pre-event RDOs (e.g., vaccination against an infectious disease) have to be applied *before* a negative event (e.g. infection) occurs. *Post-event* RDOs (*worst-case plans*) need *not* to be initiated *before* and *unless* the event happens (e.g., a medical treatment against the infectious disease).

These two types of RDOs differ in expected cost: the cost of a pre-event RDO has to be borne in any case, even if the negative event does not occur, whereas that of a post-event RDO must be borne solely if the negative event occurs. Thus, the subjectively expected utility of an alternative with a post-event RDO is larger than that of the same alternative with a pre-event RDO. Therefore, if both types of RDOs are available, decision-makers should prefer a post-event RDO, provided the occurrence of the negative event can be detected with certainty and everything else is equal. Huber and Huber (2003) confirmed this hypothesis.

5.3 Event-Prevention vs. Intervention RDOs

An *event-prevention* RDO prevents the negative event from happening. In the example with the infectious disease, wearing protective clothing that inhibits

[2] In psychological decision theory, the term "self-protection" has another connotation. It refers to a desire to escape unpleasant psychological consequences (e.g., regret) that result from a decision turning out poorly (Larrick, 1993).

any contact with a germ is an event-prevention RDO. An *intervention RDO* does not prevent the negative event from happening but interrupts the causal chain between the negative event and the negative outcome. Examples are a vaccination or a medical treatment after an infection.

Huber and Wicki (2004) confirmed the expectation that decision makers prefer an event-prevention RDO to an intervention RDO (provided everything else is equal). Note that in this experiment, the event-prevention RDOs as well as the intervention RDOs were both constructed as pre-event RDOs in all tasks.

6 Discussion

The experimental results reported in this paper contradict one traditional assumption of descriptive decision theory, that risky decisions *generally* can be modelled as gambles. The main behavioural differences between decision making in quasi-realistic tasks and choices among gambles are:

(1) In most quasi-realistic decision tasks the decision-makers are not interested actively in probability information. In gambles, on the other hand, the probability of the negative outcome is a central factor.
(2) RDOs, their search and intended application, play a central role in the decision process. In theories about decision behavior in choices among gambles, RDOs are literally nonexistent.

From the behavioral point of view, these results oppose not only the SEU model, but also all other models that have their conceptual roots in the SEU model, for example, prospect theory (Kahneman and Tversky, 1979) and cumulative prospect theory (Tversky and Kahneman, 1992), security-potential/aspiration theory (Lopes, 1995), but also regret theory (e.g., Loomes and Sugden, 1982).

From a *formal* point of view, however, the concept of RDOs can well be incorporated into the framework of normative SEU theory. Formally, the introduction of an RDO into an alternative creates a new alternative (a new gamble) consisting of the original action *and* the RDO. By using standard methods of decision analysis (e.g., a decision tree), the utility of the cost of RDO application, of the effect of an RDO and its success probability can be incorporated into the computation of the overall utility of the new alternative.

Our stable finding of most people not searching for information about the probability of the negative consequences does not mean that subjective uncertainty does not play any role in the decision process. As already mentioned, in our experiments subjects were informed explicitly about the possibility of a negative outcome. Obviously for most decision makers this coarse information is sufficient, for example, to initiate an RDO search. As discussed in the sections concerning search for and acceptance of RDOs, here too uncertainty may become relevant, namely uncertainty about the result of the search and about the effect of the RDO. One can expect that people use a coarse subjective scale also in this context.

An interesting topic for further research on RDOs is the multiperson context. For example, decision makers who have to choose independently between gambles could plan to pool their wins and losses and thus – with specific constellations of alternatives – defuse the risk for all decision makers.

Behaviour in experiments using quasi-realistic tasks is in congruence with that revealed in non-experimental research (Lipshitz and Strauss, 1997) or analyses of managerial decisions (e.g., Shapira, 1994). The findings about risk defusing behaviour fit neatly into the results of research on risk perception and risk taking. An RDO gives the decision maker at least some control over the risk. Perceived controllability has been shown to be a relevant factor in risk perception: controllable risks are evaluated as less grave than uncontrollable ones (Weinstein, 1984; Vlek and Stallen, 1981). In a study investigating information-seeking behaviour, Lion (2001) found that the more dangerous a risky situation is the more people are interested in information about controllability. Shiloh et al. (*in press*) investigated the information needs and decision-making strategies of genetic counselees in real-life decision making. Patients were interested mostly in information about the outcomes and consequences of the alternative options, and about measures to defuse the risks. They were less interested in probability information.

The concept of RDOs also fits neatly into recent normative treatments of causality. In his theory of causality, Pearl (2000) explicitly includes interventions in a causal system. An RDO is such an intervention.

The concept of RDOs has consequences also for decision analysis and decision aiding tools. Because RDOs are important for decision makers, they should be incorporated into techniques of decision analysis explicitly and in a controlled manner. It should be emphasized that the search for potential actions to defuse the risk connected with an otherwise attractive alternative is rational also from the point of view of normative decision theory. It would be rather irrational *not* to search for an RDO if there exists a genuine chance to find an effective one. If a formal structure is used for the representation of an alternative, (e.g., a decision tree), RDOs should be incorporated explicitly into that representation structure. In the process of evaluating the alternative, the decision maker has to evaluate also the RDOs, for example, in respect to their effectiveness. Decision analysis should guide the decision-maker in the incorporation of RDOs into the decision structure, in the evaluation of RDOs, and in avoiding potential evaluation biases.

In the description of the decision process in Sect. 3, two local process decisions were mentioned: the decision to search or not to search for an RDO for a specific alternative, and – if an RDO has been found – the decision to accept the RDO or not. These decisions are necessary also in the process of structuring the decision situation in the course of a decision analysis. The decision aiding procedure should guide the decision maker to avoid biased evaluations in these two decisions, like the attractiveness bias mentioned above.

A further topic, where the results of experiments with quasi-realistic tasks may be relevant for Decision analysis is probability. As reported, most decision makers do not spontaneously search for probability information. Therefore, decision makers should be alerted when and in what respect probabilities are relevant in making a decision.

As our experiments have shown, decision behaviour in quasi-realistic scenarios is quite different from choices between gambles. Furthermore, there are large differences between quasi-realistic scenarios in the search for RDOs as well as in the search for probability information. One solution would be to conclude that we need different decision theories for different types of risky tasks. This solution would, however, be quite unsatisfactory. In the long run, we need a unifying theory that enables us to understand decision behaviour in choices between gambles as well as in quasi-realistic decision situations, and that explains in which cases and why different types of decision behaviour occur.

As a first step towards such a unifying theory, Huber and Huber (2004) have investigated two types of decision maker's expectations: the expectation of finding usable probability information and the expectation to find information about the existence or non-existence of an RDO. Note that the expectation of finding information about whether an RDO exists or not is not the same as the expectation to find an RDO, and the expectation of finding usable probability information is not the same as the expectation that the probability is, say, high.

In their experiment, these expectations were varied independently by including cues in the basic description of the quasi-realistic scenarios. In addition to the scenarios, subjects had to make decisions between gambles. With the gambles also, the AIS-method was used, but no cues were included. In the quasi-realistic scenarios, cues increasing the finding expectation led to a more frequent search for probability information and RDO information, respectively, in comparison to cues that decreased finding expectation. In the gamble tasks without additional cues, the great majority of subjects searched for probability information, but not one single subject searched for RDO information.

If only high and low finding expectations for probability information and RDO information are distinguished, four types of risky decision tasks can be distinguished:

1. High Finding Expectation for Probability Information – High for RDO Information An example is the decision whether or not to travel into a country where an infectious disease rages. A decision maker may expect to get from doctors usable information about the probability of an infection as well as about the existence of a vaccination.

2. High Finding Expectation for Probability Information – Low for RDO Information An example is the situation of a patient who has to decide whether to undergo a surgery. This patient may expect to get reliable information about the chances of success of the operation, but at the same time be of the opinion the he or she could not contribute anything for the success. The classical example for this type of decisions are gambles.

3. Low Finding Expectation for Probability Information – High for RDO Information An example is the situation of a manager who has to decide whether to place a large investment into a developmental country with an unclear political and economical state of affairs. This manager may expect not being able to get usable probability information. However, he or she may presume that there are governmental guarantees for such investments, and therefore search for such information.

4. Low Finding Expectation for Probability Information – Low for RDO Information A typical example is the stock market, at least from the point of view of a small share-holder. This person may have low expectation to get reliable probability information about the value of a specific stock five years from now, and at the same time be convinced of having no influence on the stock market.

Decision behaviour in the different types of tasks should be different. For example, in situations with high expectation for finding RDO information, search for RDOs should play a prominent role. Decisions in situations of the type *high finding expectation for probability information – low for RDO information* should be most similar to decisions between gambles. This hypothesis, however, has yet to be tested.

As already mentioned, the different finding expectations can explain a substantial part of the variance in the search for probability information and RDO information, respectively. Thus, finding expectations may constitute one component of a future unifying theory for risky decisions that can also explain why behaviour in various situations is different.

Acknowledgement. I would like to thank Odilo W. Huber and an anonymous reviewer for their helpful comments on an earlier version of the manuscript.

Some of the experiments mentioned in the article were made possible by a grant from the Swiss National Science Foundations (Project Numbers 11-37315.93, 11-49188.96, 11-55474.98).

References

Bär, M. (2002), Risikoentscheidungen: Verzerrte Wahrnehmung der Risiko-Handling Operatoren. [Risky decisions: biased perception of risk-defusing operators.] Unpublished Masters-thesis (Lizentiatsarbeit), University of Fribourg, Switzerland.

Cohen, J., Hansel, C.E.M. (1959), Preference for different combinations of chance and skill in gambling. *Nature*, 183, 841–842.

Crozier, W.R., Ranyard, R. (1997), Cognitive process models and explanations of decision making. In Ranyard, R., Crozier, W.R., Svenson, O. (Eds.): *Decision Making: Cognitive Models and Explanations* (pp. 5–20). London: Routledge.

Dörner, D., Wearing, A.J. (1995), Complex problem solving: Towards a (computer-simulated) theory. In: Frensch, P.A., Funke, J. (Eds.), *Complex Problem Solving: The European Perspective.* Hillsdale, N.J.: Erlbaum, pp. 65–99.

Edwards, W. (1961), Behavioral decision theory. *Annual Review of Psychology*, 12, 473–498.

Ehrlich, I., Becker, G. (1972), Market insurance, self-insurance and self-protection. *Journal of Political Economy*, 80, 623–648.

Goldstein, W.M., Weber, E.U. (1995), Content and discontent: Indications and implications of domain specificity in preferential decision making. In J.Busemeyer, R. Hastie and D.L. Medin (Eds.), *Decision making from a cognitive perspective* (pp. 83–136). San Diego: Academic Press.

Hogarth, R.M., Kunreuther, H. (1995), Decision making under ignorance: Arguing with yourself. *Journal of Risk and Uncertainty*, 10, 15–36.

Huber, O. (1995), Ambiguity and perceived control. *Swiss Journal of Psychology*, 54, 200–210.

Huber, O. (1996), Buying protection in a multistage investment task. *Acta Psychologica*, 92, 153–167.

Huber, O. (1997), Beyond Gambles and Lotteries: Naturalistic Risky Decisions. In Ranyard, R., Crozier,W.R., and Svenson, O. (Eds.): *Decision Making: Cognitive Models and Explanations* (pp.145–162). London: Routledge.

Huber, O., Bär, A. (2005), Risky decision behavior: Effect of successful vs. unsuccessful search for risk defusing operators. Manuscript submitted for publication.

Huber, O., Bartels, L. (2005), Control belief and search for risk defusing operators in decision making. Manuscript submitted for publication.

Huber,O., Beutter, C., Montoya, J., Huber, O.W. (2001), Risk defusing behaviour: Towards an understanding of risky decision making. *European Journal of Cognitive Psychology*, 13, 409–426.

Huber, O., Huber, O.W. (2003), Detectability of the negative event: effect on the acceptance of pre- or post-event risk defusing actions. *Acta Psychologica*, 113, 1–21, a.

Huber, O., Huber, O.W. (2004), What makes gambles special? – Active search for probability and risk defusing information in gambles and quasi-realistic risky tasks. Manuscript submitted for publication.

Huber, O., Kühberger, A. (1996), Decision processes and decision trees in gambles and more natural decision tasks. *Journal of Psychology*, 130, 229–236.

Huber, O., Macho, S. (2001), Probabilistic set-up and the search for probability information in quasi-naturalistic decision tasks. *Risk Decision and Policy*, 6, 1–16.

Huber, O., Wicki, D. (2004), Risk defusing in decision making: Prevention or intervention? *Risk Decision and Policy*, 9, 359–370.

Huber, O., Wider, R., Huber, O.W. (1997), Active information search and complete information presentation in naturalistic risky decision tasks. *Acta Psychologica*, 95, 15–29.

Kahneman, D., Tversky, A. (1979), Prospect theory: An analysis of decision under risk. *Econometrica*, 47, 263–291.

Larrick, R.P. (1993), Motivational factors in decision theories: The role of self-protection. *Psychological Bulletin*, 113, 440–450.

Lipshitz, R., Strauss, O. (1997), Coping with uncertainty: A naturalistic decision-making analysis. *Organizational Behavior and Human Decision Processes*, 69, 149–163.

Lion, R. (2001), *Security or opportunity: The effects of individual and situational factors on risk information preference.* Doctoral dissertation, University of Maastricht, The Netherlands.

Loomes, G., Sugden, R. (1982), Regret theory: An alternative theory of rational choice under uncertainty. *Economic Journal*, 92, 805–824.

Lopes, L.L. (1983), Observations: Some thoughts on the psychological concept of risk. *Journal of Experimental Psychology: Human Perception and Performance*, 9, 137–144.

Lopes, L.L. (1995), Algebra and process in the modeling of risky choice. In Busemeyer, J., Hastie, R., Medin, D.L. (Eds.), *Decision making from a cognitive perspective* (pp. 177–220). San Diego: Academic Press.

Pearl, J. (2000), *Causality: models, reasoning, and inference.* Cambridge: Cambridge University Press.

Ranyard, R., Craig, G. (1995), Evaluating and budgeting with instalment credit: An interview study. *Journal of Economic Psychology*, 16, 449–467.

Ranyard, R., Hinkley, L., Williamson, J. (2001), Risk management in consumers' credit decision making. *Zeitschrift für Sozialpsychologie*, 32, 152–161.

Ranyard, R., Williamson, J., Cuthbert, L. (1999), The development of conversation-based process tracing methods for the study of naturalistic decision making. In Green, B. (ed), *Risk Behaviour and Risk Management in Business Life.* Dordrecht: Kluver.

Schulte-Mecklenbeck, M., Huber, O. (2003), Information search in the laboratory and on the web: With or without the experimenter. *Behavior Research Methods, Instruments, & Computers*, 35, 227–235.

Shapira, Z. (1994), *Managerial risk taking.* New York: Russell Sage.

Shiloh, S., Gerad, L., Goldman B. (*in press*), Patients' information needs and decision-making processes: What can be learned from genetic counseling? *Health Psychology.*

Shogren, J.F. (1990), The impact of self-protection and self-insurance on individual response to risk. *Journal of Risk and Uncertainty* 3, 191–204.

Tversky, A., Kahneman, D. (1992), Advances in prospect theory: Cumulative representation of uncertainty. *Journal of Risk and Uncertainty*, 5, 297–323.

Vlek, C., Stallen, P.J. (1981), Rational and personal aspects of risk. *Acta Psychologica*, 45, 273–300.

Weinstein, N.D. (1984), Why it won't happen to me: Perceptions of risk factors and susceptability. *Health Psychology*, 3, 431–457.

Wagenaar, W.A. (1988), *Paradoxes of gambling behavior.* Hillsdale, NJ., Erlbaum.

Williamson, J., Ranyard, R., Cuthbert, L. (2000a), Risk management in everyday insurance decisions: Evidence from a simulation study. *Risk Decision and Policy*, 5, 19–38.

Williamson, J., Ranyard, R., Cuthbert, L. (2000b), A conversation-based process tracing method for use with naturalistic decisions: an evaluation study. *British Journal of Psychology*, 91, 203–221.

THEORY AND DECISION LIBRARY

SERIES C: GAME THEORY, MATHEMATICAL PROGRAMMING
AND OPERATIONS RESEARCH

Editor: H. Peters, *Maastricht University, The Netherlands*

1. B. R. Munier and M. F. Shakun (eds.): *Compromise, Negotiation and Group Decision.* 1988 ISBN 978-90-277-2625-4
2. R. Selten: *Models of Strategic Rationality.* 1988 ISBN 978-90-277-2663-6
3. T. Driessen: *Cooperative Games, Solutions and Applications.* 1988 ISBN 978-90-277-2729-9
4. P. P. Wakker: *Additive Representations of Preferences.* A New Foundation of Decision Analysis. 1989 ISBN 978-0-7923-0050-2
5. A. Rapoport: *Experimental Studies of Interactive Decisions.* 1990 ISBN 978-0-7923-0685-6
6. K. G. Ramamurthy: *Coherent Structures and Simple Games.* 1990 ISBN 978-0-7923-0869-0
7. T. E. S. Raghavan, T. S. Ferguson, T. Parthasarathy and 0. J. Vrieze (eds.): *Stochastic Games and Related Topics. In Honor of Professor L. S. Shapley.* 1991 ISBN 978-0-7923-1016-7
8. J. Abdou and H. Keiding: *Effectivity Functions in Social Choice.* 1991 ISBN 978-0-7923-1147-8
9. H. J. M. Peters: *Axiomatic Bargaining Game Theory.* 1992 ISBN 978-0-7923-1873-6
10. D. Butnariu and E. P. Klement: *Triangular Norm-Based Measures and Games with Fuzzy Coalitions.* 1993 ISBN 978-0-7923-2369-3
11. R. P. Gilles and P. H. M. Ruys: *Imperfections and Behavior in Economic Organization.* 1994 ISBN 978-0-7923-9460-0
12. R. P. Gilles: *Economic Exchange and Social Organization.* The Edgeworthian Foundations of General Equilibrium Theory. 1996 ISBN 978-0-7923-4200-7
13. P .J.-J. Herings: *Static and Dynamic Aspects of General Disequilibrium Theory.* 1996 ISBN 978-0-7923-9813-4
14. E. van Dijk: *Social Ties and Economic Performance.* 1997 ISBN 978-0-7923-9836-3
15. W. Spanjers: *Hierarchically Structured Economies.* Models with Bilateral Exchange Institutions. 1997 ISBN 978-0-7923-4398-1
16. I. Curiel: *Cooperative Game Theory and Applications.* Cooperative Games Arising from Combinatorial Optimization Problems. 1997 ISBN 978-0-7923-4476-6
17. O.I. Larichev and H. M. Moshkovich: *Verbal Decision Analysis for Unstructured Problems.* 1997 ISBN 978-0-7923-4578-7
18. T. Parthasarathy, B. Dutta, J. A. M. Potters,T. E. S. Raghavan, D. Ray and A. Sen (eds.): *Game Theoretical Applications to Economics and Operations Research.* 1997 ISBN 978-0-7923-4712-5
19. A. M. A. Van Deemen: *Coalition Formation and Social Choice.* 1997 ISBN 978-0-7923-4750-7
20. M. 0. L. Bacharach, L.-A. Gérard-Varet, P. Mongin and H. S. Shin (eds.): *Epistemic Logic and the Theory of Games and Decisions.* 1997 ISBN 978-0-7923-4804-7
21. Z. Yang (eds.): *Computing Equilibria and Fixed Points.* 1999 ISBN 978-0-7923-8395-6

THEORY AND DECISION LIBRARY: SERIES C

22. G. Owen: *Discrete Mathematics and Game Theory.* 1999 ISBN 978-0-7923-8511-0
23. F. Patrone, I. Garcia-Jurado and S. Tijs (eds.): *Game Practice.* Contributions from Applied Game Theory. 1999 ISBN 978-0-7923-8661-2
24. J. Suijs: *Cooperative Decision-Making under Risk.* 1999 ISBN 978-0-7923-8660-5
25. J. Rosenmüller: *Game Theory: Stochastics, Information, Strategies and Cooperation.* 2000 ISBN 978-0-7923-8673-5
26. J. M. Bilbao: *Cooperative Games on Combinatorial Structures.* 2000 ISBN 978-0-7923-7782-5
27. M. Slikker and A. van den Nouweland: *Social and Economic Networks in Cooperative Game Theory.* 2000 ISBN 978-0-7923-7226-4
28. K. J. M. Huisman: *Technology Investment: A Game Theoretic Real Options Approach.* 2001 ISBN 978-0-7923-7487-9
29. A. Perea: *Rationality in Extensive Form Games.* 2001 ISBN 978-0-7923-7540-1
30. V. Buskens: *Social Networks and Trust.* 2002 ISBN 978-1-4020-7010-5
31. P. Borm and H. Peters (eds.): *Chapters in Game Theory.* In Honor of Stef Tijs. 2002 ISBN 978-1-4020-7063-1
32. H. Houba and W. Bolt: *Credible Threats in Negotiations.* A Game-theoretic Approach. 2002 ISBN 978-1-4020-7183-6
33. T. Hens and B. Pilgrim: *General Equilibrium Foundations of Finance: Structure of Incomplete Markets Models.* 2003 ISBN 978-1-4020-7337-3
34. B. Peleg and P. Sudhölter: *Introduction to the Theory of Cooperative Games.* 2003 ISBN 978-1-4020-7410-3
35. J. H. H. Thijssen: *Investment under Uncertainty, Coalition Spillovers and Market Evolution in a Game Theoretic Perspective.* 2004 ISBN 978-1-4020-7877-4
36. G. Gambarelli (ed.): *Essays on Cooperative Games.* In Honor of Guillermo Owen. 2004 ISBN 978-1-4020-2935-6
37. G. B. Asheim: *The Consistent Preferences Approach to Deductive Reasoning in Games.* 2006 ISBN 978-0-387-26235-2
38. U. Schmidt and S. Traub (eds.): *Advances in Public Economics: Utility, Choice and Welfare.* A Festschrift for Christian Seidl. 2005 ISBN 978-0-387-25705-1
39. T. S. H. Driessen, G. van der Laan, V. A. Vasil'ev and E. B. Yanovskaya (eds.): *Russian Contributions to Game Theory and Equilibrium Theory.* 2006 ISBN 978-3-540-31405-9
40. W. V. Gehrlein: *Condorcet's Paradox.* 2006 ISBN 978-3-540-33798-0
41. M. Abdellaoui, R. D. Luce, M. J. Machina, B. Munier (eds.): *Uncertainty and Risk.* 2007 ISBN 978-3-540-48934-4

Printing: Krips bv, Meppel
Binding: Stürtz, Würzburg